THE NEW GROVE
EARLY ROMANTIC MASTERS 1

THE NEW GROVE
DICTIONARY OF MUSIC AND MUSICIANS

Editor: Stanley Sadie

The Composer Biography Series

BACH FAMILY
BEETHOVEN
EARLY ROMANTIC MASTERS 1
EARLY ROMANTIC MASTERS 2
HANDEL
HAYDN
HIGH RENAISSANCE MASTERS
ITALIAN BAROQUE MASTERS
MASTERS OF ITALIAN OPERA
MODERN MASTERS
MOZART
SCHUBERT
SECOND VIENNESE SCHOOL
TURN OF THE CENTURY MASTERS
WAGNER

THE NEW GROVE®

Early Romantic Masters 1

CHOPIN   SCHUMANN
LISZT

Nicholas Temperley

Gerald Abraham

Humphrey Searle

M
MACMILLAN

First published in
The New Grove Dictionary of Music and Musicians®,
edited by Stanley Sadie, 1980

The New Grove and *The New Grove Dictionary of Music and Musicians*
are registered trademarks of Macmillian Publishers Limited, London

First published in UK in paperback with additions 1985 by
PAPERMAC
a division of Macmillan Publishers Limited
London and Basingstoke

First published in UK in hardback with additions 1985 by
MACMILLAN LONDON LIMITED
4 Little Essex Street London WC2R 3LF
and Basingstoke

British Library Cataloguing in Publication Data
Temperley, Nicholas
 Early romatic masters 1.—(The composer biography series)
 1. Music—Europe—History and criticism
 2. Composers
 I. Title II. Abraham, Gerald
 III.   Searle, Humphrey IV. The New Grove dictionary of
 music and musicians V. Series
 780'.92'2   ML240

ISBN 0-333-38545-4 (hardback)
ISBN 0-333-38546-2 (paperback)

First American edition in book form with additions 1985 by
W. W. NORTON & COMPANY
New York and London

ISBN 0-393-01691-9 (hardback)
ISBN 0-393-30095-1 (paperback)

Printed in Great Britain

Contents

List of illustrations

Cover: Liszt at the piano: painting (1840) by Joseph Danhauser; (from left to right) Alexandre Dumas, Victor Hugo, George Sand, Paganini, Rossini, Liszt and the Countess Marie d'Agoult, piano by Conrad Graf (detail, see fig.24)

Illustration acknowledgments

We are grateful to the following for permission to reproduce illustrative material: Towarzystwo imenia Fryderyka Chopina, Warsaw/photo Mansell Collection, London (fig.1); Mansell Collection, London (figs.3, 25); Heirs of Mme S. André-Maurois/photo Hachette, Paris (fig.4); Royal College of Music, London (photo, fig.5); Bibliothèque Nationale, Paris (fig.6); Musées Nationaux, Paris (photo, fig.7); Pierpont Morgan Library (D. and H. Heineman Collection), New York (fig.8); Robert-Schumann-Haus, Zwickau (figs. 9, 10. 14–16); Österreichische Nationalbibliothek, Vienna (fig.11); Deutsche Staatsbibliothek, Berlin (figs.12, 19); Richard Macnutt, Tunbridge Wells (fig.13); Gesellschaft der Musikfreunde, Vienna (fig.17a); Staatsbibliothek Preussischer Kulturbesitz, Musikabteilung, Berlin (fig.17b); Giraudon, Paris (photo, fig.18); BBC Hulton Picture Library, London (fig.21); Archiv für Kunst und Geschichte, Berlin (fig.22); Magyar Nemzeti Múzeum, Budapest (fig.23); Oberfinanz Direktion, Munich/photo Österreichische Nationalbibliothek, Vienna (fig.24, cover); British Library, London (figs.26, 29); Országos Széchényi Könyvtár, Budapest (fig.27)

General abbreviations

A	alto, contralto [voice]	frag.	fragment
acc.	accompaniment, accompanied by	Ger.	German
add, addn	addition		
ad lib	ad libitum	hn	horn
Anh.	Anhang [appendix]	Hung.	Hungarian
anon.	anonymous(ly)		
appx	appendix	inc.	incomplete
arr.	arrangement, arranged by/for	incl.	includes, including
attrib.	attribution, attributed to	Jan	January
		Jb	Jahrbuch [yearbook]
B	bass [voice]		
b	born	K	Köchel catalogue [Mozart no. after / is from 6th edn.]
Bar	baritone [voice]		
bk	book	kbd	keyboard
bn	bassoon		
BWV	Bach-Werke-Verzeichnis [Schmieder, catalogue of J. S. Bach's works]	lib	libretto
		Mez	mezzo-soprano
		movt	movement
c	circa [about]	MS	manuscript
cl	clarinet		
collab.	in collaboration with	n.d.	no date of publication
conc.	concerto	no.	number
Cz.	Czech	Nov	November
		n.p.	no place of publication
d	died		
db	double bass	ob	oboe
Dec	December	Oct	October
ded.	dedicated to	op.	opus
diss.	dissertation	orch	orchestra, orchestral
		orchd	orchestrated (by)
ed.	editor, edited by	org	organ
edn.	edition	orig.	original(ly)
Eng.	English	ov.	overture
ens.	ensemble		
esp.	especially	pf	piano
		Pol.	Polish
f., ff.,	folio, folios	pubd	published
facs.	facsimile		
Feb	February	qnt	quintet
fl	flute	qt	quartet
Fr.	French		

R	photographic reprint	transcr.	transcription, transcribed (by/for)
repr.	reprinted		
rev.	revision, revised (by/for)	U.	University
		unacc.	unaccompanied
S	soprano [voice]	unpubd	unpublished
Sept	September		
spr.	spring	v, vv	voice, voices
str	string(s)	va	viola
sum.	summer	vc	cello
sym.	symphony, symphonic	vn	violin
T	tenor [voice]		
trans.	translation, translated by	ww	woodwind

Symbols for the library sources of works, printed in *italic*, correspond to those used in *Répertoire International des Sources Musicales*, Ser. A.

Bibliographical abbreviations

ADB	*Allegemeine deutsche Biographie* (Leipzig, 1875–1912)
AMw	*Archiv für Musikwissenschaft*
AMZ	*Allgemeine musikalische Zeitung*
AnM	*Anuario musical*
BSIM	*Bulletin français de la Société International de Musique*
DJbM	*Deutsches Jahrbuch der Musikwissenschaft*
FAM	*Fontes artis musicae*
Grove 3	*Grove's Dictionary of Music and Musicians* [3rd edn.]
Grove 6	*The New Grove Dictionary of Music and Musicians*
JAMS	*Journal of the American Musicological Society*
JbMP	*Jahrbuch der Musikbibliothek Peters*
JMT	*Journal of Music Theory*
KM	*Kwartalnik muzyczny*
Mf	*Die Musikforschung*
MGG	*Die Musik in Geschichte und Gegenwart*
ML	*Music and Letters*
MMR	*The Monthly Musical Record*
MQ	*The Musical Quarterly*
MR	*The Music Review*
MT	*The Musical Times*
NRMI	*Nuova rivista musicale italiana*
NZM	*Neue Zeitschrift für Musik*
ÖMz	*Österreichische Musikzeitschrift*
PRMA	*Proceedings of the Royal Musical Association*
RBM	*Revue belge de musicologie*
RdM	*Revue de musicologie*
ReM	*La revue musicale*
RIM	*Rivista italiana di musicologia*
SM	*Studia musicologica Academiae scientarium hungaricae*
SMA	*Studies in Music* [Australia]
SMz	*Schweizerische Musikzeitung/Revue musicale suisse*
SovM	*Sovetskaya muzika*
ZIMG	*Zeitschrift der Internationalen Musik-Gesellschaft*

Preface

This volume is one of a series of collections of biographies derived from *The New Grove Dictionary of Music and Musicians* (London, 1980). In its original form, the text was written in the mid-1970s, and finalized at the end of that decade. For this reprint, the texts have been re-read and modified. That on Schumann has been amended by its original authors; that on Liszt (originally by the late Humphrey Searle) has been substantially revised by Sharon Winklhofer; and in the case of Chopin the biographical section has been freshly written by Nicholas Temperley, author of the original musical discussion. In particular, an effort has been made to bring the bibliographies up to date and to incorporate the findings of recent research.

The fact that the texts of the books in this series originated as dictionary articles inevitably gives them a character somewhat different from that of books conceived as such. They are designed, first of all, to accommodate a very great deal of information in a manner that makes reference quick and easy. Their first concern is with fact rather than opinion, and this leads to a larger than usual proportion of the texts being devoted to biography than to critical discussion. The nature of a reference work gives it a particular obligation to convey received knowledge and to treat of composers' lives and works in an encyclopedic fashion, with proper acknowledgment of sources and due care to reflect different standpoints, rather than to embody imaginative or speculative writing about a

composer's character or his music. It is hoped that the comprehensive work-lists and extended bibliographies, indicative of the origins of the books in a reference work, will be valuable to the reader who is eager for full and accurate reference information and who may not have ready access to *The New Grove Dictionary* or who may prefer to have it in this more compact form.

S.S.

FRYDERYK CHOPIN

Nicholas Temperley

CHAPTER ONE

Birth and ancestry

The date now generally accepted for Chopin's birth, 1 March 1810, is based chiefly on his own assertions and the traditions of his family; it lacks documentary confirmation. The baptismal certificate states that he was born a week earlier, on 22 February. He was baptized on 23 April as Fryderyk Franciszek, but later used the French form of his first name.

His father, Nicolas Chopin (*b* Marainville, Vosges, 15 April 1771; *d* Warsaw, 3 May 1844) was the son of a vine-grower and wheelwright, François Chopin (*b* 1738), who came of a well-established peasant family. Nicolas departed from the locality and occupation of his forebears, probably because he was offered a chance to improve his prospects. A Polish count, Michal Pac, owned land at Marainville, which was managed by his Polish steward, Jan Adam Weydlich. When Weydlich offered to take Nicolas back to Poland with him in 1787, the boy accepted. He may have first intended only a temporary visit; the one surviving letter to his parents, dated 15 September 1790, shows that he kept away from France to avoid conscription in the Revolutionary army. From this time, however, he seems to have cut himself off completely from his family, leaving them behind as he rose in the social scale. When his son later came to be an eminent Parisian, he made no attempt to contact his humble French relations; he may not even have

1

known that he had two aunts living at Marainville.

Nicolas's first job in Poland was that of clerk in a tobacco factory run by a Frenchman. The firm went bankrupt during the revolt that led to the final partition of Poland (1794–5). Nicolas joined the Polish Guard and rose to the rank of captain. Later he became tutor to the children of aristocratic Polish families, where knowledge of French was a much prized accomplishment. In 1802 he was engaged as tutor to the children of Count Skarbek in Żelazowa Wola, near Warsaw. While in this employment he met Tekla-Justyna Krzyżanowska (*b* 1782), a relative of the Skarbeks, whom he married on 2 June 1806. Their two elder children, Ludwika (1807) and Fryderyk, were born there. Shortly after Fryderyk's birth the family moved to Warsaw, where Nicolas secured a position as teacher of French language and literature in the new high school. Two more daughters were born in Warsaw: Isabella in 1811 and Emilia in 1813. Emilia died of consumption when she was 14. The other two sisters married and Ludwika had children – three boys and a girl.

It is not clear how Nicolas acquired enough education to teach French literature in a Warsaw high school, even if the standards there were not very high. He may well have learnt from his wife, whose social and educational background were higher than his own. This may also have been the main reason for his ready adoption of Polish language, culture and patriotism. Justyna brought him little wealth and he had to supplement his modest income with additional teaching. But she did supply a certain distinction and knowledge of the social world, as well as a good education. She was able to pass on these attributes to her husband and, more thor-

oughly, to her son. Because of Nicolas's wholehearted adoption of Polish nationality, the dual national inheritance was not a source of conflict but proved a most happy and fertile union in the son's career.

After the defeat of Napoleon, the Chopin family came to terms with the fact that Warsaw was to be permanently under the oppressive rule of Russia. The liberal constitution proclaimed by the tsar on 24 May 1815 was soon undermined. Yet they could not openly oppose the authorities, whose goodwill was needed for Chopin's future success.

CHAPTER TWO

Childhood and youth, 1809–31

Chopin was sensitive to music from early years; he would often dissolve in tears when his mother played the piano or sang to him. It soon became clear that he was unusually gifted, and as early as 1816 his parents sent him to Wojciech (Adalbert) Żywny for music lessons. Żywny was a violinist of moderate accomplishments who gave Chopin a thorough grounding in the music of J. S. Bach (then a relatively unknown figure) and the Viennese Classics. Chopin seems to have taught himself how to play the piano. In these early years Żywny would write down the boy's keyboard improvisations for him. One of them, a polonaise in G minor, was even published in 1817. It was dedicated to Countess Wiktoria Skarbek, one of his aristocratic cousins and the daughter of his godparents. This perhaps indicates Chopin's entry, even at this tender age, into the world of the aristocratic salon, where French, his father's native tongue, was the language of cultivation. On 23 February 1818 he had made his first appearance in public, playing a concerto by Gyrowetz at a fashionable charity concert. The Grand Duke Konstantyn Pavlovich, brother of the tsar and military commander of Poland, began to patronize Chopin, inviting him to play at the palace and sending his carriage to fetch him. In 1817 he had a march by Chopin orchestrated and played by his band.

Meanwhile the boy's general education had not been neglected, and from 1823 to 1826 he attended the high school where his father taught. He began to take lessons with Józef Elsner (1769–1854), a Silesian composer who was the founder and first director of the Warsaw Conservatory. It was Elsner who taught him how to write out his own compositions, and his Rondo op.1 was published in June 1825. The tsar, when visiting Warsaw to open the nominal Polish Diet on 13 May 1825, heard him perform on a new keyboard instrument, the aeromelodicon, and gave him a diamond ring.

Chopin's health was already delicate, for he wrote to Jan Białobłocki on 2 October 1826:

Wouldn't it be the height of absurdity to sit still for six hours a day when the doctors, German and Polish, have told me to walk as much as possible? . . . I go to bed at nine o'clock. Tea parties, soirées, and dances are all forbidden on Malcz's orders. I drink emetic waters, and feed myself on oatmeal – like a horse.

He obtained his high-school diploma on 27 July 1826. After a summer holiday at the Silesian spa of Duszniki (Reinertz) he entered the conservatory as a full-time student, continuing his studies with Elsner, from whom he took two years in harmony and counterpoint and one year of composition exercises. One result was his Sonata op.4, written in 1828; but it is a stiff and uncharacteristic piece, and clearly his real interest was in the many polonaises and mazurkas he had already completed, or in such works as the *Rondo à la Mazur* op.5, and the Nocturne published as op.72 no.1. Elsner was not a martinet, however, and must be given credit for having recognized the exceptional qualities of his pupil. He wrote on one of his reports: 'Extraordinary ability. Musical genius'.

1. Fryderyk Chopin: drawing (1826) by E. Radziwiłł

Chopin's summer holidays were generally spent at resorts in Poland, but in 1828 a zoology professor, Dr Jarocki, who was a friend of his father's, offered to take him to Berlin, where he was going to attend a scientific congress. They set off on the five-day coach journey on 9 September and stayed for two weeks at the Kronprinz Hotel. Chopin saw Spontini and Mendelssohn, but was too shy to introduce himself. He heard music that he could not hear in Warsaw: although this included Weber's *Freischütz*, Chopin wrote that the work that 'came nearest to the idea I have formed of great music' was Handel's *Ode for St Cecilia's Day*. On the way home he and Jarocki stayed with Prince Antonin Radziwiłł, a cellist, who tried out the newly written first movement of the Trio op.8, later dedicated to him. It was in 1828 too that he composed the Rondos opp.14 and 73 and the first two studies of op.10.

Warsaw under Russian rule was a provincial place. In spite of visits from Hummel and Paganini, it offered few opportunities for Chopin to hear up-to-date music well performed, or – more important – to be heard by leading musicians and patrons. After completing his final exams at the conservatory with distinction, he set off with three friends for Vienna in July 1829; the travel money was presumably found by his father, for the government had turned down an application for a travelling scholarship. A prime purpose in Vienna was to see his publisher, Tobias Haslinger, about the publication of the Variations op.2, the sonata and other early works. Haslinger was instrumental in arranging for Chopin to give two concerts at the Kärntnertor-Theater on 11 and 18 August. He played the variations, but the greatest successes were the 'exotic' pieces: his im-

provisation on the Polish wedding song *Chmiel* and his concert rondo *Krakowiak*, op.14 (postponed until the second concert because the orchestral parts were not ready). Reviews were enthusiastic: he was described as 'a master of the first rank, with exquisite delicacy of touch, unequalled dexterity of finger, and deep feeling shown by his command of expression'. The only criticism of his playing was that it was too delicate for a large audience and did not sufficiently accent the first note of each phrase. Chopin left Vienna on 19 August and returned to Warsaw via Prague and Dresden.

He was increasingly restless at home, but political troubles kept him in Warsaw for over a year, apart from short visits to friends. On 17 March 1830 he gave his first public concert in Warsaw, where his F minor Piano Concerto and Fantasia on Polish Airs op.13 scored a resounding success: the concert was repeated five days later, with *Krakowiak* replacing the fantasia. At the first concert he played his own piano but at the second he was persuaded to use a more powerful Graf instrument, which he did not like. The notices show that Chopin was beginning to be regarded as a national figure. The nationalist colouring of his style was appreciated in Warsaw as much as in Vienna, if for different reasons. He gave two more concerts in Warsaw: at the final one, on 11 October, he played his Concerto in E minor, which he had been composing during the preceding months.

On 2 November 1830 Chopin left Warsaw – as it turned out, for ever. Accompanied by his friend Tytus Woyciechowski, having made intermediate stops at Breslau and Dresden, he arrived at Vienna on 23 November for an eight-month stay which was one of

the most frustrating periods of his career. His two concerts, on 4 April and 11 June 1831, made little impression, and Haslinger would not publish any more of his music. He soon became very short of money. Above all, he was depressed by increasingly bad news from home. A revolt against Russian rule had broken out shortly after his departure; Woyciechowski promptly returned to Poland. The Poles, who had a well-equipped army, at first made headway, and Grand Duke Konstantyn was forced to leave the country; but the Russians soon prevailed and reduced Poland to the status of a Russian province, with the object of stamping out all national feeling. The other European powers ignored Polish appeals for help. Chopin's extreme bitterness was understandable and was coupled with anxiety for his family. These feelings were expressed in his despairing letters, and (some would say) in the mood of some of the compositions of this period, such as the G minor Ballade, the B minor Scherzo and the C minor Study op.10 no.12. Other compositions of this fertile period include the Mazurkas opp.6 and 7, the Grand Polonaise op.22, waltzes and songs.

Chopin's move to Paris was an understandable one both in personal and political terms. Vienna had turned the cold shoulder on this second visit; he was conscious of his French heritage; Napoleonic France and Poland had been natural allies against Russia and the other reactionary powers, and he had hopes that the new French government would come to Poland's assistance. The Russian ambassador in Vienna would not allow travel to France, which had harboured so many Polish exiles, so Chopin said he was going to London and had his passport stamped 'Munich', where he did in fact

stay, playing his E minor Concerto on 28 August. A week later, the news reached him in Stuttgart that Warsaw had fallen to the Russians, and he poured out his extreme distress in a short-lived 'journal'. Pressing on, he arrived at Paris in the third week of September.

CHAPTER THREE

Personal characteristics

Chopin's character was well formed when, at the age of 21, he left his native country for the last time; and it did not change in any fundamental way during the rest of his life. He possessed exceptional inner self-confidence, which allowed him to maintain a quiet and modest demeanour in his dealings with others. His early training prepared him well for social intercourse with the upper classes of European society. His appearance was distinguished – he was always correctly and fashionably dressed – and his manner was impeccably polite and reserved. Few of his acquaintances were privileged to know his real feelings on sensitive matters. In letters to his Polish contemporaries, he revealed strong views on many subjects, often expressed with a degree of poetic feeling; and in describing his musical experiences he was humorous and realistic about his own strengths and weaknesses and about his effect on others. He showed himself to be tactful, considerate and generous in spirit.

Chopin's emotional life has been the subject of endless speculation by romantic biographers. He was certainly attractive to women, but his ability to reciprocate in a physical sense must be regarded as doubtful. The letters he wrote to Tytus Woyciechowski contain unequivocal expressions of homosexual feeling. Many such passages were excised by Hedley in his *Selected Correspondence*,

but they can be read in Sydow's edition of the letters. At the same time, Chopin was toying with undeclared love for the singer Constance Gladkowska, which he confided to Tytus in a letter of 3 October 1829. But later (27 March 1830) he told Tytus he would send him his portrait: 'You want it, you will have it, and no-one besides you, with the exception of one person [presumably Constance] – and not before you, because you are dearest to me'. Publicly he let it be known that he was attached to a childhood friend, Alexandrine de Moriolles, for, as he explained to Woyciechowski, 'one must conform and keep one's real feelings hidden. I would never have thought myself capable of dissembling so far when my heart is not inclined to reveal what is tormenting it' (15 May 1830). On another occasion he wrote that it was only in letters that he could 'give free reign to his exaltation'; but he would have to close, because Mlle de Moriolles was waiting to see him: 'I must confess to you that I have often allowed people to think that she is the cause of my melancholy. . . Many think so, and I am content, at least on the surface' (4 September). He ended the same letter: 'Don't embrace me, because I haven't yet washed! You? Even if I was imbued with Byzantine perfumes, you would only embrace me if I compelled you by magnetism; [but] there are natural forces. To-day you will dream that I embrace you! I must avenge myself for the terrible dream that you gave me last night'. The meaning of these passages seems clear. Their anguished tone is quite distinct from the romantic warmth that was more generally adopted between young men in this period. They show a passionate nature rarely revealed to any of Chopin's everyday acquaintances.

Personal characteristics

Politically Chopin was no radical. Not only his musical career, but the rapid upward mobility of his family, completed in his own person, required acceptance of the existing social order. The Polish nationalists who agitated for greater independence made no attempt to alleviate the terrible plight of their own peasant population, and we must assume that Chopin's views coincided with theirs. There is no evidence that he was actively engaged in the nationalist movement and when the insurrection came in 1830 he accepted the advice of his mother and Woyciechowski by staying away from Poland. But his romantic nostalgia for his homeland and grief for its sufferings grew to tragic proportions during his long exile. He tended to remain aloof from the French Revolution of 1830 and he was at first disgusted with France for failing to help Poland; but he quickly settled into a happy relationship with French society under Louis-Philippe, which in fact suited him very well with its bourgeois capitalism and relative political freedom. He seems to have had little or no interest in religion.

CHAPTER FOUR

Paris, 1831–8

Arriving in the French capital in September 1831, Chopin established himself in rooms on the fifth floor of no.27 boulevard Poissonnière, and lost little time in presenting a letter of recommendation from Beethoven's physician, Dr Malfatti, to the composer Ferdinando Paer, who in turn introduced him to Rossini, Cherubini, Kalkbrenner and other leading figures of Parisian musical life. The Romantic movement had recently burst upon Paris with a sudden intensity. But Chopin was unsympathetic to the music of Liszt and Berlioz, and considered that only Kalkbrenner, of all the virtuoso pianists who were dazzling Parisian society, was worth his emulation. He even contemplated taking a three-year course of piano lessons from Kalkbrenner but gave up the idea when his friends impressed on him that he was already a better pianist than his proposed master.

Another important introduction was to Camille Pleyel, one of the leading publishers and piano manufacturers. Chopin soon came to prefer Pleyel's pianos to all others. It was Kalkbrenner and Pleyel who helped to arrange Chopin's first concert in Paris, in the Salle Pleyel on 26 February 1832. Although not a financial success, it was enthusiastically praised by Liszt – always conspicuously generous to other musicians – and by Fétis in the *Revue musicale*. Mendelssohn was also present and added his approval. Meanwhile Schumann's

famous recognition of Chopin's genius had appeared in the *Allgemeine musikalische Zeitung* in December 1831, in a review of the published Variations op.2. Within a few months of his arrival in Paris, therefore, Chopin had reached the heights of the musical landscape. It is no wonder that, in spite of financial difficulties and growing anxiety about his family and friends at home, he felt exhilarated by his new life.

He made few appearances as a pianist in these years, playing in only about two concerts a year and never (after the first time) as principal performer. On the other hand, he was dissatisfied with the reactions to his playing (as he made ferociously clear in some of his letters home) and found that he could not compete in the eyes of the larger public with spectacular virtuosos like Herz, Thalberg and Liszt. However, he learnt that it was possible to acquire both fame and fortune without appearing in public. His early supporters had been largely drawn from the growing throng of Polish émigrés in Paris, leavened by a few fellow musicians. But an early meeting with Prince Valentin Radziwiłł led to an introduction to the Baron and Baroness James de Rothschild and his entry into high society. In the climate of the 1830s this was a mixture of older aristocratic families and *nouveaux riches*. 'I sit with ambassadors, princes, ministers', he wrote to a Polish friend in 1832, 'and don't even know how it happened, because I did not try for it.' He was able to charge the highest fees for the lessons he gave to society ladies. He moved to more spacious apartments, employed a manservant and formed friendships on terms of equality with such romantic hostesses as Princess Belgiojoso and Countess Delfina Potocka. His relationship with Mme Potocka was a very warm

2. *Programme for Chopin's first Paris concert: it was postponed because of Kalkbrenner's illness and finally held on 26 February 1832*

one. The erotic letters he is said to have written to her may or may not be authentic; the originals have never been produced for scholarly scrutiny.

Chopin had reached a social milieu in which his financial security was assured. The adulation he attracted would have satisfied the vainest of men – which he certainly was not. He had won this position quite as much by his personal qualities as by his musical genius. Few of the lady pupils who admired him can have understood the unique originality of his gifts, but they all appreciated his romantic looks, his beautifully sensitive playing and his sympathetic, courteous manner. There was an added piquancy attaching to an exile from a troubled and distant country who kept his feelings hidden behind a mask of *politesse*. The idea of a composer who was above and beyond the pressures of the crowd was one that many Romantics aspired to; Chopin achieved it with little effort. As the *Musical World* grudgingly admitted:

it is impossible to deny that he occupies a foremost place among the piano-forte composers of the present day. . . In Paris. . .his admirers regard him as a species of musical Wordsworth, inasmuch as he scorns popularity and writes entirely up to his own standard of excellence.

This is not entirely true, however, of the compositions of the early 1830s, for they were mostly on the popular side, coming as near as he ever did to the showy virtuosity that was the rage in Paris at the time (the Rondo op.16, the Grand Duo on themes from *Robert le diable*, the Variations op.12 and the Bolero). His other publications were mostly of music he had brought with him from Poland: op.9 was dedicated to Marie Pleyel, op.10 to Liszt, op.11 to Kalkbrenner. He began a third con-

17

certo (the *Allegro de concert* op.46) and completed the
G minor Ballade.

But as the pattern of Chopin's career in Paris became
clear (teaching and playing in private salons rather than
in public) he developed a style of composition more in
line with it. He gave up writing concertos and concert
pieces. He resumed the composition of teaching pieces:
three of the Studies op.10 and all those of op.25 date
from 1832–6 and the Preludes op.28 were begun in 1836;
and he also returned to mazurkas, nocturnes and the
occasional waltz.

An old Polish friend of Chopin's and the recipient of
many of his letters was Jan Matuszynski. In spring 1834
he went to Paris as a medical student and Chopin gave
him lodgings for two years. In May 1834 Chopin set off
with Ferdinand Hiller on a German tour that included
the Aachen Festival, Cologne, Koblenz and Düsseldorf;
at Aachen he met Mendelssohn. In July 1835 he was the
guest of the Marquis de Custine at Enghien, and the
following month he met his parents for the last time at
Karlsbad. After they had left to return to Poland on 14
September, Chopin made his way to Dresden, where he
joined the Wodziński family, whose three sons had been
old schoolfriends of his. One of their sisters, Maria, now
16, was much attracted to Chopin and he was greatly
touched by her youth and beauty. He gave her his Waltz
'L'adieu' (op.69 no.1) shortly before his departure. At
Leipzig, his next stopping place, Chopin was introduced
by Mendelssohn to Schumann and to Clara Wieck,
whom he regarded as 'the only woman in Germany who
can play my music'. Next he travelled to Heidelberg to
stay with his 19-year-old pupil Adolf Gutmann. Becoming ill, he tried to conceal the fact from his parents, but

rumours of his death began to appear in the Warsaw newspapers.

He returned to the pursuit of Maria Wodzińska the following summer, visiting her with her family at Marienbad and returning with them to Dresden. He proposed marriage on 7 September 1836, two days before going to Leipzig for another visit to Schumann. The Countess Wodzińska, Maria's mother, continued to write to Chopin for some time, Maria adding rather coquettish postscripts; but opposition to the marriage on the part of the Wodziński family gradually strengthened, perhaps on the grounds of Chopin's health. The engagement was broken off in summer 1837. Chopin was saddened by this episode: a bundle of letters from Maria and her mother was found after his death in an envelope marked 'Moja Biéda' ('my sorrow'). It was not long, however, before his relationship with George Sand had absorbed his emotional energies.

In July 1837 Chopin travelled with Pleyel to London, where he stayed privately for two weeks, visiting his London publishers, Wessel & Co., and the piano manufacturer John Broadwood, who induced him to play to a small gathering.

CHAPTER FIVE

Years with George Sand, 1838–47

Late in 1836 Chopin was introduced by Liszt to the novelist Baroness Aurore Dudevant (née Dupin), whose pen name was George Sand (1804–76). She had been brought up as a boy and was masculine in appearance and manner, but had nevertheless had numerous lovers; she left her husband, Baron Casimir Dudevant, in 1831, after having two children by him, Maurice and Solange. In 1821 she had inherited from her grandmother an estate and château at Nohant, in Berry. She was a woman of extraordinary character and independence and wrote voluminously about Chopin in her autobiography, letters and the novel *Lucrezia Floriani*; but her testimony is quite untrustworthy on many points as she was an assiduous cultivator of her own image and reputation.

Sand was strongly attracted to Chopin and in 1837 invited him to Nohant, but Chopin did not at first find her appealing and refused. In a long letter (? June 1838) to Adalbert Grzymala, a friend of Chopin's, she spoke of their wonderful relationship, but complained that Chopin persisted in his refusal to have full sexual relations. Shortly after this, however, it seems that she succeeded in seducing him, and they had a short but passionate affair. Chopin himself was most reticent about the affair and hardly mentioned it in his letters to

friends. It was much discussed by others, notably in the circle of Liszt and the Countess d'Agoult; there was an abundance of gossip, much of it coloured by jealousy.

After spending summer 1838 together in Paris, Chopin and Sand left for Palma, Majorca, where they arrived on 8 November. The purpose of the visit was to take Sand's 15-year-old son Maurice to a warmer climate to cure his rheumatic fever. Chopin told only a few close friends that he was going to Majorca with her and hid the news from his parents. He took with him some scores of Bach; the Preludes op.28, on which he was working; and large amounts of paper, pens and ink. He sent the finished manuscript of the Preludes to Pleyel in Paris in January 1839. At about that time the piano he had ordered before departure arrived at Valldemosa, the monastery to which the travellers had been obliged to move on 15 December. Shortly after this Chopin, who had been suffering for some years from latent tuberculosis, became seriously ill, and they left Majorca on 13 February. It had been a disastrous trip from Chopin's point of view; but his illness seems to have brought out some of George Sand's best qualities. Their relationship cooled rapidly to one of warm friendship, with a strong element of the maternal on her side. She spoke of him almost condescendingly and treated him as a pampered pet. Her nickname for him was 'Chip-Chip'.

The doctor Chopin consulted at Marseilles insisted that he must stay in a southern climate, so he and Sand remained at Marseilles for some three months, with one expedition to Genoa. Then they went to Nohant late in May 1839. Chopin, who was seeing the place for the first time, was captivated; it was the only country house in which he ever made a permanent home. He had his

own bedroom with a piano and with a beautiful view of the peaceful countryside; and there in tranquillity he composed much of his best music. On this first visit he soon recovered his strength and resumed correspondence. In October he returned to Paris, taking up new lodgings at 5 rue Tronchet. Sand now decided to live in Paris too, but in a separate house. In October 1841, however, Chopin moved to the smaller of the two houses she had taken in rue Pigalle.

The early 1840s was a happy and productive period in Chopin's life in spite of his physical deterioration. His time was divided between teaching and composing. He was emphatically not a member of the 'Parisian School' of virtuosos, who vied with one another in feats of spectacular pianism. He stood aloof and above the contest, becoming an almost legendary figure whom few could approach. Charles Hallé expressed his sense of privilege and awe when he secured an introduction to one of the 'réunions intimes' at which Chopin played:

With greater familiarity my admiration increased, for I learned to appreciate what before had principally dazzled me. In personal appearance he was also most striking, his clear-cut features, diaphanous complexion, beautiful brown waving hair, the fragility of his frame, his aristocratic bearing, and his princelike manners, singling him out, and making one feel in the presence of a superior man... It is impossible at the present day [c1890], when Chopin's music has become the property of every schoolgirl, when there is hardly a concert-programme without his name, to realise the impression which these works produced upon musicians when they first appeared, and especially when they were played by himself. I can confidently assert that nobody has ever been able to reproduce them as they sounded under his magical fingers. In listening to him you lost all power of analysis; you did not for a moment think how perfect was his execution of this or that difficulty; you listened, as it were, to the improvisation of a poem and were under the charm as long as it lasted.

One of his few public concerts at this time was at the

3. Fryderyk Chopin: sketch (1841) by George Sand, destroyed in World War II

Salle Pleyel on 26 April 1841, when he played the two Polonaises op.40, the Ballade op.38, the third scherzo and several preludes. According to Sand's letter of 18 April, he only agreed to give the concert because his friends had plagued him; he hoped that it would prove impossible to arrange. But

scarcely had he uttered the fatal *yes* than everything was settled as if by a miracle, and three-quarters of the tickets were taken before it was even announced. Then he awoke as if from a dream; and there is nothing more amusing than to see our meticulous and irresolute Chip-Chip compelled to keep his promise. He hoped you [Pauline Viardot-Garcia] would come and sing for him. When I received your letter and he could no longer hope for that, he wanted to cancel his concert. But there was no way – he had gone too far. . . He doesn't want posters, or programmes, or a large audience.

It would be difficult to imagine an attitude to performing publicly more diametrically opposed to that of Liszt.

The concert was widely acclaimed; *La France musicale* compared Chopin with Schubert: 'One has done for the pianoforte what the other has done for the voice. . . Chopin is a pianist apart, who should not and cannot be compared with anyone'. An even more successful concert was given at the Salle Pleyel in February 1842 when Chopin introduced the third Ballade, the Impromptu op.51, three mazurkas and four nocturnes. He had also performed with Moscheles before the French royal family in October 1839. In general, however, he clearly preferred the plutocratic salons of the Rothschilds and the Stockhausens, where he was called on to repeat his most popular pieces, to improvise and even to give comic shows in which his talents as a mimic were much admired. He was reproached for this curious preference by Polish compatriots, including the great

poet Adam Mickiewicz (1798–1855), a fellow exile in Paris. But he seems to have known what was best for himself. There is no escaping the fact that this period of pleasing the wealthy, teaching their daughters and retiring to Nohant for intervals of poetic seclusion was the one in which most of his greatest masterpieces were composed.

In September 1842 Chopin and Sand moved to luxurious lodgings at 9 square d'Orléans where he remained for seven years. The death of his father on 3 May 1844 caused him deep and lasting sadness. It was followed in August by a visit from his sister Ludwika and her husband Josef Jędrzejewicz; they were entertained at Nohant by George Sand, who wrote in advance to Ludwika:

You will find my dear child quite puny and changed since you last saw him! But do not be too alarmed about his health. It has remained much the same for the last six years, during which time I have seen him every day. A rather violent fit of coughing every morning; two or three more serious attacks only in winter, each lasting only two or three days; some spells of neuralgia from time to time: such is his usual condition. . .I am sure that with a regular life and care, [his constitution] will last as long as anyone else's.

In fact, Chopin was able to rally himself to better health many times during these years and had long spells of successful creative effort.

Chopin had no difficulty now in having his music published on excellent terms, but his relations with publishers were not easy. Because of the absence of effective international copyright the simultaneous publication of his works in Paris, London and Leipzig was necessary. He used various Paris publishers but he was obliged to keep the same firm in London and Leipzig, in spite of frequent causes of irritation. Wessel of London annoyed

him by adding silly titles to his works; Breitkopf of Leipzig failed to reply to his letters and paid scant attention to his wishes. An additional problem was the need to provide three copies of each new work. Chopin hated writing out his music. He employed copyists but was anxious to be sure they understood his wishes; whenever possible he used his old friend from Warsaw days, Julian Fontana (1810–65), and also frequently entrusted him with negotiations with his publishers. After Fontana left for America in 1841, other friends of Chopin, such as Grzymala or the cellist Auguste Franchomme, were saddled with this unenviable task: in this aspect of his career Chopin could be surprisingly ruthless. Hedley considered, however, that his fall in productivity after 1841 was a direct result of Fontana's departure.

Chopin's way of life with George Sand was brought to an end by a family quarrel that erupted in 1846. Sand's children and Augustine Brault, whom she adopted as a daughter in 1846, were increasingly resentful of Chopin and began to slight him and his friends. The crisis came over the affairs of Solange. She was engaged to marry a wealthy landowner, Fernand de Preaulx, and Chopin, who was fond of the girl and acted as her confidant, supported the match in spite of her mother's opposition. Under pressure from her mother, Solange broke off the match and instead married the sculptor Auguste Clésinger, by whom she was probably pregnant. The marriage took place at Nohant in May 1847 and Chopin was not told about it until too late. A great family quarrel was brewing at Nohant: Solange's affairs were only one element. Sand invited Chopin there but he refused, probably for a variety of reasons: he

4. *Chopin giving Pauline Viardot a piano lesson at Nohant: caricature (1844) by Maurice Sand; although Madame Viardot was a frequent visitor to Nohant, there is no record of a visit that year*

was bored with the relationship; he was hurt by his treatment at her hands; he wished to avoid further involvement in the rather sordid quarrels of the Dudevant family. At any rate the breach with Sand was permanent and was sealed by her famous letter of adieu, probably written near the end of July 1847.

CHAPTER SIX

Last years, 1847–9

Chopin's career as a composer was now virtually at an end. Loss of the tranquil atmosphere of the earlier years at Nohant, combined with rapidly deteriorating health, took away the will to create large-scale compositions. His last major work, the Sonata for cello and piano, had been sent to the publishers before the final separation. Apart from a song and a few mazurkas, he wrote nothing more.

Remarkably, however, there was a brief revival of his activity as a concert pianist. On 16 February 1848 he gave his first public concert for six years at the Salle Pleyel, playing the Cello Sonata with Franchomme, the Berceuse, the Barcarolle and Mozart's E major Piano Trio. The hall was packed and the critics enthralled, though Chopin himself thought that he had played 'worse than ever'.

A week later, revolution broke out in Paris, bringing Chopin's teaching to a halt. He now took up a long-standing invitation from his devoted Scottish pupil, Jane Stirling, to visit Britain. With great courage he gave several public concerts in London and submitted to being dragged round by Stirling and her sister, Mrs Erskine, to meet leading ladies of London society. Queen Victoria heard him play at a party given by the Duchess of Sutherland. After the season he went to Scotland for three months and was allowed some rest

5. Fryderyk Chopin: daguerreotype, 1849

and recuperation. He travelled to Edinburgh on the recently completed railway line from Euston and stayed in some luxury at several Scottish castles. He played at Manchester, Glasgow and Edinburgh. Writing to Grzymala from Keir House on 1 October he praised his view of Stirling Castle but lamented his poor health. Jane Stirling was in love with him, but he was totally bored by her constant attentions, though he acknowledged her kindness. By the time he returned to London his relations with the Scottish ladies had become strained. He stayed in bed for two weeks, rallied for one last concert at the Guildhall in aid of Polish émigrés (16 November) and left for Paris on 23 November.

The last year of his life was a melancholy time indeed. His two sources of income, teaching and composing, were now beyond his powers and he would have had to make a drastic change in his luxurious style of living if Jane Stirling had not come to the rescue with a generous gift of money. On doctors' advice he spent the summer at Chaillot, on the heights of Passy, where he was visited by Jenny Lind. He sent for his sister Ludwika, who arrived with her husband and daughter on 9 August 1849 and nursed him to the end. Tytus Woyciechowski tried to reach him and got as far as Ostend, but Russian subjects were not allowed into France. At the end of September, Chopin moved to 12 place Vendôme, where he died on 17 October after taking the last sacraments from his friend Alexander Jelowicki. It is generally supposed that tuberculosis was the cause of his death at the age of 39.

The funeral took place at the Madeleine on 30 October and was attended by nearly 3000 people. Pauline Viardot and Luigi Lablache took part in a performance

of Mozart's Requiem, and Chopin's own funeral march from the B flat minor Sonata was played, for the first time, in an orchestral version by Napoléon-Henri Reber. Chopin was buried at the cemetery of Père-Lachaise, where a monument by Clésinger was unveiled on 17 October 1850. A posthumous edition of unpublished works (op.66–77) was brought out by Fontana in 1855 and 1859.

CHAPTER SEVEN

Approach to composition

Most composers since the 17th century have been keyboard musicians, but few have devoted themselves so exclusively to the keyboard as Chopin. Every one of his compositions includes the piano: even in the 19th century, the heyday of the instrument, there are few others of whom this may be said (John Field is one). For Chopin there was a peculiarly intimate relation between playing and composing. Not for him the careful reworkings and revisions to establish a final inviolate text of his music. From all accounts he invariably composed at the piano, and carried the essence of the music in his head; at every performance there would be a slightly different version, so that there is no clear distinction with Chopin between improvisation and composition. As a contemporary, Karl Filtsch, put it in a letter of 8 March 1842:

The other day I heard Chopin improvise at George Sand's house. It is marvellous to hear Chopin compose in this way: his inspiration is so immediate and complete that he plays without hesitation as if it could not be otherwise. But when it comes to writing it down, and recapturing the original thought in all its details, he spends days of nervous strain and almost terrible despair.

Arguing from this and from Chopin's legendary image as an impulsive romantic artist, some critics have per-

suaded themselves that he was merely a musical dreamer, unable to concentrate his mind to the working out of large structures.

But improvisation is not an introspective art, nor is it a vague one. It is designed for an audience, and its starting-point is that audience's expectations, which include the current conventions of musical form. So it was with Chopin. His development as a composer can be seen as an ever more sophisticated improvisation on a fundamentally simple, classical formal principle – the principle of departure and return. That this was sufficient to support complex large-scale structures is abundantly demonstrated by such works as the ballades, the scherzos, the Barcarolle, the F minor Fantasie and the Polonaise-fantaisie, not to mention the sonatas in B♭ minor and B minor.

Chopin's output as a composer clearly reflects the circumstances of his life. For the concert-giving years 1828–32 he wrote his two concertos and other music for piano and orchestra (this was before the time of the solo piano recital) and such brilliant virtuoso pieces as the rondos and the Bolero. The Andante spianato was originally an introduction to the Polonaise in E♭ for piano and orchestra; for the *Allegro de concert*, another work in the 'concert' manner, an orchestral accompaniment was also planned. These works are conventional in plan. A prominent feature of most of them is the grand winding-up in the last pages, for which the most brilliant virtuosity is reserved.

The teaching side of Chopin's career is reflected not only in the pieces designed in part to satisfy a pedagogical purpose – the studies – but, even more, in the pieces composed for his pupils to play. These are

generally of only moderate difficulty, but make the most of an amateur's technical capacity both in the brilliance of rapid finger passages and in the possibilities for sentimental expression. They include the preludes, nocturnes, waltzes, impromptus and mazurkas, as well as a number of single pieces such as the Berceuse and the Tarantelle, and the earlier polonaises. Many of these pieces were dedicated to Chopin's pupils. Their style is shaped by the tastes of the aristrocratic Parisian salon, expressed with superb polish and inventive detail; in the mazurkas and polonaises a strong national colouring is added. All these pieces served in part to display and advertise Chopin's quality as a teacher – but as a teacher of musicianship and artistic expression as well as of technique.

In the third main category of his works Chopin wrote for himself, and for a small circle of like-minded musicians and admirers. Here also the mazurkas may be included. Their nationalistic colouring, though on the one hand it satisfied his pupils' taste for novelty and piquancy and nurtured their romantic image of him as an exiled patriot, was at the same time for Chopin himself a sincere expression of deeply felt emotion. Here, too, should be placed those large-scale works, too profound for the salon, in which Chopin stretched his musical invention to its greatest capacity. They include some of the later polonaises, the Polonaise-fantaisie, the F minor Fantasie, the scherzos, ballades and the Barcarolle, the sonatas in B♭ minor and B minor, and the Cello Sonata. With the exception of the sonatas, many of these pieces are probably examples of the kind of improvisation that evoked so much astonishment in Chopin's admirers: certainly he did not always play

them in the form in which they were eventually published.

Almost alone among the Romantics of his generation, Chopin showed little interest in linking his music to pictorial, literary or autobiographical subjects. Chopin himself apparently never attached a story to any of his works (though this did not inhibit his friends and biographers from foisting their own stories on the music, sometimes pretending that they had the composer's authority). Even the titles to his pieces rarely suggest anything beyond a musical genre, the Berceuse and the *Marche funèbre* from the Bb minor Sonata being perhaps the only exceptions. Such romantic titles as 'Raindrop Prelude', 'Revolutionary Study' and 'Winter Wind Study' are spurious.

On the other hand Chopin was obviously interested in evoking nationalistic feeling by means of the Polish dances, the polonaise and the mazurka. In his songs he went further by allying his music to Polish words. Some of them are like mazurkas, some are in other simple, folklike forms. Two, whose words bear direct reference to the sufferings of Poland (*Melodya* op.74 no.9, and *Śpiew grobowy* ('Hymn from the tomb') op.74 no.17), drew from Chopin some of his most poignantly emotional music; both are characterized by major–minor alternations, and both are through-composed.

CHAPTER EIGHT

Pianistic style

Chopin is generally admired above all for his great originality in the artistic exploitation of the piano. Though the instrument was more than a century old when he began to play and compose, it was undergoing a series of transformations in the early decades of the 19th century that produced what was essentially a new instrument. The 18th-century piano, with its hard hammers, had had a tone not far removed from that of the harpsichord, but the recent tendency had been to cover the hammers with more and more layers of felt or soft leather. The new, soft tone that resulted was a vehicle for a type of romantic expression that fascinated a number of composers of the time. Of these Field and Hummel had the most striking influence on Chopin.

The improvements made by rival manufacturers had produced wide variety: there was as yet no standard piano tone. Hummel in his piano method of 1828 described the difference between the two fundamental types of action, the English and the Viennese. A third type was the Erard action, with the new double escapement (invented in 1821) which allowed more rapid trills and repeated notes than before. According to Liszt, Chopin preferred Pleyel's pianos, basically of the older Viennese type, 'particularly on account of their silvery and somewhat veiled sonority, and ... easy touch'. His second choice was Erard, though he used a

37

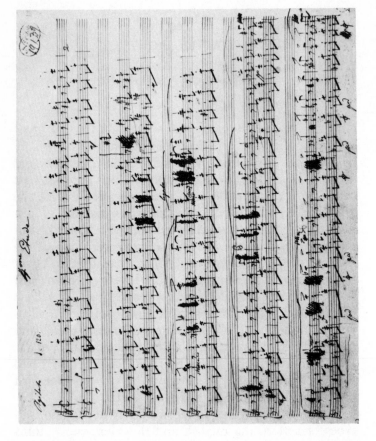

6. Autograph MS
of the opening of
Chopin's Study in
A minor op.25 no.4,
composed 1832–4

Broadwood for his concerts in England in 1848. In a period when spectacular and forceful playing was greatly admired, Chopin, though capable of great power when it was called for, gained most of his effects through a more subtle and restrained style of playing, as was suggested in the *Daily News* (10 July 1848):

He accomplishes enormous difficulties, but so quietly, so smoothly and with such constant delicacy and refinement that the listener is not sensible of their real magnitude. It is the exquisite delicacy, with the liquid mellowness of his tone, and the pearly roundness of his passages of rapid articulation which are the peculiar features of his execution.

Of his predecessors in Paris, Kalkbrenner was the pianist most noted for this type of playing, and he was the one whom Chopin particularly admired.

The fundamental texture of Chopin's music is accompanied melody, and much of its fascination comes from the limitless resource with which he varied it. Sometimes a tune is harmonized in plain chords, generally with a thickness of spacing unknown to earlier generations; sometimes the accompaniment carries the rhythm of a dance, but with many subtle variations. Often he varied the articulation and relative emphasis of the melody and accompaniment without changing the figure in use (Study op.25 no.4).

The 'nocturne' type of texture, clearly derived from Field, isolates a right-hand melody, allowing the fullest possible expression in its playing, while the left hand, assisted by the pedal, provides the entire rhythmic and harmonic background in broken chord accompaniments. The ancestor of this texture was the Alberti bass of Classical music, but in the hands of Field and Dussek it had grown from a routine method of sustaining har-

mony into an expressive counterfoil to the melody, having in itself some melodic interest. Chopin greatly expanded these possibilities, notably by regularly enlarging the compass of the left-hand figures well beyond the octave and by using the sustaining pedal. He was capable, indeed, of covering almost the entire six-and-a-half-octave range of his piano with a piece of melodic figuration that also generates a vast spread chord (ex.1). He did not mind letting the pedal catch

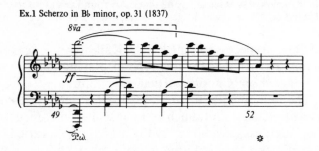

Ex.1 Scherzo in B♭ minor, op.31 (1837)

passing notes in the upper octaves, where, on the type of piano he preferred, the tone died away too quickly to blur the harmonies severely. Even in the tenor register some blurring was acceptable (Prelude op.28 no.3), so that in many cases the 'accompanying' figure is no longer merely a broken chord but a freely moving melodic pattern.

Although the 'nocturne' texture does not lend itself easily to textural inversion, Chopin developed for left-hand melodies a kind of setting possibly derived from Weber (e.g. *Aufforderung zum Tanze*). The melody is devised so as to need few bass notes beneath it, and can thus be played with maximum expressiveness: sometimes it acts as its own bass, sometimes the true bass is

understood or is lightly played by the little finger. The right hand plays repeated chords or chordal figures, sometimes including a counter-melody (ex.2), and the

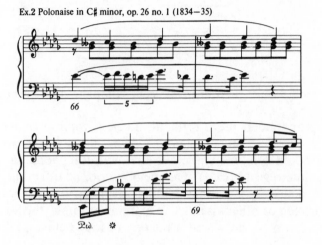

Ex.2 Polonaise in C♯ minor, op. 26 no. 1 (1834—35)

liberal use of the pedal helps to enrich the texture. A famous example of left-hand melody is the Study op.25 no.7; in the Prelude op.28 no.22 the left hand plays the tune in octaves throughout.

Many of Chopin's most beautiful effects derive from combining melodies of equal prominence and expressive content. Sometimes both tunes are clearly shown to be equal in the notation of the music (Nocturne op.55 no.2); more often, one of them emerges from the accompaniment figures (ex.3). Chopin was not at his best in formal counterpoint, though there is an effective strict canon here and there in the mazurkas; but in this kind of keyboard counterpoint he was supreme, bringing together in perfect union and balance two melodies

41

Ex.3 Prelude in F♯ major, op. 28 no. 13 (1836−39)

which would each in itself seem complete and spontaneous.

For louder, more brilliant effects Chopin invented a variety of accompaniments. By careful use of the pedal he could combine an octave bass with full chords in the middle register (ex.4). The melody, too, he often played in octaves, sometimes filling them in with chords (broken or otherwise). In an extreme example, the Study op.25 no.10, he scored a melody in four octaves, filling

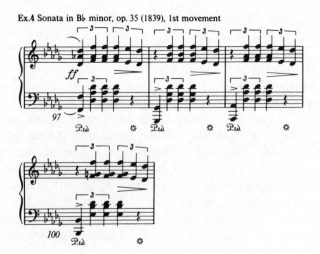

Ex.4 Sonata in B♭ minor, op. 35 (1839), 1st movement

in the harmonies as best he could. In a section of the Nocturne op.48 no.1, on the other hand, it is the accompaniment which is played in four octaves, while the melody is represented by an occasional full chord. By contrast, the last movement of the B♭ minor Sonata consists of entirely unaccompanied melody in two octaves, the harmonies being implied but never stated; similarly, on a smaller scale, does the Prelude op.28 no.11.

Chopin's most characteristic melodies can never be thought of without their accompaniment, which makes them pianistic. Nevertheless, they have much in common with the vocal melodies of contemporary Italian opera. The similarity of Chopin's nocturnes to Bellini's cavatinas (such as 'Casta diva' from *Norma*) has often been noticed, though there is little evidence of direct influence in either direction. The influence of Rossini's and Weber's operatic melodies is apparent. The structure of Chopin's melodies often relies heavily on the prevailing four- and eight-bar phrase, with conventional cadences on dominant and tonic. But there are many exceptions, including what some consider his greatest melody, that of the Study op.10 no.3, which is built in units five, three, five and seven and a half bars long. In the mazurkas there are many melodies of unusual construction, some imitating the 'modular' structure characteristic of the folk mazurka (ex.5). These melodies

Ex.5 Mazurka in G minor, op. 24 no. 1 (1834—35)

also use modal and chromatic scales foreign to Western art music of the time, and other features such as crushing notes and cross-rhythms in imitation of Polish folk music.

CHAPTER NINE

Harmony

Chopin is one of the great 19th-century harmonic in-
novators. Although, like every other aspect of his music,
his harmony is inseparable from the character of the
piano, it had a profound influence on later music of all
kinds, above all on Wagner's music dramas. On occasion
Chopin's harmonic innovations were revolutionary, but
the more important aspect of his originality was in
pushing the accepted procedures of chromatic dissonance and modulation into previously unexplored ter-
ritory. Novel harmonic effects frequently result from
the combination of ordinary appoggiaturas or passing
notes with 'melodic' figures of accompaniment of the
kind already described (Nocturne op.9 no.3, bars 1ff);
but instead of leaving such clashes to pass by as incidental details, Chopin frequently dwelt on them for expressive significance (ex.6). He often used chords am-
biguously, particularly the diminished 7th, with its
unlimited possibilities for modulation. He would often
delay a cadence by extending the Neapolitan 6th into a
passage in the key of the flat supertonic, returning by
way of an ambiguous diminished 7th (ex.7). Other
remote keys could be used in the same way as a colourful digression, which sometimes amounted to a whole
section in one of the dance forms, for example the B
major episode in the C minor Mazurka op.56 no.3.
Elsewhere he used similar remote modulations in a more

Ex.6 Etude in E♭ minor, op. 10 no. 6 (1830)

Ex.7 Nocturne in B♭ minor, op. 9 no. 1 (1830–31)

purposeful way, to intensify the return to the tonic; a good example is in the Study in C minor op.10 no.2, where the harmony passes through G♭, F♭ and E♭ in a climactic six-bar sequence before reaching a quiet cadence (with Neapolitan 6th) in C minor – here again the vehicle of modulation is the diminished 7th. Sometimes Chopin modulated deliberately through a variety of keys, so that little sense of overall tonality remained (Nocturne op.37 no.2). More typically, he would glide through a sequence of keys so rapidly that, although the modulations were technically complete, the overall effect was one of kaleidoscopic colour, with no real feeling of departure from the main key (ex.8). On

Ex.8 Nocturne in E♭ major, op. 9 no. 2 (1830–31)

occasion the process passes into a realm of pure chromaticism without any clear tonal base (ex.9). Such passages could indeed be called revolutionary. Their precedent is not to be found in Classical music, but in certain passages from the keyboard works of Bach (Partita no.6: Toccata, bars 106–8; Partita no.1: Gigue, bars 34–40). A chain of unresolved dominant 7ths seems to point straight to Debussy (ex.10). Sometimes in introductions Chopin stayed adrift from any clear

Ex.9 Etude in G♯ minor, op. 25 no. 6 (1832–4)

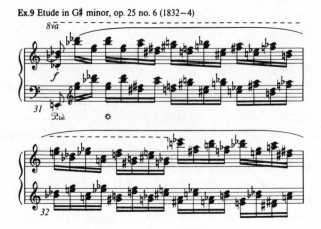

Ex.10 Mazurka in C♯ minor, op. 30 no. 4 (1836)

tonal centre. Tonal instability in slow introductions was a tradition of the Classical period, but Chopin went beyond it in the introductions to the G minor Ballade, the Polonaise-fantaisie, and above all in the first 24 bars of the C♯ minor Scherzo, where the listener is given no clue whatever to the main key of the work. Of course such radical experiments were only incidental.

Chopin extended harmonic thought in other, less spectacular ways. There was nothing new in a piece in a minor key ending in the major, but Chopin found new ways – different from Schubert's – of making this device emotionally intense, as in one of his best nocturnes (op.27 no.1) and one of his best mazurkas (op.24 no.4). In both cases the change to the tonic major is extended with new melodic material and reinforced by a surprising cadence. Chopin's cadences are conspicuously original. Only in conventional 'concert' pieces did he habitually use the plain perfect cadence. The plagal cadence was a favourite of his, as was a cadence that can be analysed as both perfect and plagal, with a diminished 7th on the leading note over a tonic pedal (ex.11).

Ex.11 Nocturne in E minor, op. 72 no. 1 (1827)

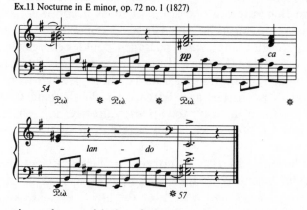

Sometimes he avoided a formal cadence altogether by leaving a melody unaccompanied (Mazurka op.24 no.4) or simply ending in the middle of a phrase (Mazurka op.41 no.4). And who can explain the final cadences of the B♭ minor Sonata, the Prelude op.28 no.11, or the mazurkas op.17 no. 4 and op.30 no.1? Or explain the

famous E♭ in the last chord of the F major Prelude of op.28?

Another innovation of great importance for the future was the 'harmonic daydream', as it might be called. The music seems to go into a daze, cut off from the world of reality – in musical terms, from the business of continuing the harmonic, thematic and structural development of the piece. Usually the harmony is completely static in these passages, or it repeats, with hypnotic monotony, a series of two or three chords. The earliest such daydream occurs in the Nocturne in B♭ minor op.9 no.1, before the return of the opening melody: there are 17 slow bars on the chord of D♭ major. Other examples are in the Prelude op.28 no.21, bars 17–32, the Study op.25 no.7, bars 29–38 and the Barcarolle, bars 78–82. The Berceuse may be thought of as a continuous daydream. In terms of the Classical conceptions of form and development such passages could be regarded as flaws or weaknesses, but they exercised important influence on the impressionists, through such works as Fauré's nocturnes and barcarolles. With them Chopin had shown a way to appeal directly to sensation, suspending for a while the stimulation of the listener's intellect.

Chopin was also a pioneer in the consistent use of modal harmony as the Romantics conceived it. In certain passages he would deliberately forsake all the richness of chromatic dissonance for a harmonic scheme consisting entirely of common chords in root position and first inversion. The Nocturne op.15 no.3 contains a 32-bar passage of this sort, marked 'religioso', as the middle section in what at first appears to be ternary form. To enhance the 'purity' of this passage Chopin deliberately refrained from using the

sustaining pedal. The expected return to the opening, however, is replaced by a new idea, also somewhat modal in character. This seems to approach a cadence in D minor, but the concluding chords bring the music back to G, with an archaic 4–3 suspension and Picardy 3rd. Similar excursions into modal purity are found in the Nocturne op.37 no.1 and the Prelude op.28 no.2. The harmonies are very similar to those that were recommended for accompanying plainsong by French church musicians from Niedermeyer onwards, and were used in the religious scenes in many Romantic operas. It is doubtful whether any consistent example of such harmony can be found of earlier date than op.15 no.3 (1833), unless the third movement, 'in the Lydian mode', of Beethoven's String Quartet op.132 is included.

Modality also characterizes many of Chopin's mazurkas, and here it is especially the sharpened 4th of the Lydian mode that is used to evoke the idiom of the Polish folkdance. There are many examples of its use, the earliest being in op.68 no.2 (1827). At times it is treated merely as incidental melodic colour, with the harmonies of the ordinary major or minor scale – the latter producing a colourful augmented 2nd. Elsewhere bucolic drone 5ths are found in imitation of the bagpipe (ex.12). Flattened 2nds and 7ths are also used from time

Ex.12 Mazurka in F major, op.68 no.3 (1829)

to time. A sustained effort at modal harmony is found in op.24 no.2, one of Chopin's most original experiments. Clinging rigidly to the white notes, he nevertheless contrived to effect 'modulations' in and out of the 'keys' of C major, G major, A minor and F major – the latter with, of course, a sharpened 4th. As if in impatience after two pages of this, he plunged into D♭ for the middle section, but returned to strict modality at the close.

The modal principle is extended, in several mazurkas, to the choice of key for the middle or contrasting section. Tovey, in his article 'Harmony' for the 11th edition of the *Encyclopaedia Britannica*, labelled as 'contradictory' the key that is a whole tone below the tonic, which he said had the effect of undermining the tonality of a piece. Yet this relationship is selected by Chopin in four mazurkas (op.30 no.4, op.50 no.3, op.56 no.3 and op.63 no.1). Other unusual choices occur: in op.33 no.4 in B minor, for instance, the contrasting section is in B♭ major. Normally however the choice falls on the dominant, the subdominant, the parallel major or minor, or a key a 3rd away from the tonic.

The basically modal and diatonic nature of the folklike mazurka did not inhibit Chopin from employing the full resources of chromatic harmony in many of his mazurkas. There are even a number of examples of the rapidly sliding chromatic modulations already noted in other works of Chopin (see ex.10). His last work, the Mazurka in F minor op.68 no.4, is an extreme example of chromaticism both in melody and harmony; its main tonal excursion is into the extremely remote key of A.

CHAPTER TEN

Form

I Piano works

In the matter of musical form Chopin has always been rather underestimated. Like other Romantic composers he has been accused of inattention to musical structure or of the inability to develop his materials on a large scale. This verdict seems to be based chiefly on the fact that Chopin did not often apply strict sonata form in his larger works. Even the three mature sonatas opp.35, 58 and 65 depart from Classical first-movement form by not recapitulating the first subject in the tonic, and the last two movements of op.35 are, in particular, unconventional as sonata movements. It is curious that critics have persisted in equating lack of interest in sonata form with inability to develop large forms of any kind. Hedley (in *Grove 5*) quoted with approval the verdict of a writer in the *Musical World* of 17 August 1843: 'The entire work is not a consequence of the first idea. . . . Therefore Chopin is incapable of a large and profound work of art'. And Abraham (1939) remarked that 'his sense of form is primitive, being limited almost exclusively to the possibilities of more or less modified ternary form'.

It is true that many of the dances – mazurkas, polonaises, waltzes, and others less numerous – are in a strict ternary form, so strict in some cases that the words 'da capo' are sufficient to indicate the reprise. Others (such

7. *Fryderyk Chopin: portrait (1838) by Eugène Delacroix*

as the Tarantelle and many of the waltzes) have a more variable arrangement of sections, but with an almost inevitable return of the opening section, often followed by a coda. But some of the more important mazurkas, though basically ternary, involve modifications and a substantial coda that takes them a stage beyond the simple dance form. Typically the return of the opening melody, when it comes, is shortened, and leads on into an extended coda – the emotional climax of the piece.

In pieces independent of dance forms, a similar principle is adopted. They are seldom strictly ternary, but they do usually bring back the opening idea after a section that provides a contrast of key and theme. They show great resource and subtlety in the way the return is varied, delayed, foreshortened, or extended. Nothing could be less 'primitive', for instance, than the reprise in the F♯ Impromptu. Two melodies form the *A* section, both in the tonic key. After the melodramatic *B* section in D major, the expected return to the tonic is postponed by a diversion to F major – where the first melody appears with an enriched accompaniment. Before it concludes, a characteristic twist of harmony returns to the tonic, where modification of the same first melody is heard. Then a totally unexpected section of brilliant scale passages intervenes before the second melody closes the piece. Tovey, discussing this work, pointed out that Chopin is 'usually most classical in free forms, because these have shaped themselves from the new material, just as the old classical forms shaped themselves from their own'.

In the nocturnes and other pieces of medium length the reprise is sometimes a point of repose after a turbulent middle section, sometimes a climactic restate-

55

ment. Well aware of the audience's expectations, Chopin frequently introduced some element of surprise at this point, and occasionally omitted the reprise altogether (Nocturnes op.15 no.3 and op.32 no.1). In many of these pieces there is no contrast in the middle section: the reprise forms part of a continuous musical statement, which modulates freely between the two occurrences of the main idea. It is hardly enough to describe such a structure as 'ternary'.

The principle of departure and return, far from being inadequate, lends itself well to a large structure, as Chopin and many other 19th-century composers discovered. The more the middle section is extended, and the further it departs in key, mood and theme from the opening idea, the more important and dramatic is the reprise when at last it comes. Usually the reprise is the climax of the piece, and it is a strength, not a weakness, that it comes nearer the end than it would in strict sonata form. In mature examples it is not played out in full, but leads directly into a coda based largely on new and more brilliant material. Chopin showed in his larger forms that this principle is enough in itself to provide direction and meaning. The listener needs no thematic 'unity' for his satisfaction, no matter how much this may be admired by theorists as the *sine qua non* of great music.

This method is followed, with immense variety of mood, thematic material, and structural detail, in the first, second and fourth scherzos, the first, third and fourth ballades, the Barcarolle, the Bolero, the Polonaise op.44, and the Polonaise-fantaisie. By 'scherzo' Chopin at first meant a piece in strict ternary form, like a Classical scherzo and trio; but after the first he treated

the form more flexibly. The third is in a remarkable modified sonata form. The term 'ballade' was an invention of Chopin's in this connection. All the ballades are in compound time, but their similarity goes no further. Though the title vaguely suggests a narrative, the pieces are free of specific 'programmes', published or otherwise, though some have said that they were influenced by the poems of Mickiewicz. The second has a unique structure with alternating sections of Andantino and Presto con fuoco: the contrast between them is violent and is never fully resolved. The other three ballades are supreme examples of Chopin's command of large-scale form: the fourth is often regarded as his greatest work.

The three solo works that include the word 'fantasia' in their titles are all based on the principle of departure and return, with the added feature that the middle section is in slow tempo. The most important of them is the Polonaise-fantaisie in A♭ op.61. This is a highly satisfying, intensely dramatic structure. Chopin's starting-point was a stereotyped dance form, the polonaise, but he broke down its conventional ternary structure completely (a process foreshadowed in the Polonaise op.44). He used thematic development to a degree unusual with him, but in ways that owe little to the methods of the Classical composers. The climactic reprise of the main tune in the tonic is surmounted by an even more exciting statement, also in the tonic, of the contrasting tune (first heard in B major and marked 'poco piu lento'). The melodious accompaniment figure for this tune takes over as a melody in its own right and forms the basis of the coda. Similarly, in the Barcarolle the shortened reprise and coda recapitulate the themes of both the *A* and the *B* sections.

8. Autograph MS
of the opening of
Chopin's
Polonaise
in Ab op.53,
composed 1842

In the larger works Chopin indulged in an unfettered exploration of key relationships, both in large-scale contrasts of tonal centres and for incidental colour. Three of them end in keys other than that of the principal melody (Scherzo op.31, Ballade op.38, Fantasie op.49).

In the studies and preludes, though the process of departure and return may be at work, another structural principle generally prevails. A small motif or pattern, often chosen to exemplify a technical problem, is worked out consistently through a series of harmonies and keys to the end of the piece. Sometimes the result is pure passage-work, lacking the melody–accompaniment texture that dominates most of Chopin's music; but a great deal of harmonic interest may derive from the ruthless pursuit of the chosen formula. (In the Study op.10 no.2 it produces simultaneous triads of A major and A minor.) This type of piece derives from various piano methods from Cramer (1804) onwards; the Studies op.10 were apparently inspired by Paganini's Caprices, which Chopin heard at Warsaw in 1829. Another influence, however, was Bach's *Das wohltemperirte Clavier* (the '48') which he had studied with his first teacher, Żywny. The Study in C minor op.25 no.12 bears a striking resemblance to the prelude in the same key from book 1 of the '48' and it is not the only one of Chopin's studies that recalls Bach: there is a strong flavour of the E♭ minor Prelude from book 1 in the bass-melody C♯ minor Study op.25 no.7.

The idea of writing a set of preludes in all the keys obviously recalls the '48'. At Prague in 1829 Klengel had played Chopin his set of 48 Canons and Fugues, also inspired by Bach, and this may have sparked off Chopin's later decision to produce his own set of pieces,

the Preludes op.28. The order of keys is different, but the great variety of form and texture recalls Bach's preludes, and the 'continuous' form without intermediate cadences found in many of Bach's preludes is transferred to some of Chopin's, most notably nos.1, 4, 8, 14 and 23. Hedley (1947) said: 'the title *Preludes* is not a particularly happy one. These pieces have obviously nothing of the character of introductions to more substantial works; they are not preludes in the sense in which Bach's "Preludes" are understood'. This is an error, however. There is evidence that the art of 'preluding' was very much alive, and was used in the type of salon concert that Chopin occasionally gave. Kalkbrenner's *Traité d'harmonie du pianiste* (Leipzig, c1848) set out 'rational principles of modulation for learning to prelude and to improvise'. He asked: 'How many of our best pianists are there who can play a prelude that is even moderately satisfying?'. He listed Chopin among 'the most distinguished improvisers who have ever lived', and provided a 'modèle des préludes' – a set of examples of preludes in various keys and of various styles and lengths. Clearly Chopin's op.28 can be regarded as another such set of examples. To provide a prelude in every key was a convenience as well as an artistic challenge. At the same time Chopin so arranged his preludes that they could also be played as a cycle, with appropriate contrast, tension and release, and with careful attention to key transitions from one piece to the next.

II   Orchestral and chamber music

Chopin was really no master of orchestration, though Berlioz might be considered to have been perhaps a

little harsh when he said that Chopin's orchestra was 'nothing but a cold and almost useless accompaniment'. His concertos and other works for piano and orchestra are merely vehicles for brilliant piano playing. Formally they are longwinded and extremely conservative. In chamber music Chopin showed real interest in only one instrument other than his own: the cello, for which he wrote several pieces culminating in the late G minor Sonata. He sometimes gave the cello the same type of rich, dark melody that he used for left-hand melodies on the piano. The opening movement shows the same peculiarity as the two mature piano sonatas (shared with Berlioz's *Symphonie fantastique*): the first subject is not recapitulated in the tonic. The early piano trio has an even more unusual first-movement structure: the exposition ends firmly in the tonic, but the second-subject material is recapitulated in the dominant before the coda. All Chopin's sonatas have admirably terse scherzos.

CHAPTER ELEVEN

Historical position

During the early years of the 19th century there was a
radical change in the composer's function in society.
Instead of providing music for an aristocratic employer,
he was now expected to please a multitude. The
Romantics responded in widely varying ways to this
new situation. Many composers, now largely despised
or forgotten, simply catered to the crowd and provided
music that would sell easily. A few heroic spirits – most
outstandingly Berlioz, Liszt and Wagner – grasped and
held the attention of the huge new public and sought to
lead it on to higher and greater musical experience. A
third group turned its back on the 'philistines', hoping to
preserve the values of true art from attack by withdraw-
ing into a small circle of sensitive aesthetes. Schumann
was the most articulate spokesman of this philosophy;
Chopin was its most extreme practitioner. Not only did
he withdraw himself at the earliest possible date from a
public eager to hear him play. In his compositions he
declined to make any concession to the gallery. He
provided music for the exclusive milieu of the Parisian
salon.

Even more than the other major composers of this
'escapist' group, Chopin was happiest with small forces
and on a small scale, though it has been shown above
how in the end he succeeded in evolving satisfactory
large forms. Like them, too, he felt the importance of

preserving musical tradition against the attacks of the philistines, and hence was drawn to the music of the past. Bach and Mozart were his two favourite masters. In spite of this he pushed forward into unknown realms of music, especially in his harmony. He did so not from any desire to lead a revolution, but from an intellectual curiosity to explore beyond the boundaries of a style that was second nature to him. Others more ambitious than he took their cue from some of his innovations. His influence on Liszt and Wagner is too obvious to need emphasis; but this influence can be traced to the very boundaries of 20th-century tonality, in Schoenberg's early piano music. Another strong line of his influence is found in French music – Franck, Saint-Saëns, Fauré and Debussy all owed much to him. The nationalist composers – Grieg, Smetana, Dvořák, Albéniz, Falla, as well as the Russians – followed him in bringing unaccompanied modal folksongs into the system of Western art music. And to the more Westernized Russians – Rubinstein, Serov, Tchaikovsky, Moszkowski, Rakhmaninov – he was the starting-point and model. Even Brahms occasionally succumbed to his spell, especially in some of the later songs and piano pieces.

CHAPTER TWELVE

Editions

Chopin's musical texts have been subjected to more interference from editors than usual. The ultimate reason for this can be traced to the composer's own dislike, already referred to, of writing out his music. Because much of it was published simultaneously by Brandus (later Schlesinger) in Paris, Wessel in London, and Breitkopf & Härtel in Leipzig, three manuscript copies were often required for each work. For this reason Chopin frequently employed a copyist, one of whom, Julian Fontana, even went out of his way to imitate Chopin's handwriting. Fontana also wrote out and published several works after Chopin's death, claiming that he was incorporating the composer's revisions.

Three German collected editions appearing about 1880 (Breitkopf & Härtel, Kistner and Peters) all suffered from this uncertainty as to the sources. Each made use of manuscripts and editions not available to the others, but none was based on a full examination of all the sources. Most of the immense number of more recent editions can be traced to one of these, often with gratuitous alterations and additions by later revisers. The Kalmus and Lea editions are direct reprints of Breitkopf & Härtel, while the Schirmer edition is largely from Kistner.

The first serious effort to provide a scholarly edition

was made by the Oxford University Press in 1932. The complete piano music was issued in three large volumes edited by Edouard Ganche, who used early French editions that formerly belonged to Chopin's pupil Jane Stirling, with corrections in the composer's hand. Unfortunately Ganche attached far too much weight to this source, neglecting others of greater importance.

The only complete edition of all Chopin's works was published between 1937 and 1966 by the Chopin Institute in Warsaw and Polskie Wydawnictwo Muzyczne in Kraków. It is in 20 volumes of scores, with a further six of instrumental parts, and is fully annotated. The editorial aims are of the highest, but they were frustrated in some cases by the fact that important sources in Western libraries were not available to the editors. Where sources differed, the editors tended to choose a reading on aesthetic grounds alone – a shaky principle for a scholarly edition.

The best available, though as yet incomplete, edition is Ewald Zimmermann's, published by the Henle Verlag in Munich. For each work a primary 'source layer', consisting of the autograph (if extant) and the contemporary edition that was most probably based on it, is established; all other sources are then related to this. When there are two substantially different sources, both emanating from Chopin himself, both are printed in full.

WORKS

Edition: F. F. Chopin: *Dzieła wszystkie* [Complete works], ed. I. J. Paderewski (Warsaw and Kraków, 1949–61) [P]
Catalogues: M. J. E. Brown: *Chopin: an Index of his Works in Chronological Order* (London, 1960, rev. 2/1972) [B]
    K. Kobylańska: *Rękopisy utworów Chopina: Katalog* (Kraków, 1977; Ger.trans., 1979) [KK]
    A. Koptiajew: 'Najdiennyi sbornik junoszeskich proizwiedienij Szopiena' [Newly discovered collection of Chopin's early works], *Birżewïe Wiedomosti*, no.12320 (St Petersburg, 1911) [Koptiajew]

Collections of photographs of MSS are in the Chopin Institute, Warsaw, and *A-Wn*. *PL-K* indicates the Kraków National Museum in cases where it is not known to which of its sections – Biblioteka Czartoryskich (*Kc*) or Biblioteka Czapskich (*Kcz*) – an MS belongs. The number in the German translation of Kobylańska's catalogue (KK), with a roman-numeral prefix, is given for works without opus number.
* – *autograph* CI – *Chopin Institute, Warsaw* Ferra – *collection of A. M. Ferra, Valldemosa, Majorca* US-NYlehman – *private collection of R. O. Lehman, New York* US-LApiatigorsky – *private collection of G. Piatigorsky, Los Angeles*
Numbers in the right-hand column denote references in the text.

PIANO SOLO

B	op. or KK	Key	Composition	MSS	Publication; dedication; remarks	P	
				mazurkas			
7	7/4a	Ab	1824	*PL-Wtm	facs. in K. Kobylańska: *Chopin w Kraju* (Kraków, 1955; Eng. trans., 1955); orig. version of op.7 no.4	—	5, 18, 24, 29, 35, 36, 41, 43–4, 51–2, 53, 55
8	17/4a	a	1824		?orig. version of op.17 no.4	—	
16	IIa/2, 3	G, Bb	1826	Wtm, F-Pn	Warsaw, 1826 (2nd versions); Poznań, 1875 (1st versions)	x	
18	68/2	a	1827	—	Berlin, 1855	x	32, 51
34	68/3	F	1829	—	Berlin, 1855	x	32, 51
38	68/1	C	1829	—	Berlin, 1855	x	32
45	7/2a	a	1829	—	Leipzig and Warsaw, 1902; orig. version of op.7 no.2	—	
31	IVa/7	D	1829	—	Poznań, 1875; 1st version of B 71	—	
60	6	f♯, c♯, E, e♭	1830	*Cologne, Stadtarchiv (no.1), *F-Ppo (no.2, sketch), *USSR-Lsc (no.4, sketch)	Leipzig; Paris and London, 1833; ded. Countess Pauline Plater	x	9, 300
61	7	Bb, a, f, Ab, C	1830–31	*Basle, Flörsheim Collection (nos.1, 3), *S-Smf (no.3), *PL-Kj (no.4, sketch), Schloss Kórnik, Poland (no.2)	Leipzig, 1832; Paris and London, 1833; no.1, Warsaw, 1835	x	9

73	IVb/1	B♭	24 June 1832	*PL-Kcz	Lamus, ii (Lwów, 1909); ded. Alexandrine Wolowska	x	
71	IVb/2	D	1832	—	Leipzig, 1880		49
77	17	B♭, e, A♭, a	1832–3	*Kj (no.2, sketch)	Leipzig, Paris and London, 1834; ded. Lina Freppa	x	
82	IVb/3	C	1833	D-MZsch	Warsaw and Mainz, 1870	x	
85	IVb/4	A♭	July 1834	*F-Ppo	ed. M. Mirska, Warsaw, 1930	x	
89	24	g, C, A♭, b♭	1834–5	*PL-Wn	Leipzig, Paris and London, 1836; ded. Count de Perthuis	x	43, 49, 52
93	67/1	G	1835	—	Berlin and Paris, 1855; ded. Anna Młokosiewicz	x	32
93	67/3	C	1835	—	Berlin and Paris, 1855; ded. Mme Hoffman	x	32
105	30	c, b, D♭, c♯	1836–7	Wn	London, 1837; Leipzig and Paris, 1838; ded. Princess Maria Czartoryska Württemberg	x	47, 48, 49, 52
115	33	g♯, D, C, b	1837–8	*Wn	Leipzig, Paris and London, 1838; ded. Countess Róża Mostowska; order varies	x	52
122	41/2	e	28 Nov 1838	*US-LApiatigorsky, *A-Wgm, *F-Pn	Leipzig, Paris and London, 1840; ded. Stefan Witwicki; numbering varies	x	
126	41/1, 3, 4	c♯, B, A♭	1839	PL-Wn	same as B 122; numbering varies	x	49
134	IIb/4	a	1840	—	Six morceaux de salon, Paris, 1841; Notre temps (Mainz, 1842)	x	
140	IIb/5	a	1840		Album de pianistes polonais, i (Paris, 1841); ded. Emile Gaillard	x	
145	50	G, A♭, c♯	1841–2	*US-NYpm, *PL-Kj (no.3, 1st version)	Vienna, Paris and London, 1842; ded. Leon Szmitkowski	x	52
153	56	B, C, c	1843	*Wn, *GB-Lbl (no.2, sketch)	Leipzig and Paris, 1844; London, advertised 1845; ded. Catherine Maberly	x	45, 52
157	59	a, A♭, f♯	1845	*D-MZsch (nos.1, 2), *F-Po (no.2), *GB-Ob (no.2), *US-NYpm (no.3)	Berlin and London, 1845; Paris, 1846	x	
162	63	B, f, c♯	1846	*F-Pn (no.1), *private collection of Mme Y. Faure (La Croix en Touraine) (no.2, sketch)	Leipzig and London, 1847; Paris, 1848; ded. Countess Laura Czosnowska	x	52
163	67/4	a	1846	*A-Wgm, *private collection of Mme K. H. Strauss (Paris)	Berlin, 1855; Paris, 1856; 3 versions	x	32

B	op. or KK	Key	Composition	MSS	Publication; dedication; remarks	P	
167	67/2	g	1849	—	Berlin, 1855; Paris, 1856	x	32
168	68/4	f	1849	*C (sketch)	Berlin, 1855; Paris, 1856; realized 1852 by A. Franchomme from sketches	x	32, 52

Lost: several early mazurkas, KK Vf, see Koptiajew; Mazurka, D, KK Ve/5, ded. ?W. Kolberg, mentioned in 3 Polonaises 1817–21, ed. Z. Jachimecki (Kraków, 1947); KK Vc/2, 1832, mentioned in letter from Chopin to J. K. Jędrzejewicz, 10 Sept 1832; KK Ve/7, 14 Sept 1832, listed in auction catalogue, Paris, 1906; ?Mazurka, Bb, KK Ve/4, 1835, sold in Paris, 20 June 1977; Mazurka, KK Vc/4, 1846, mentioned in letter from Chopin to W. Grzymała, Dec 1846; KK Ve/8, mentioned in letters from Breitkopf & Härtel to Izabela Barcińska, 1878; KK Ve/6, ded. Mme Nicolai, mentioned in note from Augener to C. A. Spina, 21 May 1884
Doubtful: Mazurka in D, B 4, KK Anh. Ia/1, ?1820, ed. in Kurier (Warsaw, 20 Feb 1910), P x; in bb, KK Anh.Ib, MS with J. T. Stopnicki, Warsaw
Spurious: in F♯, KK Anh.II/1 (by Charles Mayer)

nocturnes

B	op. or KK	Key	Composition	MSS	Publication; dedication; remarks	P	
19	72/1	e	1827	*Ferra	Berlin, 1855; Paris, 1856	vii	24, 35, 43, 55–6
49	IVa/16	c♯	1830		Poznań, 1875; ded. Ludwika Chopin; 'Lento con gran espressione'	xviii	5, 18, 49
54	9	bb, Eb, B	1830–31		Leipzig, 1832; Paris and London, 1833; ded. Marie Pleyel	vii	17, 45, 46, 47, 50
55	15/1–2	F, F♯	1830–31		Leipzig, 1833; Paris and London, 1834; ded. Ferdinand Hiller	vii	
79	15/3	g	1833	*US-NYlehman (frag.)	same as B 55	vii	50–51, 56, 118
91	27/1	c♯	1835		Leipzig, Paris and London, 1836; ded. Countess Thérèse d'Appony	vii	49
96	27/2	Db	1835	*PL-Wn	same as B 91	vii	
106	32	B, Ab	1836–7		Berlin, Paris and London, 1837; ded. Baroness Camille de Billing	vii	56
108	IVb/8	c	1837	*F-Pn (sketch), *C (sketch)	ed. L. Bronarski (Warsaw, 1938); sketches not entitled Nocturne	xviii	
119	37/1	g	1837	*Ferra (frag.) PL-Wn	Leipzig, Paris and London, 1840	vii	51
127	37/2	G	1839	Wn	same as B 119	vii	47
142	48	c, f♯	1841	CI	Paris, 1841; Leipzig and London, 1842; ded. Laure Duperré	vii	43
152	55	f, Eb	1843	*PL-Wn *F-Pn (no.1)	Leipzig and Paris, 1844; London, advertised 1845; ded. Jane Stirling	—	41

B	op. or KK	Key	Composition	MSS	Publication; dedication; remarks	P	
62	18	E♭	1831	*B-MA, *US-NH, *private collection of Vicomte P. de la Panouse (Yvelines)	Leipzig, Paris and London, 1834; ded. Laura Horsford	ix	
64	34/2	a	1831	A-Wn (frag.)	Leipzig and London, 1838; Paris, 1839; ded. Baroness G. d'Ivry	ix	
92	70/1	G♭	1833	*private collection of Vicomte P. de la Panouse (Yvelines), *US-NH	Berlin, 1855	ix	32
94	34/1	A♭	15 Sept 1835	*PL-Wm	Leipzig and London, 1838; Paris, 1839; ded. Josefine von Thun-Hohenstein	ix	
95	69/1	A♭	1835	*F-Pc, *Harvard U., Dumbarton Oaks Research Library	Berlin, 1855; autographs ded. Charlotte de Rothschild Mme Peruzzi, Maria Wodzińska; 'L'adieu'	ix	18, 32
118	34/3	F	1838	—	Leipzig and London, 1838; Paris, 1839; ded. Mlle A. d'Eichtal	ix	
131	42	A♭	1840		Leipzig, Paris and London, 1840	ix	
138	70/2	f	1841	*F-Pn; *private collection of J. Samuel (Vienna); *private collection of F. Lang (Abbaye de Royaumont)	Kraków, 1852; autographs ded. Marie de Krudner, Mme Oury, Elise Gavard, Countess Esterházy	ix	32
150	IVb/11	a	?1843	*F-Pn (incl. sketch)	facs. in ReM (1955), no.225, p.13	—	
164	64/1	D♭	1846–7	*Pn (incl. sketch); *GB-Lcm	Leipzig, 1847; Paris and London, 1848; Paris edn. ded. Countess Delfina Potocka; 'Minute'	ix	
164	64/2	c♯	1846–7	*F-Pn, *Po (sketch)	Leipzig, 1847; Paris and London, 1848; Paris edn. ded. Baroness Charlotte de Rothschild	ix	

no.	op./KK	title, key	date	other	first publication, remarks	vol	pp.
164	64/3	A♭	1846–7	*Po (sketch)	Leipzig, 1847; Paris and London, 1848; Paris edn. ded. Countess Katarzyna Branicka	ix	
166	Va/3	B	12 Oct 1848	—	ded. Mrs Erskine; MS not available, formerly owned by W. Westley Mannings (London)	—	

Lost: several early waltzes, KK Vf, incl. waltz, a, 1824, ded. Countess Łubieńska, see Koptiajew; in C, KK Vb/8, ?1824; in C, B p.11, KK Vb/3, 1826; in A♭, KK Vb/4, 1827; in d (*'La partenza'*), B p.23, KKVb/6, 1828; in A♭, KK Vb/5, B p.62, 1829 or 1830; in E♭, KK Vb/7, 1829 or 1830, listed in catalogue by Ludwika Jędrzejewicz; KK Ve/10, listed in auction catalogue, Paris, March 1906; KK Ve/11, mentioned in letters from Breitkopf & Härtel to Izabela Barcińska, 1878; KK Ve/12, mentioned in diary of L. Niedzwiecki, 1845

Doubtful: 'Valse mélancolique', E♭, KK Ib/7, Cl (New York, 1932)

no.	op./KK	title, key	date	other	first publication, remarks	vol	pp.
10	1	Rondo, c	1825	—	Warsaw, 1825; arr. pf 4 hands, Leipzig, 1834; ded. Mme Bogumił Linde	xii	34–5, 36, 53, 55–60; 5
12	72/3	Three Ecossaises, D, G, D♭	1826	Stanford U., Memorial Library of Music (nos.2,3)	Berlin, 1855	xviii	32
14	IVa/4	Introduction and Variations, E, on a German air ('Der Schweizerbub')	1826	*PL-Kp	Paris, 1856; Vienna and Paris, ?1851; ded. Katarzyna Sowińska	xiii	
15	5	Rondo à la Mazur, F	1826	—	Warsaw, 1828; ded. Countess Alexandrine de Moriolles	xii	5
20	72/2	Funeral March, c	1827	Cl, *F-Pn*, Schloss Kórnik, Poland; *US-NYlehman*	Berlin, 1855; version in Cl, ed. E. Ganche (London, 1932)	xviii	32
23	4	Sonata, c	1828	*A-Wgm*	Vienna and Paris, 1851; ded. Józef Elsner	vi	5, 7
26	73a	Rondo, C	1828		P; ded. Aloys Fuchs; orig. version of Rondo, C, 2 pf, op.73	xii	7, 32
37	IVa/10	Variations, A, 'Souvenir de Paganini'	1829	—	ed. in *Echo muzyczne i teatralne*, v (Warsaw, 1881); on Venetian air 'Le carneval de Venise'	xiii	

B	op. or KK	Key	Composition	MSS	Publication; dedication; remarks	P	
42	10/8–11	Four Studies, F, f, Ab, Eb	Oct–Nov 1829	*C (nos.8, 10), *US-NYlehman (no.9), *S-Smf (no.11)	Leipzig, Paris and London, 1833; ded. Franz Liszt	ii	17, 59
57	10/5–6	Two Studies, Gb, eb	?sum. 1830	*CI	same as B 42	ii	17, 45, 46, 59
59	10/1–2	Two Studies, C, a	late aut. 1830	*CI (MSS, no.1, ?no.2), *S-Smf (no.2)	same as B 42	ii	7, 17, 47, 59
67	10/12	Study, c	? Sept 1831	*Smf	same as B 42	ii	9, 17, 59
65	20	Scherzo, b	1831–2	—	Leipzig, Paris and London, 1835; ded. Thomas Albrecht	v	9, 56
68	10/7	Study, C	spr. 1832	*US-NYpm	same as B 42	ii	17, 18, 59
74	10/3	Study, E	25 Aug 1832	*CI, *US-NYlehman	same as B 42	ii	17, 18, 43, 59
75	10/4	Study, c♯	Aug 1832	*Basle, Flörsheim Collection	same as B 42	ii	17, 18, 48, 59
76	16	Introduction, c, and Rondo, Eb	1832	—	Leipzig, Paris and London, 1834; ded. Caroline Hartmann	xii	17, 32
80	12	Introduction and Variations, Bb, on 'Je vends des scapulaires' from Hérold's Ludovic	1833	—	Leipzig, 1833; Paris and London, 1834; ded. Emma Horsford	xiii	17
81	19	Introduction, C, and Bolero, a–A	1833	—	Leipzig, 1834; Paris and London, 1835; ded. Countess Emilie de Flahault	xviii	17, 34, 56
78	25/4–6, 8–10	Six Studies, a, e, g♯, Db, Gb, b	1832–4	*F-Po (no.4), *PL-Wn (no.8), Wn (nos.4,5, 6, 9,10)	Leipzig, Paris and London, 1837; ded. Countess Marie d'Agoult	ii	18, 38, 39, 42, 47, 48

83	25/11	Study, a	1834	*Wn*	Leipzig, Paris and London, 1837; ded. Countess Marie d'Agoult	ii	18
84	IVb/6	Cantabile, Bb	1834	—	ed. L. Bronarski in *Muzyka* (Warsaw, 1931), nos. 4-6	xviii	
86	IVb/7	Prelude, Ab	July 1834	*Spokane Conservatory, Washington	ed. in *Pages d'art* (Geneva, Aug 1918); ded. Pierre Wolff	i	
88	22	Andante spianato, G	1834	—	Leipzig, Paris and London, 1836; introduction to Grand polonaise, pf, orch, B 58	viii	34
66	23	Ballade, g	1831–5	*US-LApiatigorsky*	Leipzig, Paris and London, 1836; ded. Baron de Stockhausen	iii	9, 48, 56, 57
87	66	Fantaisie-impromptu, c♯	1835	*private collection of A. Rubinstein (Paris), CI, *F-Pn* CI, *F-Pn*	Berlin and Paris, 1855; London, n.d.; ded. Baroness d'Este	iv	32
97	25/2	Study, f	Jan 1836	*Ferra (frag.), *PL-Wn*, *K*	Leipzig, Paris and London, 1837; ded. Countess Marie d'Agoult	ii	18
98	25/7	Study, c♯	early 1836	*Wn*	same as B 97	ii	18, 41, 50, 59
99	25/3, 12	Two studies, F, c	1836	*Wn*	same as B 97	ii	18, 59
104	25/1	Study, Ab	Sept 1836	*Wn*	same as B 97	ii	18
110	29	Impromptu, Ab	1837	*CI*	Paris and London, 1837; Leipzig, 1838; ded. Countess Caroline de Lobau	iv	
111	31	Scherzo, bb	1837	*F-Pn*	Paris and London, 1837; Leipzig, 1838; ded. Countess Adèle de Fürstenstein	v	40, 56, 59
113	IIb/2	Variation no.6, E, in Hexaméron (Variations on the March from Bellini's I puritani)	1837	*PL-Kj*	Vienna and London, 1839; Paris, ?1839; composed for Princess Christina Belgiojoso-Trivulzio; collab. Liszt, Thalberg, Pixis, H. Herz and Czerny	xiii	270
114	—	Funeral March, bb	1837	*A-Wn*	pubd as 3rd movt of Sonata, B 128	vi	32, 36

B	op. or KK	Key	Composition	MSS	Publication; dedication; remarks	P	
109 117	IVb/5 —	Largo, Eb; Andantino, g	?1837 1838	*F-Pn A-Wgm, PL-WRol	ed. L. Bronarski (Warsaw, 1938) ed. A. Orga (London, 1968); pf part of song, B 116	xviii —	
102	38	Ballade, F/a	1836-9	*F-Pn	Leipzig, Paris and London, 1840; ded. Robert Schumann	iii	24, 57, 59
100, 107, 123-4	28	Twenty-four Preludes	1831-9	*PL-Wn, *US-LApiatigorsky (no.2; no.4, sketch), *F-Pn (no.20), *USSR-MI (no.20)	Leipzig, Paris and London, 1839; autograph and Leipzig edn. ded. J. C. Kessler, London and Paris edns. ded. Camille Pleyel	i	18, 21, 24, 35, 40, 41, 42, 43, 49, 50, 51, 59-60
125	39	Scherzo, c#	1839	PL-Wn	Leipzig, Paris and London, 1840; ded. Adolf Gutmann	v	24, 57
128	35	Sonata, bb	1839	*CI (frag.), PL-Wn	Leipzig, Paris and London, 1840; 3rd movt composed 1837 as Funeral March, B 114	vi	32, 34, 35, 36, 42, 43, 49, 53, 61
129	36	Impromptu, F#	1839	*CI (sketch)	Leipzig, Paris and London, 1840	iv	55
130	IIb/3	Trois nouvelles études, f, Ab, Db	1839	*PL-Kc (sketch), *Ferra	Berlin and Paris, 1840; London, 1841; for Moscheles's 'Méthode'	ii	
129b	IVc/1	Canon, f	?1839	*private collection of M. Uzielli (Liestal) (frag.)	ed. L. Bronarski, Annles Chopin, ii (Warsaw, 1948); inc.	—	
133	IVb/10	Sostenuto, Eb	1840	*F-Pc	ed. M. J. E. Brown (London, 1955)	—	
72	46	Allegro de concert, A	1832-41	*PL-Wn	Leipzig and Paris, 1841; London, 1842; ded. Friederike Müller-Streicher; incl. material originally intended for a 3rd pf conc.	xiii	18, 34
136	47	Ballade, Ab	1840-41	F-Pn	Paris, 1841; Leipzig and London, 1842; ded. Pauline de Noailles	iii	24, 56, 57
137	49	Fantasie, f/Ab	1841	*PL-Wn	Paris, 1841; Leipzig and London, 1842; ded. Princess Catherine de Souzzo	xi	34, 35, 59
139	43	Tarantelle, Ab	1841	F-Pn	Hamburg, Paris and London, 1841	xviii	35, 55
141	45	Prelude, c#	1841	—	Vienna and Paris, 1841; London, 1842; ded. Princess Elisabeth Tschernischeff; composed for Mechetti's 'Beethoven-Album'	i	

144 146	IVc/2 52	Fugue, a Ballade, f	1841–2 1842	*Ferra *GB-Ob, *private collection of R. F. Kallir (New York) (frag.) *PL-Kj	Leipzig, 1898 Leipzig and Paris, 1843; London, advertised 1845; ded. Baroness Charlotte de Rothschild	xviii iii	56, 57
148	54	Scherzo, E	1842	*PL-Kj	Leipzig and Paris, 1843; London, advertised 1845; ded. Jeanne de Caraman	v	56
149	51	Impromptu, G♭	aut. 1842	private collection of Lady Gwynne-Evans (Wales) (1st version), *private collection of Mrs G. Selden-Goth (Florence) (2nd version)	Leipzig, Paris and London, 1843; ded. Countess Jeanne Batthyany-Esterházy	iv	24
151	IVb/12	Moderato, E	1843	USSR-Mcl	ed. H. Pachulski in Świat (Warsaw, 1910), no.23; ded. Countess de Cheremetieff; 'Albumblatt'	xviii	
154	57	Berceuse, D♭	1843–4	*F-Pn, *US-NYlehman (sketch)	Leipzig, Paris and London, 1845; ded. Élise Gavard	xi	29, 35, 36, 50, 273
155	58	Sonata, b	sum. 1844	*PL-Wn, *CI (sketch), *private collection of J. Reande (Paris) (sketch)	Leipzig, London and Paris, 1845; ded. Countess Emilie de Perthuis	vi	34, 35, 53, 61
158	60	Barcarolle, F♯	1845–6	*CI (sketch), *PL-Kj	Leipzig, Paris and London, 1846; ded. Baroness de Stockhausen	xi	29, 34, 35, 50, 56, 57
159	61	Polonaise-fantaisie, A♭	1845–6	*Wn, *CI (sketch)	Leipzig, Paris and London, 1846; ded. Mme A. Veyret	viii	34, 35, 48, 56, 57
—	IVc/13	Galopp, A♭	1846	*private collection of Mme K. H. Strauss, Paris	—	—	4

Lost: Military March, B 2, KK Vd/4, 1817 (1817), ded. Grand Duke Konstantyn Pavlovich; other early marches, KK Vf, see Koptiajew; Variations, KK Ve/9, 1818, see Pamiętnik Warszawski, Jan 1818; Andante dolente, bb, KK Vb/1, B p.19, and Ecossaise, KK Vb/9, B p.19, both 1827, listed in catalogue by Ludwika Jędrzejewicz; three marches, KK Vd/1–3, mentioned in letter from J. Fontana to Ludwika Jędrzejewicz, 14 March 1854 (perhaps incl. Funeral March, c, B 10); Ecossaise, KK Ve/3, mentioned in letter from O. Kolberg to M. A. Szulc, 15 Dec 1874; two lost works for Äolopantaleon, KK Ve/1–2, mentioned in Gazeta Warszawska (23 Sept 1873)
Doubtful: Contredanse, G♭, B 17, KK Anh.Ia/4, ?1827 (Warsaw, 1843), P xviii; Prélude and Andantino, F, KK Anh.Ia/2–3, 1845, ed. Z. Mycielski in Muzyka (1930), no.1

PIANO FOUR HANDS

B	op. or KK	Title, key	Composition	MSS	Publication; remarks	P	
12a	IVa/6	Introduction, Theme and Variations, D	1826	PL-Kj (frag.)	ed. J. Ekier (Kraków, 1965); on a Venetian air	—	

Lost: Variations, F. KK Vb/2, 1826, ded. Tytus Woyciechowski, mentioned in catalogue by Ludwika Jędrzejewicz, see B p.11; Sonata, op. 28. KK Vc/5, mentioned in letter from Chopin to Breitkopf & Härtel, 1835

TWO PIANOS

B	op. or KK	Title, key	Composition	MSS	Publication; remarks	P	
27	73	Rondo, C	1828	—	Berlin, 1855; orig. for pf solo, B 26	xii	7, 32
							60-61

CHAMBER

B	op. or KK	Title, key, scoring	Composition	MSS	Publication; dedication; remarks	P	
25	8	Piano Trio, g	1828-9	*CI	Leipzig, 1832; London, 1833; Paris, 1934; ded. Prince Antoni Radziwiłł	xvi	7, 61
52, 41	3	Introduction and Polonaise, C, vc, pf	1829-30	—	Vienna, 1831; Berlin, 1832; Paris, 1835; London, 1836; ded. Joseph Merk	xvi	
70	IIb/1	Grand Duo, E, on themes from Meyerbeer's Robert le Diable, vc, pf	1832	*F-Pn	Berlin, Paris and London, 1833; ded. Adèle Forest; collab. Auguste Franchomme	—	17
160	65	Cello sonata, g	1845-6	*CI (sketches), *F-Pn (sketch), *US-NYlehman (sketch), F-Pn	Leipzig, 1847; Paris, 1848; ded. Auguste Franchomme	xvi	29, 35, 53, 61

Lost: several works for vn, pf, see Koptiajew

Doubtful: Variations, E, on 'Non più mesta' from Rossini's La Cenerentola, fl, pf, B 9, KK Anh. Ia/5, ?1824, ed. in K. Kobylańska: *Chopin w kraju* (Kraków, 1955; Eng. trans., 1955); P xvi

34, 60-61

B	op. or KK	Title, key	Composition	MSS	Publication; dedication; remarks	P	
22	2	Variations, B♭, on 'La ci darem' from Mozart's Don Giovanni	1827	*A-Wn, *US-NYlehman (sketch)	Vienna, 1830; Paris and London, 1833; ded. Tytus Woyciechowski	xv, xxi (score)	7, 15, 110
28	13	Fantasia on Polish Airs, A	1828	*Geneva, Bibliothèque Bodmeriana (sketch)	Leipzig, Paris and London, 1834; arrs. for pf solo and pf, str qt, Leipzig, ?1834; ded. Johann Peter Pixis	xv, xxi (score)	8
29	14	Krakowiak, rondo, F	1828	*PL-Kc, *US-NYlehman (sketch)	Leipzig, Paris and London, 1834; arr. for pf solo, Leipzig, ?1834, pf, str qt and 2 pf, Paris, 1834; ded. Princess Anna Czartoryska	xv, xxi (score)	7, 8
43	21	Concerto no.2, f	1829-30	*PL-Wn, *CI (sketch)	Leipzig, Paris and London, 1836; arrs. for pf solo, pf qt and pf qnt, Leipzig, ?1836; ded. Countess Delfina Potocka	xiv, xx (score)	8, 34, 257
53	11	Concerto no.1, e	1830	F-Pn	Leipzig and Paris, 1833; London, 1834; arrs. for pf solo, 2 pf, pf qt and pf qnt, Leipzig, ?1833; ded. Friedrich Kalkbrenner	xiv, xix (score)	8, 10, 17, 34, 257
58	22	Grand polonaise, E♭	1830-31	—	Leipzig, Paris and London, 1836; arrs. for pf solo and pf qt, Paris, ?1836, pf solo, London, 1838; ded. Baroness d'Este; see also Adante spianato, pf solo, B 88	viii, xxi (score)	9, 34

SOLO SONGS
(all with pf acc.)

B	op. or KK	Title (text)	Composition	MSS	Publication; dedication; remarks	P	
32	74/5	Gdzie lubi [There where she loves] (S. Witwicki)	1829	A-Wn, PL-Wn	Berlin, 1857	xvii	32
33	74/1	Życzenie [The wish] (Witwicki)	1829	A-Wn, PL-Wn, CS-Pnm	Kiev, 1837	xvii	32
39	IVa/9	Jakież kwiaty [Which flowers] (I. Maciejowski)	1829	*CS-Pnm	ed. in F. Wójcicki: Cmentarz Powązkowski pod Warzawa, ii (Warsaw, 1856); ded. Vaclav Hanka; no pf part known	—	
47	74/10	Wojak [The warrior] (Witwicki)	1830	*CI, A-Wn, PL-Wn,	Kiev, 1837	xvii	32
48	74/6	Precz z moich oczu! [Out of my sight!] (A. Mickiewicz)	1830	*CI (sketches), A-Wn, F-Pn	Berlin, 1857	xvii	32
50	74/4	Hulanka [Merrymaking] (Witwicki)	1830	A-Wn, F-Pn, PL-Wn	Berlin, 1857; as Patryot Piesn [Patriotic song] (S. Hernisz), B 168b, 1831	xvii	32
50	74/7	Poseł [The envoy] (Witwicki)	1830	*private collection of G. Mecklenburg (Marburg), A-Wn, F-Pn, PL-Wn	Berlin, 1857	xvii	32
51	IVa/11	Czary [Charms] (Witwicki)	1830	—	facs., Leipzig, 1910	xvii	
63	74/3, 15, 16	Smutna rzeka [The sad stream] (Witwicki), Narzeczony [The bridegroom] (Witwicki), Piosnka litewska [Lithuanian song] (L. Osiński)	1831	*F-Ppo (no.16) A-Wn	Berlin, 1857	xvii	32
101	74/17	Śpiew grobowy [Hymn from the tomb] (W. Pol)	1836	PL-Wtm, A-Wn	Berlin, 1857	xvii	32, 36
103	74/14	Pierścień [The ring] (Witwicki)	8 Sept 1836	*CI (sketch), A-Wn, F-Pn	Berlin, 1857	xvii	32
112	74/12	Moja pieszczotka [My darling] (Mickiewicz)	1837	*PL-K, F-Pn	Berlin, 1857	xvii	32
116	74/2	Wiosna [Spring] (Witwicki)	1838	*GB-Cfm	Berlin, 1857; ? version for pf solo, see Andantino, B 117	—	32

B	KK		Date				
132	IVb/9	Dumka [Reverie] (B. Zaleski)	1840	—	A-Wn, F-Pn	ed. S. Lam in *Słowo Polskie* (Lwów, 22 Oct 1910)	xvii
143	74/8	Śliczny chłopiec [Handsome lad] (Zaleski)	1841		A-Wn, F-Pn	Berlin, 1857	xvii 32
156	74/11, 13	Two Dumkas (Zaleski): Dwojaki koniec [The double end], Nie ma czego trzeba [I want what I have not]	1845		A-Wn (nos.11, 13), F-Pn (no.13)	Berlin, 1857	xvii 32
165	74/9	Melodya (Z. Krasiński)	1847		A-Wn	Berlin, 1857	xvii 32, 36

Lost: song, KK Vc/10, mentioned in letter from Ludwika Jędrzejewicz to Chopin, 9 Jan 1841; Piotno [Linen], KK Vd/5, mentioned in letter from J. Fontana to Ludwika Jędrzejewicz, 2 July 1852; three songs, KK Vd/6–8, listed in letter from J. W. Stirling to Ludwika Jędrzejewicz, July 1852

Doubtful: Dumka na Wygnaniu [Song of the exile] (M. Gosławski), KK Anh.Ic/1; Tam na błoniu [There on the green], KK Anh.Ic/2; Trzeci maj [The third of May] (S. Starzeński), KK Anh.Ic/3; O wiem, że Polska [Oh, I know that Poland] (Z. Krasiński), KK Anh.Ic/4; Pytasz się, czemu [You ask why] (Krasiński), KK Anh.Ic/5; Pieśni pielgrzyma polskiego [Songs of a Polish pilgrim] (K. Gaszyński), KK Anh.Ic/6

Two sacred works, KK Va/1–2, incl. Veni creator, before 1846, ded. Bohdan Zaleski and Zofia Rosengardt, not available

ARRANGEMENTS, TRANSCRIPTIONS

B	KK	Work arranged	Forces	Date	MS: publication
—	VIIa/1	'Casta diva' from Bellini's Norma	pf acc.	?c1831	US-LApiatigorsky
—	VIIa/2	Three Fugues, a, F, d, from Cherubini's Cours de contrepoint et de fugue	pf		private collection (Paris)
160b	VIIb/1, 2	Two Bourrées, G, A	pf	?1846	ed. A. Orga (London, 1968)
—	VIIb/7, 8	Allegretto, A, Mazurka, d	pf		MS sold in Paris, 21 Nov 1974

Frags. (v part only): Czułe serca [Tender hearts], B 140b (3–4), KK VIIb/3; Dawniej Polak [Previously a Pole], B 140b(2), KK VIIb/4, CI; Doyna Vallacha [Romanian folksong], KK VIIb/5, CI; song, a, B, 140b(1)

Lost: Variations, v, pf, on a Ukrainian Dumka by Antoni Radziwiłł, completed by Chopin, KK VIIa/3, mentioned in letter from Chopin to T. Woyciechowski, 5 June 1830; songs, KK VIIb/6, mentioned in letter from Chopin to Ludwika Jędrzejewicz, 18 Sept 1844

OTHER WORKS

Exercises, theoretical works, KK VIIc/1–7
Unidentified sketches, KK VI/1–12
Further lost works, see KK

BIBLIOGRAPHY

CATALOGUES

Thematisches Verzeichnis der im Druck erschienenen Compositionen von Friedrich Chopin (Leipzig, 1852, rev. 2/1888)

A. Koptiajew: 'Najdiennyji sbornik junoszeskich proizwiedienij Szopiena' [Newly discovered collections of Chopin's early works], *Birżewje Wiedomosti*, no. 12320 (St Petersburg, 1911)

M. J. E. Brown: *Chopin: an Index of his Works in Chronological Order* (London, 1960, rev. 2/1972)

Towarzystwo imienia Fryderyka Chopina: katalog zbiorów [The Chopin Society in Warsaw: a catalogue of its collections], i–ix (Warsaw, 1969–71)

T. D. Turło: 'Registers of Chopin's Manuscripts in Polish Collections', *Studies in Chopin*, ed. D. Żebrowski (Warsaw, 1973), 90

K. Kobylańska: *Rękopisy utworów Chopina: Katalog* [Manuscripts of Chopin's works: catalogue] (Kraków, 1977)

——: *Frédéric Chopin: Thematisch-bibliographisches Werkverzeichnis* (Munich, 1979)

DISCOGRAPHIES

A. Parigel: *L'oeuvre de Frédéric Chopin: discographie générale* (Paris, 1949)

H. C. Schonberg: *The Collector's Chopin and Schumann* (Philadelphia, 1959/R1978)

D. Melville: *Chopin: a Biography, with a Survey of Books, Editions and Recordings* (London, 1977)

G. Mannoni: 'Discographie comparée: les 24 études de Frédéric Chopin', *Harmonie*, cxlv (1979), 96

J. Methuen-Campbell: *Chopin Playing from the Composer to the Present Day* (London, 1981)

BIBLIOGRAPHIES

B. E. Sydow: *Bibliografia F. F. Chopina* (Warsaw, 1949; suppl., 1954)

K. Michałowski: *Bibliografia chopinowska/A Chopin Bibliography 1849–1969* (Kraków, 1970)

——: 'Bibliografia chopinowska 1970–1973', *Rocznik chopinowski*, ix (1975), 121–75

ICONOGRAPHIES

L. Binental: *Chopin: dokumenty i pamiątki* [Chopin: documents and souvenirs] (Warsaw, 1930; Ger. trans., 1932)

R. Bory: *La vie de Frédéric Chopin par l'image* (Geneva, 1951)

Bibliography

M. Idzikowski and B. E. Sydow: *Portret Fryderyka Chopina* [Portraits of Chopin] (Kraków, 1952, enlarged 2/1963; Fr. trans., 1953)

K. Kobylańska: *Chopin w kraju: dokumenty i pamiątki* [Chopin in his homeland: documents and souvenirs] (Kraków, 1955; Eng. trans., 1955)

A. Boucourechliev: *Chopin: a Pictorial Biography* (London, 1963)

M. Mirska and W. Hordyński: *Chopin na obczyźnie: dokumenty i pamiątki* [Chopin abroad: documents and souvenirs] (Kraków, 1965)

A. Murgia: *The Life and Times of Chopin* (London, 1967)

W. Dulęba: *Chopin* (Kraków, 1975)

H. Wróblewska and M. Lewkowicz, eds.: *Portret Fryderyka Chopina* [Portraits of Chopin] (1975) [exhibition catalogue]

A. Zborski and J. Kański: *Chopin i jego ziemia/Chopin and the Land of his Birth* (Warsaw, 1975)

A. Orga: *Chopin: his Life and Times* (Tunbridge Wells, 1976)

LETTERS

M. Karłowicz, ed.: *Nie wydane dotychczas pamiątki po Chopinie* [New unpublished souvenirs of Chopin] (Warsaw, 1904; Fr. trans., 1904)

B. Scharlitt, ed.: *Friedrich Chopins gesammelte Briefe* (Leipzig, 1911)

H. Opieński, ed.: *Chopin's Letters* (New York, 1931/*R*1971; Pol. orig., 1936)

B. E. Sydow, ed.: *Korespondencja Fryderyka Chopina*, i–ii (Warsaw, 1955; Fr. trans., 1953–60)

A. Hedley, ed.: *Selected Correspondence of Fryderyk Chopin* (London, 1962/*R*1979)

Z. Lissa: 'Chopins Briefe an Delfine Potocka', *Mf*, xv (1962), 341

J. M. Smoter: *Spór o 'listy' Chopina do Delfiny Potockiej* [The controversy over Chopin's 'letters' to Delfina Potocka] (Kraków, 1967)

M. Gliński, ed.: *Chopin: listy do Delfiny* [Chopin: letters to Delfina] (New York, 1972)

K. Kobylańska, ed.: *Korespondencja Fryderyka Chopina z rodziną* [Chopin's correspondence with his family] (Warsaw, 1972)

——: *Korespondencya Fryderyka Chopina z George Sand i z jej dziećmi* [Chopin's correspondence with George Sand and her children] (Warsaw, 1981)

COLLECTED ESSAYS, COLLECTIVE PUBLICATIONS

R. Schumann: *Gesammelte Schriften über Musik und Musiker* (Leipzig, 1854, 4/1891/*R*1968, 5/1914; Eng. trans., 1877)

E. Ganche: *Dans le souvenir de Frédéric Chopin* (Paris, 1925)

ReM (1931), no.121 [special Chopin issue]

B. Wójcik-Keuprulian: *Chopin: studia, krytyki, szkice* [Chopin: studies, criticisms, sketches] (Warsaw, 1933)

81

E. Ganche: *Voyages avec Frédéric Chopin* (Paris, 1934)

L. Bronarski: *Etudes sur Chopin*, i–ii (Lausanne, 1944–6)
——: *Chopin et l'Italie* (Lausanne, 1946)

A. Gide: *Notes sur Chopin* (Paris, 1948; Eng. trans., 1949)

A. Cortot: *Aspects de Chopin* (Paris, 1949; Eng. trans., 1951/R1974, as *In Search of Chopin*)

M. Mirska: *Szlakiem Chopina* [Along Chopin's paths] (Warsaw, 1949)

'Autour de Frédéric Chopin, sa correspondance, ses portraits', *ReM* (1955), no.229

Rocznik chopinowski/Annales Chopin, i–xii (Warsaw, 1956–80)

Chopin-Jahrbuch, ed. F. Zagiba, i–iii (Vienna, 1956, 1963, 1970)

M. Tomaszewski, ed.: *Kompozytorzy polscy o Fryderyku Chopinie* [Polish composers on Chopin] (Kraków, 1959, 2/1964)

G. Edelman, ed.: *Friderik Shopen: stat'i i issledovaniya sovetskikh muzïkovedov* [Articles and research by Soviet musicologists] (Moscow, 1960)

Z. Lissa, ed.: *F. F. Chopin* (Warsaw, 1960)

L. Bronarski: *Szkice chopinowskie* [Sketches of Chopin] (Kraków, 1961)

Z. Lissa, ed.: *The Book of the First International Musicological Congress devoted to the Works of Frederick Chopin* (Warsaw, 1963) [*Chopin Congress: Warszawa 1960*]

Chopin, Génies et réalités, xxv (Paris, 1965, 3/1973)

A. Walker, ed.: *Frédéric Chopin: Profiles of the Man and the Musician* (London, 1966, 2/1973 as *The Chopin Companion*, 3/1979)

A. Harasowski: *The Skein of Legends around Chopin* (Glasgow, 1967)

S. M. Chentova, ed.: *Shopen, kakim mï evo slïshim* (Moscow, 1970)

Z. Lissa: *Studia nad twórczością Fryderyka Chopina* [Studies on the works of Chopin] (Kraków, 1970)

L. A. Mazel': *Issledovaniya o Shopene* [Research on Chopin] (Moscow, 1971)

D. Żebrowski, ed.: *Studies in Chopin* (Warsaw, 1973)

Piano Quarterly, cxiii (1981), spr. [special Chopin issue]

BIOGRAPHY: MEMOIRS, LIFE

G. Sand: *Un hiver à Majorque* (Paris, 1842, 5/1929)
——: *Histoire de ma vie* (Paris, 1854, 7/1928)

W. von Lenz: *Die grossen Pianoforte-Virtuosen unserer Zeit aus persönlicher Bekanntschaft* (Berlin, 1872; Eng. trans., 1899/R1972), 19–50

A. Wodziński: *Les trois romans de Frédéric Chopin* (Paris, 1886, 6/1927)

C. Willeby: *Frédéric François Chopin: a Biography* (London, 1892)

Bibliography

E. Delacroix: *Journal*, i–iii (Paris, 1893–5, 4/1960)

G. de Pourtalès: *Chopin ou le poète* (Paris, 1927, 10/1963; Eng. trans., 1927, as *Frederick Chopin: a Man of Solitude*, 2/1933 as *Polonaise: the Life of Chopin*)

E. Vuillermoz: *La vie amoureuse de Chopin* (Paris, 1927, enlarged 2/1960)

W. D. Murdoch: *Chopin: his Life* (London, 1934/R1971)

N. Salvaneschi: *Il tormento di Chopin* (Milan, 1934, 9/1946)

A. Gronowicz: *Chopin* (New York, 1943, 2/1946)

K. Stromenger: 'Kronika życia' [Chronicle of his life], *Almanach chopinowski* (Warsaw, 1949), 7–79

K. Wierzyński: *The Life and Death of Chopin* (New York, 1949, 2/1972; Pol. orig., 1953)

S. Tenand: *Portraits de Chopin* (Paris, 1950)

J. Iwaszkiewicz: *Chopin* (Kraków, 1955, 3/1965; Ger. trans., 1958, 2/1964; Fr. trans., 1966)

C. Bourniquel: Chopin (Paris, 1957; Eng. trans., 1960)

J. Rousselot: *La vie passionnée de Frédéric Chopin* (Paris, 1957)

A. Czartkowski and Z. Jeżewska: *Fryderyk Chopin* (Warsaw, 1958, 5/1975)

L. Fábián: *Wenn Chopin ein Tagebuch geführt hätte* ... (Budapest, 1964, 2/1965)

V. Seroff: *Frederic Chopin* (New York, 1964)

G. Belotti: *Chopin, l'uomo*, i–iii (Milan, 1974)

B. Gavoty: *Frédéric Chopin* (Paris, 1974; Eng. trans., 1977)

J. M. Smoter, ed.: *Album Chopina/L'album de Chopin: 1829–1831* (Kraków, 1975)

W. Dulęba: *Chopin* (Kraków, 1975)

G. R. Marek and M. Gordon-Smith: *Chopin* (New York, 1978)

D. Melville: *Chopin: a Biography, with a Survey of Books, Editions and Recordings* (London, 1977)

A. Zamoyski: *Chopin* (London, 1979)

LIFE AND WORKS

J. W. Davison: *Essay on the Works of Frederick Chopin* (London, 1843)

J. Sikorski: 'Wspomnienie Chopina' [In memory of Chopin], *Biblioteka warszawska*, xxxvi (1849), 510–59

F. Liszt: *F. Chopin* (Paris, 1852, 14/1959; Eng. trans., 1877, 2/1963)

H. Barbedette: *Chopin* (Paris, 1861, enlarged 2/1869)

M. A. Szulc: *Fryderyk Chopin i utwory jego muzyczne* [Chopin and his musical works] (Poznań, 1873)

M. Karasowski: *Friedrich Chopin: sein Leben, seine Werke und Briefe*, i–ii (Dresden, 1877, 4/1914; Eng. trans., 1879/R1970, 3/1938)

F. Niecks: *Frederick Chopin as a Man and Musician*, i–ii (London, 1888, 3/1902/*R*1973)

J. G. Huneker: *Chopin: the Man and his Music* (New York, 1900/*R*1972)

——: 'The Greater Chopin', *Mezzotints in Modern Music* (London, 1900/*R*1972)

J. C. Hadden: *Chopin* (London, 1903, rev. 2/1935/*R*1975)

E. Ganche: *Frédéric Chopin: sa vie et ses oeuvres* (Paris, 1909/*R*1972, 7/1949; Eng. trans., 1922)

F. Hoesick: *Chopin: życie i twórczość* [Chopin: life and work], i–iii (Warsaw, 1910–11, 3/1962–8)

B. Scharlitt: *Chopin* (Leipzig, 1919)

Z. Jachimecki: *Fryderyk Chopin* (Kraków, 1926, 2/1949; Fr. trans., 1930)

L. Binental: *Chopin* (Paris, 1934)

A. Hedley: *Chopin* (London, 1947, rev. 3/1974)

H. Weinstock: *Chopin: the Man and his Music* (New York, 1949, 2/1959)

I. F. Belza: *Friderik Frantsishek Shopen* (Moscow, 1960; Pol. trans., 1969)

J. M. Chominski: *Fryderyk Chopin* [Ger. trans.] (London, 1980)

SPECIAL BIOGRAPHICAL STUDIES

H. Volkmann: *Chopin in Dresden* (Dresden, 1933; suppl., 1936)

B. Ferra: *Chopin and George Sand in Majorca* (Palma de Mallorca, 1936/*R*1974)

S. Brookshaw: *Concerning Chopin in Manchester* (Manchester, 1938, enlarged 2/1951)

S. Łobaczewska: 'Fryderyk Chopin w świetle nauki o typach ludzkich' [Chopin in the light of the science of human typology], *Ruch muzyczny*, vii (1947), no.9, p.8; no.10, p.5; no.11, p.11; no.12, p.9; nos.13–14, p.21; nos.15–16, p.13

J. Prosnak: 'Środowisko warzawskie w życiu Fryderyka Chopina' [The Warsaw society in Chopin's life], *KM* (1949), no.28, pp.7–126

B. E. Sydow: 'Um Chopins Geburtsdatum', *Mf*, iii (1950), 246

S. and D. Chainaye: *De quoi vivait Chopin?* (Paris, 1951)

S. Szuman: 'Dowcip i ironia Chopina' [Chopin's wit and irony], *Muzyka*, ii/2 (1951), 23

F. Zagiba: *Chopin und Wien* (Vienna, 1951)

J. Miketta: 'Aus der Genealogie der lothringischen bäuerlichen Vorfahren Chopins', *Chopin-Jb*, i (1956), 156

M. Godeau: *Le voyage à Majorque de George Sand et Frédéric Chopin* (Paris, 1959)

Bibliography

F. German: *Chopin i literaci warszawscy* [Chopin and Warsaw's men of letters] (Kraków, 1960)

A. Molnár: 'Die Persönlichkeit Chopins', *Chopin Congress: Warszawa 1960*, 701

T. Frączyk: *Warszawa młodości Chopina* [Warsaw in Chopin's youth] (Kraków, 1961)

M. Gliński: *Chopin the Unknown* (Windsor, Ont., 1963)

A. Hedley: 'Chopin – the Man', *Frédéric Chopin*, ed. A. Walker (London, 1966), 1

J. Procházka: *Chopin and Bohemia* (Prague, 1969)

L. Ripoll: *The Majorcan Episode of Chopin and George Sand 1838–1839* (Palma de Mallorca, 1969)

A. Zborski: *Chopin and the Land of his Birth* (Warsaw, 1975)

F. Ziejka: 'Chopin w Marsylii' [Chopin in Marseilles], *Kultura i spoleczenstwo*, xix/4 (1975), 113

B. Jaeger: 'Quelques nouveaux noms d'élèves de Chopin', *RdM*, lxiv (1978), 76

W. G. Atwood: *The Lioness and the Little One: the Liaison of George Sand and Frédéric Chopin* (New York, 1980)

S. Jarocinski: 'Alcune osservazioni sull'impegno ideologico di Chopin', *Pagine*, iv (1980), 191

J. J. Eigeldinger: 'Un concert inconnu de Chopin à Paris', *Revue musicale de Suisse romande*, xxxiv (1981), 2

R. Fiske: *Scotland in Music: a European Enthusiasm* (Cambridge, 1983), 149

HEALTH

K. Barry: *Chopin and his Fourteen Doctors* (Sydney, 1934)

E. Ganche: *Souffrances de Frédéric Chopin: essai de médicine et de psychologie* (Paris, 1935)

E. R. Long: *A History of the Therapy of Tuberculosis and the Case of Frederic Chopin* (Lawrence, Kansas, 1956)

F. H. Franken: *Das Leben grosser Musiker im Spiegel der Medizin: Schubert–Chopin–Mendelssohn* (Stuttgart, 1959)

D. Kerner: 'Frédéric Chopin', *Krankheiten grosser Musiker* (Stuttgart, 1963), 127; (3/1973)

S. Szpilczyński: 'War Frédéric Chopin Allergiker?', *Mickiewicz-Blätter* (1964), no.26, p.102

C. Sielużycki: 'Czy Chopin mógl żyć dłużej? rozważania lekarza' [Could Chopin have lived longer? considerations of a doctor], *Ruch muzyczny*, xix/1 (1975), 15

——: 'W sprawie "alergicznego podłoża" chorób Chopina' [The 'allergic basis' of Chopin's diseases], *Polski tygodnik lekarski*, xxx (1975), 229

——: 'Lekarze Chopina' [Chopin's doctors], *Archiwum historii medy-cyny*, xxxix (1976), 305

G. Böhme: *Medizinische Porträts berühmter Komponisten: Mozart, Beethoven, Weber, Chopin, Tschaikowski, Bartók* (Stuttgart, 1979)

CONTEMPORARIES

M. Doüel: 'Chopin and Jenny Lind', *MQ*, xviii (1932), 423

C. Chamfray: *Musset–Chopin: Confrontation* (Paris, 1934)

M. Wöss: 'Chopin–Lenau–Schumann', *ÖMz*, i (1949), 279

A. Hedley: 'Chopin and Countess Delphine', *ML*, xxxii (1951), 399

S. Pugliatti: *Chopin e Bellini* (Messina, 1952)

F. German: 'Chopin i Mickiewicz' [Chopin and Mickiewicz], *Rocznik chopinowski*, i (1956), 227

M. J. E. Brown: 'Chopin and his English Publishers', *ML*, xxxix (1958), 363

A. E. Bone: *Jane Wilhelmina Stirling 1804–1859* (Chipstead, 1960)

W. Eggert: 'Chopins Lieblingsschüler Adolf Gutmann', *Mickiewicz-Blätter* (1960), no.15, p.221

Z. Lissa: 'Chopin im Lichte des Briefwechsels von Verlegern seiner Zeit', *FAM*, vii (1960), 46

D. V. Zhitomirsky: 'Shopen i Shuman' [Chopin and Schumann], *Friderik Shopen*, ed. G. Edelman (Moscow, 1960), 296

A. Melbechowska: 'Teofil Kwiatkowski: malarz Chopina' [Kwiatkowski: Chopin's painter], *Biuletyn historii sztuki*, xxiii/2 (1961), 118

J. Starzyński: *Delacroix et Chopin* (Warsaw, 1962)

P. A. Gaillard: 'Jugements portés sur Chopin par Mickiewicz', *SMz*, ciii (1963), 289

S. A. Semenovsky: 'Russkiye druz'ya i znakomïye Shopena' [Russian friends and acquaintances of Chopin], *Russko-polskiye muzïkal'nïye svyazi* (Moscow, 1963), 119

C. Colombati: 'Chopin a Bellini', *Ruch muzyczny*, xix/21 (1975), 3

T. Higgins: 'Delphine Potocka and Frédéric Chopin', *Journal of the American Liszt Society*, viii (1980), 64; ix (1981), 73

PIANIST AND TEACHER

W. Landowska: 'How Chopin Played Chopin', *Musical Standard*, i (1913), 10

E. J. Hipkins: *How Chopin Played* (London, 1937)

S. Stookes: 'Chopin the Teacher', *MMR*, lxxxiv (1954), 119

L. Bronarski: 'Les élèves de Chopin', *Annales Chopin*, vi (1965), 7

J. J. Eigeldinger: *Chopin vu par ses élèves* (Neuchâtel, 1970)

J. Holland: *Chopin's Teaching and his Students* (diss., U. of North Carolina, 1973)

Bibliography

J. Ekier: 'Chopin jako pedagog' [Chopin as a pedagogue], *Ruch muzyczny*, xviii/10–11 (1974)

J. Methuen-Campbell: *Chopin Playing from the Composer to the Present Day* (London, 1981)

WORKS: MUSICAL ELEMENTS

E. S. Kelley: *Chopin the Composer* (New York, 1913)

J. P. Dunn: *Ornamentation in the Works of Frederick Chopin* (London, 1921, 2/1930/*R*1971)

H. Leichtentritt: *Analyse der Chopinschen Klavierwerke*, i–ii (Berlin, 1921)

B. Wójcik-Keuprulian: *Melodyka Chopina* (Lwów, 1930)

L. Bronarski: *Harmonika Chopina* (Warsaw, 1935)

M. Ottich: *Die Bedeutung des Ornaments im Schaffen Friedrich Chopins* (Berlin, 1937)

J. Miketta: 'Ze studiów nad melodyką Chopina' [Studies on Chopin's melody], *KM* (1949), nos.26–7, pp.289–359

F. Eibner: 'Chopins kontrapunktisches Denken', *Chopin-Jb*, i (1956), 103

H. Zelzer: 'Zur Satztechnik Chopins: eine musiktheoretische Untersuchung ihrer Kriterien', *Chopin-Jb*, i (1956), 85

Z. Lissa: 'Die Chopinsche Harmonik aus der Perspektive der Klangtechnik des 20. Jahrhunderts', *DJbM*, ii (1957), 68; iii (1958), 74

K. Hławiczka: 'Reihende polymetrische Erscheinungen in Chopins Musik', *Annales Chopin*, iii (1958), 68–99

J. M. Chomiński: 'Harmonika a faktura fortepianowa Chopina' [Chopin's harmony and piano texture], *Muzyka*, iv/4 (1959), 3

S. Borris: 'Chopins Bedeutung für den Chromatismus des XIX. Jahrhunderts', *Chopin Congress: Warszawa 1960*, 107

J. Chailley: 'L'importance de Chopin dans l'évolution du langage harmonique', *Chopin Congress: Warszawa 1960*, 30

J. M. Chomiński: 'Z zagadnień faktury fortepianowej Chopina' [Problems of Chopin's piano texture], *F. F. Chopin*, ed. Z. Lissa (Warsaw, 1960), 150

V. A. Cukkerman: 'De l'emploi des genres et des formes dans l'oeuvre de Chopin', *Chopin Congress: Warszawa 1960*, 114

E. J. Dreyer: 'Melodisches Formelgut bei Chopin', *Chopin Congress: Warszawa 1960*, 132

K. Hławiczka: 'Chopin – Meister der rhythmischen Gestaltung', *Annales Chopin*, v (1960), 31–81

——: 'Eigentümliche Merkmale von Chopins Rhythmik', *Chopin Congress: Warszawa 1960*, 185

Yu. A. Kremlyov: 'La place historique de l'harmonie de Chopin', *Chopin Congress: Warszawa 1960*, 202

87

Z. Lissa: 'Die Formenkreuzung bei Chopin', *Chopin Congress: Warszawa 1960*, 207

L. A. Mazel': 'Nekotorïye chertï kompozitsii v svobodnïkh formakh Shopena' [Some patterns of composition in the free forms of Chopin], *Friderik Shopen*, ed. G. Edelman (Moscow, 1960), 182–231

Ya. I. Mil'shteyn: 'Fortepiannaya faktura Shopena i Lista' [The piano texture of Chopin and Liszt], *Chopin Congress: Warszawa 1960*, 341

W. W. Protopopow: 'Variacionnyi metod razvitiya tematizma v muzyke Shopena' [The variational method of thematic process in Chopin's works], *Friderik Shopen*, ed. G. Edelman (Moscow, 1960), 232–95

K. Reinhardt: 'Zur Frage des Tempos bei Chopin', *Chopin Congress: Warszawa 1960*, 449

D. T. Turło: 'Formotwórcza rola dynamiki w utworach Chopina' [The formal role of dynamics in Chopin's works], *Muzyka*, vi (1961), no.2, p.3; no.3, p.29

J. Scherpereel: 'Chopin et l'élargissement de la tonalité', *Revue d'esthétique*, xviii (1965), 73

D. T. Turło: 'The Evolution of Dynamics as an Element of Construction in Chopin's Works', *Annales Chopin*, vi (1965), 90

A. Walker: 'Chopin and Musical Structure: an Analytical Approach', *Frédéric Chopin* (London, 1966), 227–57

R. Klein: 'Chopins Sonatentechnik', *ÖMz*, xix (1967), 389

G. Abraham: 'Chopin and the Orchestra', *Slavonic and Romantic Music* (New York, 1968), 23

F. F. McGinnis: *Chopin: Aspects of Melodic Style* (diss., Indiana U., 1968)

Z. Chechlińska: 'Rodzaje tempa w utworach Chopina' [The kinds of tempo in Chopin's works], *Muzyka*, xiv/2 (1969), 45

F. Eibner: 'Über die Akkorde im Satz Chopins', *Chopin-Jb*, iii (1970), 3

Z. Chechlińska: 'Studies on the Chopin Melodic Design', *Studies in Chopin*, ed. D. Żebrowski (Warsaw, 1973), 62

U. Dammeier-Kirpal: *Der Sonatensatz bei Frédéric Chopin* (Wiesbaden, 1973)

Z. Helman: 'Chopin's Harmonic Devices in 20th-century Theoretical Thought', *Studies in Chopin*, ed. D. Żebrowski (Warsaw, 1973), 49

T. Higgins: 'Tempo and Character in Chopin', *MQ*, lix (1973), 106

R. S. Parks: 'Voice Leading and Chromatic Harmony in the Music of Chopin', *JMT*, xx (1976), 189

K. M. Moore: *Linearity of Voice Structure in Selected Works of Frédéric Chopin and its Implications in Performance* (diss., New York U., 1978)

Z. Chechlińska: 'Italianismi e caratteri vocale della melodia di Chopin', *Pagine*, iv (1980), 181

Bibliography

H. Krebs: 'Alternatives to Monotonality in Early Nineteenth-century Music', *JMT*, xxv (1981), 1

WORKS: STYLE, INFLUENCE, RELATIONSHIPS

K. Szymanowski: 'Frédéric Chopin et la musique polonaise moderne', *ReM* (1931), no.121, p.398

G. Abraham: *Chopin's Musical Style* (London, 1939, rev. 4/1960)

V. V. Pashchalov: *Shopen i pol'skaya narodnaya muzïka* [Chopin and Polish folk music] (Moscow, 1941, 2/1949)

W. L. Landowski: *Frédéric Chopin et Gabriel Fauré* (Paris, 1946)

S. Borris: 'Chopin und die deutsche Romantik', *Chopin Almanach* (Potsdam, 1949), 34; repr. in *ÖMz*, vi (1954), 314

Z. Lissa: 'Der Einfluss Chopins auf die russische Musik', *Chopin-Almanach* (Potsdam, 1949), 88

F. Zagiba: 'Chopin als Mozart-Verehrer', *Chopin-Jb*, i (1956), 177–207

J. M. Chomiński: 'La maîtrise de Chopin compositeur', *Annales Chopin*, ii (1957), 179–237

Z. Lissa: 'Le style national des oeuvres de Chopin', *Annales Chopin*, ii (1957), 100–78

S. Łobaczewska: 'L'apport de Chopin au romantisme européen', *Annales Chopin*, ii (1957), 7–99

I. F. Belza: 'Natsional'nïye istoki tvorchestva Shopena' [The folk sources of Chopin's music], *Chopin Congress: Warszawa 1960*, 23

K. Biegański: 'Evolution de l'attitude de Chopin à l'égard du folklore (suivant ses mazurkas)', *Chopin Congress: Warszawa 1960*, 95

J. M. Chomiński: 'Die Evolution des Chopinschen Stils', *Chopin Congress: Warszawa 1960*, 44

R. L. Henderson: 'Chopin and the Expressionists', *ML*, xli (1960), 38

J. A. Kremlev: 'L'importance mondiale de l'esthétique de Chopin', *Chopin Congress: Warszawa 1960*, 696

M. Ladmanová: 'Chopin und Smetana', *Chopin Congress: Warszawa 1960*, 324

S. Łobaczewska: 'La culture musicale en Pologne au début du XIXe siècle et ses relations avec la musique de Chopin', *Chopin Congress: Warszawa 1960*, 63

H. J. Moser: 'Chopin stilkundlich betrachtet', *Chopin Congress: Warszawa 1960*, 707

M. Ottich: 'Chopin und die Komponisten der nachfolgenden Generationen', *Chopin Congress: Warszawa 1960*, 350

W. Siegmund-Schultze: 'Chopin und Brahms', *Chopin Congress: Warszawa 1960*, 388

A. Silbermann: 'Sozialpsychologische Aspekte im Wandel des Chopin-Idols', *Chopin Congress: Warszawa 1960*, 604

V. A. Tsukkerman: 'Zametki o muzïkal'nom yazïke Shopena' [Remarks

on Chopin's musical language], *Friderik Shopen*, ed. G. Edelman (Moscow, 1960), 44–181

W. Wiora: 'Chopins Préludes und Etudes und Bachs Wohltemperiertes Klavier', *Chopin Congress: Warszawa 1960*, 73

H. Hollander: 'Chopin als Vorläufer des Impressionismus', *NZM*, cxxxii (1961), 400

P. Badura-Skoda: 'Chopin und Liszt', *ÖMz*, xiv (1962), 60

Z. Lissa: 'Shopen i Skryabin', *Russko-polskiye muzïkal'nïye svyazi* (Moscow, 1963), 293–374

E. H. Mueller von Asow: 'Fryderyk Chopin in Deutschland', *Chopin-Jb*, ii (1963), 161–215

Yu. N. Tyulin: *O programnosti v proizvedeniyakh Shopena* [The programmatic in Chopin's works] (Leningrad, 1963, 2/1968)

I. F. Belza: 'Problemï izucheniya stilya Shopena' [Problems of examining Chopin's style], *Annales Chopin*, vi (1965), 39–72

J. Kłobukowska: 'Chopin et la musique d'autrefois', *Annales Chopin*, vi (1965), 112

J. Racek: 'Les idées de Leoš Janáček sur la structure de composition des oeuvres pour piano de Frédéric Chopin', *Annales Chopin*, vi (1965), 13

J. Starzyński: *O romantycznej syntezie sztuk: Delacroix–Chopin–Baudelaire* [The Romantic synthesis of the arts: Delacroix–Chopin–Baudelaire] (Warsaw, 1965)

K. Wilkowska-Chomińska: 'Chopin et Paganini', *Annales Chopin*, vi (1965), 104

P. Badura-Skoda: 'Chopin's Influence', *Frédéric Chopin*, ed. A. Walker (London, 1966), 258

Z. Lissa: 'Max Regers Metamorphosen der Berceuse op.57 von Frédéric Chopin', *FAM*, xiii (1966), 79; repr. in *Mf*, xxiii (1970), 277

——: 'Der Einfluss Chopins auf A. K. Lyadow', *DJbM*, xiii (1968), 5–42

J. M. Chomiński: 'Szymanowski a Chopin', *Rocznik chopinowski*, vii (1969), 52

S. Lazarov: 'Chopin–Wagner–Mahler', *Rocznik chopinowski*, vii (1969), 100

Z. Lissa: 'Beethovens Stilelemente im Schaffen Fryderyk Chopins', *II. Internationales musikologisches Symposium: Piešt'any 1970*, 75

D. Branson: *John Field and Chopin* (London, 1972)

Z. Chechlińska: 'Chopin a impresjonizm' [Chopin and Impressionism], *Szkice o kulturze muzycznej XIX wieku*, ii (Warsaw, 1973), 21

E. Dziębowska: 'Chopin – romantyk, klasyk, modernista' [Chopin: Romantic, Classic, modernist], *Szkice o kulturze muzycznej XIX wieku*, ed. Z. Chechlińska, ii (Warsaw, 1973), 7

C. Colombati: 'Chopin, Scarlatti i muzyka włoska' [Chopin, Scar-

Bibliography

latti and Italian music], *Ruch muzyczny*, xx/4 (1976), 3

H. Kinzler: *Frédéric Chopin: über den Zusammenhang von Satztechnik und Klavierspiel* (Munich, 1977)

C. Colombati: 'Federico Chopin e il melodramma italiano', *Pagine*, iv (1980), 167

C. Rosen: 'Influence: Plagiarism and Inspiration', *19th Century Music*, iv (1980–81), 87

EDITORIAL QUESTIONS, INTERPRETATION

J. Kleczyński: *O wykonywaniu dzieł Szopena* (Warsaw, 1879; Eng. trans., 1896, 6/1913, rev. 7/1970 as *How to Play Chopin*)
——: *Chopin w celniejszych swoich utworach* (Warsaw, 1886; Eng. trans., 1896, 2/1898 as *Chopin's Greater Works: how they Should Be Understood*)

G. C. A. Jonson: *A Handbook to Chopin's Works* (London, 1905, rev. 2/1908/R1972)

R. Koczalski: *Chopin-Zyklus: vier Klaviervorträge* (Leipzig, 1909)

L. Kamieński: 'Zum Tempo rubato', *AMw*, i (1918–19), 108

J. B. McEwen: *Tempo rubato or Time Variation in Musical Performance* (London, 1928)

J. F. Porte: *Chopin: the Composer and his Music: an Analytical Critique of Famous Traditions and Interpretations* (London, 1935)

R. Koczalski: *Frédéric Chopin: Betrachtungen, Skizzen, Analysen* (Cologne, 1936)

R. Caporali: 'La scrittura pianistica chopiniana e la sua interpretazione', *RaM*, xix (1949), 286

B. von Poźniak: *Chopin: praktische Anweisungen für das Studium der Chopin-Werke* (Halle, 1949)

D. N. Ferguson: *Piano Interpretation: Studies in the Music of Six Great Composers* (London, 1950), 183–249

M. J. E. Brown: 'The Posthumous Publications of Chopin's Songs', *MQ*, xlii (1956), 51

O. Jonas: 'On the Study of Chopin's Manuscripts', *Chopin-Jb*, i (1956), 142

Z. Drzewiecki: 'Der zeitgenössische polnische Aufführungsstil der Werke von Fryderyk Chopin', *Annales Chopin*, ii (1957), 243

P. Badura-Skoda: 'Um den Chopinschen Urtext', *NZM*, cxxi (1960), 82

M. J. E. Brown: 'First Editions of Chopin in Periodicals and Serial Publications', *Annales Chopin*, v (1960), 7

Z. Drzewiecki: 'Le style d'interprétation de Chopin dans la pédagogie polonaise contemporaine', *Chopin Congress: Warszawa 1960*, 430

A. Hedley: 'Some Observations on the Autograph Sources of Chopin's Works', *Chopin Congress: Warszawa 1960*, 474

W. Heinitz: 'Essentielle Erkenntnisse zur Werk-Ästhetik Frédéric Chopins', *Annales Chopin*, v (1960), 13

Ya. I. Mil'shteyn: 'K istorii izdanii sochinenii Shopena' [History of the editions of Chopin's works], *Friderik Shopen*, ed. G. Edelman (Moscow, 1960), 323–60

M. Sobieski and J. Sobieska: 'Das Tempo rubato bei Chopin und in der polnischen Volksmusik', *Chopin Congress: Warszawa 1960*, 247

E. Zimmermann: 'Probleme der Chopin-Edition', *Mf*, xiv (1961), 155

H. Keller: 'Zur Textkritik der Préludes und Etudes von Chopin', *Chopin-Jb*, ii (1963), 80

Z. Drzewiecki: 'How to Play Chopin', *Polish Perspectives*, viii/2 (1965), 14

T. Higgins: *Chopin Interpretation: a Study of Performance Directions in Selected Autographs and other Sources* (diss., U. of Iowa, 1966)

Ya. I. Mil'shteyn: *Soveti̇̆ Shopena pianistam* [Chopin's advice to pianists] (Moscow, 1967)

G. Belotti: *Le origini italiane del 'rubato' chopiniano* (Wrocław, 1968)

J. Ekier: 'The National Edition of the Works of Frédéric Chopin', *Polish Music*, iii/1 (1968), 3

K. Kobylańska: 'Prace Chopina nad zbiorowym wydaniem dzieł własnych' [Chopin's work on the collected edition of his own works], *Ruch muzyczny*, xii/14 (1968), 3

E. Hutcheson: *The Literature of the Piano* (London, rev. 2/1969), 206–50

D. T. Turło: 'Dziedzictwo formy artystycznej Chopina' [The heritage of Chopin's artistic manner of interpretation], *Annales Chopin*, viii (1969), 7–43

J. Ekier: *Wydanie narodowe dzieł Fryderyka Chopina: komentarze źródłowe: ballady* [Source commentary to the (Polish) national edition of Chopin's ballades] (Kraków, 1970)

——: 'The Purpose, Scope and Character of the National Edition', *Polish Music*, ix/3 (1974), 4

——: *Wstęp do Wydania narodowego dzieł Fryderyka Chopina*, i: *Zagadnienia edytorskie* [Introduction to the (Polish) national edition of Chopin's works: editorial problems] (Warsaw, 1974)

R. Nowacki: *Ocena wykonań konkursowych na przykładzie konkursów chopinowskich* [The value of competitive performances as revealed at the Chopin piano competitions in Warsaw] (Warsaw, 1974)

N. Temperley: 'Scorography: the Music of Chopin', *Musical Newsletter*, iv/1 (1974), 11

J. J. Eigeldinger: 'Un autographe musical inédit de Chopin', *SMz*, cxv (1975), 18

E. M. Frederick: 'The "Romantic" Sound in Four Pianos of Chopin's Era', *19th Century Music*, iii (1979–80), 150

Bibliography

T. Higgins: 'Whose Chopin?', *19th Century Music*, v (1981–2), 67

J. Kallberg: *The Chopin Sources: Variants and Versions in Later Manuscripts and Printed Editions* (diss., U. of Chicago, 1982)

BALLADES

K. Wilkowska: 'Środki wyrazu emocjonalnego w balladach Chopina' [Methods of emotional expression in Chopin's ballades], *KM* (1949), no.28, pp.167–239

F. Eibner: 'Über die Form der Ballade op.23 von Fr.Chopin', *Annales Chopin*, iii (1958), 107

W. W. Protopopow: 'Polifonia w Balladzie f-moll Chopina' [Polyphony in Chopin's F minor Ballade], *Rocznik chopinowski*, vii (1969), 34

N. D. Witten: *The Chopin Ballades: an Analytical Study* (diss., Boston U., 1979)

MAZURKAS

J. Miketta: *Mazurki* (Kraków, 1949)

G. Belotti: 'L'assimetria ritmica nella mazurca chopiniana', *NRMI*, v (1971), 657, 827

W. Nowik: 'Chopin's Mazurka F-moll, op.68, Nr.4: "die letzte Inspiration des Meisters" ', *AMw*, xxx (1973), 109

PRELUDES

A. Eaglefield-Hull: 'Chopin's Preludes', *MMR*, lvi (1926), 662

J. M. Chomiński: 'Problem formy w preludiach Chopina' [Problems of form in Chopin's Préludes], *KM* (1949), nos.26–7, pp.183–288; (1949), no.28, pp.240–395

——: *Preludia* (Kraków, 1950)

M. J. E. Brown: 'The Chronology of Chopin's Préludes', *MT*, xcviii (1957), 423

Z. Chechlińska: 'Das Problem der Form und die reelle Klanggestalt in Chopins Präludien', *Chopin Congress: Warszawa 1960*, 425

W. Wiora: 'Über den geistigen Zusammenhang der Präludien und Etüden Chopins', *Musik des Ostens*, i (1962), 76

G. Belotti: 'Il problema delle date dei preludi di Chopin', *RIM*, v (1970), 159–215

T. Higgins: *Frederic Chopin: Preludes, opus 28: an Authoritative Score, Historical Background, Analysis, Views and Comments* (New York, 1973)

V. Toncitch: 'Regards sur les préludes de Chopin', *SMz*, cxiv/2 (1974),78

——: 'Regards sur les préludes de Chopin', *AnM*, xxxiii–xxxv (1978–81), 161

SONGS

S. Barbag: *Studium o pieśniach Chopina* [A study on Chopin's songs] (Lwów, 1927)

S. Stookes: 'Chopin, the Song-writer', *MMR*, lxxx (1950), 96

R. Prilisauer: 'Frédéric Chopins "Polnische Lieder" ', *Chopin-Jb*, ii (1963), 117

WALTZES

A. Koszewski: 'Melodyka walców Chopina' [Melody in Chopin's waltzes], *Studia muzykologiczne*, ii (1953), 276–341

——: 'Problemy rytmiczne i agogiczne w walcach Chopina' [Problems of rhythm and tempo in Chopin's waltzes], *Annales Chopin*, iii (1958), 113

——: 'Das Walzerelement im Schaffen Chopins', *DJbM*, v (1960), 58

——: 'Das Wienerische in Chopins Walzern', *Chopin-Jb*, ii (1963), 27

OTHER WORKS

L. A. Mazel': *Fantaziya f-moll Shopena: opït analiza* [Towards an analysis of Chopin's F minor Fantasia] (Moscow, 1937; repr. in *Issledovaniya o Shopene* (Moscow, 1971), 5–141

D. F. Tovey: Programme Notes on Works by Chopin, *Essays in Musical Analysis: Chamber Music* (London, 1944/*R*1972), 155

H. Feicht: 'Ronda Fr. Chopina' [Chopin's rondos], *KM* (1948), nos.21–2, p.35; no.23, pp.23–62; no.24, pp.7–54

K. Wilkowska: 'Impromptus Chopina', *KM* (1949), nos.26–7, pp.102–82

J. Prosnak: 'Wariacje fletowe Chopina' [Chopin's Variations for flute and piano], *Studia muzykologiczne*, i (1953), 267–307

M. J. E. Brown: 'Chopin's Lento con gran espressione', *MMR*, lxxxvi (1956), 207

A. Frączkiewicz: 'Faktura fortepianowa koncertów Fryderyka Chopina' [Piano texture in Chopin's concertos], *Annales Chopin*, iii (1958), 133

J. M. Chomiński: 'Structure à motifs des études de Chopin en tant que problème d'éxécution', *Annales Chopin*, iv (1959), 75–122

Z. Chechlińska: 'Ze studiów nad źródłami do Scherz F. Chopina' [A source study of Chopin's scherzos], *Annales Chopin*, v (1960), 82–199

J. M. Chomiński: *Sonaty Chopina* [Chopin's sonatas] (Kraków, 1960)

A. Whiteside: *Mastering the Chopin Etudes and other Essays* (New York, 1969), 26–127

W. S. Newman: *The Sonata since Beethoven* (Chapel Hill, 1969, rev. 2/1972)

Bibliography

I. Barbag-Drexler: 'Die Impromptus von Fryderyk Chopin', *Chopin-Jb*, iii (1970), 25–108

G. Belotti: 'Le prime composizioni di Chopin: problemi e osservazioni', *RIM*, vii (1972), 230–91

J. Ekier: 'Das "Impromptu" cis-moll von Frédéric Chopin', *Melos*, iv (1978), 201

J. T. Lam (Fang): *Chopin's Approach to Form in his Four Piano Scherzos* (diss., Michigan State U., 1979)

J. Falenciak: 'Variazioni di Federico Chopin su temi italiani', *Pagine*, iv (1980), 233

W. Nowik: 'Chopin e Bellini: "Casta Diva", manoscritto di F. Chopin', *Pagine*, iv (1980), 241

W. R. YaDeau: *Tonal and Formal Structure in Selected Larger Works of Chopin* (diss., U. of Illinois, 1980)

J. S. Bollinger: *An Integrative and Schenkerian Analysis of the B-flat Minor Sonata of Frédéric Chopin* (diss., Washington U., 1981)

OTHER STUDIES

I. J. Paderewski: *Chopin: a Discourse* (London, 1911)

K. Szymanowski: *Fryderyk Chopin* (Warsaw, 1925, 2/1949)

N. Slonimsky: 'Chopiniana: some Material for a Biography', *MQ*, xxxiv (1948), 467

J. C. Romero: *Chopin en México* (Mexico City, 1950)

J. Holcman: *The Legacy of Chopin* (New York, 1954)

I. F. Belza: 'Shopen i russkaya muzykal'naya kul'tura' [Chopin and the Russian musical culture], *Annales Chopin*, ii (1957), 254

G. B. Bernandt: 'Shopen v Rossii' [Chopin in Russia], *SovM* (1960), no.2, p.29

B. Johnsson: 'Chopin og Danmark', *Dansk musiktidskrift*, xxxv (1960), 33

G. Ohlhoff: 'Die Tragik der Chopinforschung', *NZM*, cxxi (1960), 89

Ting Shan-teh: 'What Makes the Chinese People Accept and Appreciate Chopin's Music', *Chopin Congress: Warszawa 1960*, 399

A. Zakin: 'Chopin and the Organ', *MT*, ci (1960), 780

W. Brennecke: 'Internationaler musikwissenschaftlicher Chopin-Kongress in Warschau', *Mf*, xiv (1961), 68

E. N. Waters: 'Chopin by Liszt', *MQ*, xlvii (1961), 170

R. Steglich: 'Chopins Klaviere', *Chopin-Jb*, ii (1963), 139

Z. Lissa: 'I-er Congrès international de musicologie à Varsovie consacré à l'oeuvre de Frédéric Chopin', *Annales Chopin*, vi (1965), 194

G. Belotti: 'Okoliczności powstania pierwszej monografii o Chopinie' [The circumstances of the origin of the first monograph on Chopin (by Liszt)], *Annales Chopin*, vii (1969), 7

L. Sługocka: 'Fryderyk Chopin in der deutschen Lyrik', *Mickiewicz-Blätter* (1969), nos.40–41, p.8

J. Prosnak: *The Frederic Chopin International Piano Competitions, Warsaw 1927–1970* (Warsaw, 1970)

K. Kobylańska: 'Chopin's Biography: Contemporary Research and History', *Studies in Chopin*, ed. D. Żebrowski (Warsaw, 1973), 116

W. Nowik: 'The Receptive-informational Role of Chopin's Musical Autographs', *Studies in Chopin*, ed. D. Żebrowski (Warsaw, 1973), 77

O. Pisarenko: 'Chopin and his Contemporaries: Paris, 1832–1860', *Studies in Chopin*, ed. D. Żebrowski (Warsaw, 1973), 30

Chopin in Silesia: Lectures from the Symposium 'Chopin in Silesia' (Katowice, 1974)

E. Galińska: 'Zastosowanie muzyki Chopina w terapii nerwic' [The application of Chopin's music in the treatment of neuroses], *Rocznik chopinowski*, ix (1975), 81–120

G. Belotti and W. Sandelewski: *Chopin in Italia: conferenzi e studi* (Wrocław, 1977)

G. Belotti: 'La fortina in Italia dell'opera di Chopin durante la vita del compositore', *Pagine*, iv (1980), 137

J. Opalski: *Chopin i Szymanowski w literaturze dwudziestolecia miedzywojennego* [Chopin and Szymanowski in literature of the inter-war period, 1919–39], i (Kraków, 1980)

J. Kallberg: 'Chopin in the Marketplace – Aspects of the International Music Publishing Industry', *Notes*, xxxix (1982–3), 535, 795

ROBERT SCHUMANN

Gerald Abraham

Eric Sams

CHAPTER ONE

Early years, 1810–27

Robert Alexander Schumann was born in Zwickau, Saxony, on 8 June 1810. He was the fifth and youngest child of August Schumann, bookseller, publisher and author (*b* 1773, son of a Saxon clergyman), and his wife Johanna Christiane (*b* 1771, daughter of a surgeon named Schnabel). The parents were married in 1795 and lived for more than 11 years at Zeitz; in the spring of 1807 they moved to Zwickau, where Schumann founded a publishing house. In the year of Roberts's birth his father was attacked by a 'nervous disorder' which affected his remaining years. The boy's education began in 1816, when he was sent to a local private school, where he showed no special gifts; at about the same time he seems to have had his first piano lessons from J. G. Kuntzsch (1775–1855), organist of St Mary's Church, a somewhat pedantic musician of limited ability. In August 1819 he heard Moscheles play at Karlsbad, and the impression was indelible; the piano style of Moscheles is still easily perceptible in Schumann's earlier published compositions. Either the same year or the year before the boy was taken to hear his first opera, *Die Zauberflöte*, at Leipzig. On 6 November 1821 he took part, at the piano, in Kuntzsch's performance of Friedrich Schneider's oratorio *Weltgericht* in St Mary's and, possibly under the influence of this work, then enormously popular, he composed in January 1822 a

9. Robert Schumann: miniature by an unknown artist

setting of Psalm cl for soprano, contralto, piano and
orchestra, inscribed 'Oeuv. I' – 26 pages of 13-line score
on plain paper ruled by himself – which was performed
by his fellow pupils and other young friends. For them
also he composed in the same year an *Ouverture et Chor
fürs grosse Orchester, Oeuv. 1/No. 3* – a nine-page overture followed by a brief chorus on words beginning 'Wie
reizend ist der schöne Morgen' – apparently suggested
by a vocal score of Paer's *Achille* which had fallen into

his hands. At Easter 1820 he had entered the Zwickau Lyceum, where he remained for eight years, and the record of his public appearances at the 'evening entertainments' arranged from time to time by the head of the Lyceum shows his considerable ability as a pianist.

At the same time the boy showed equal if not greater literary ability. Side by side with his regular education at the Lyceum he largely educated himself by promiscuous reading in his father's shop and library. This was encouraged by his father, who allowed him at the age of 13 to contribute some short articles to one of his publications, the *Bildergalerie der berühmtesten Menschen aller Völker und Zeiten*. At about the same time the boy compiled an anthology of album-verses, poems (partly his own, including a scene from a five-act tragedy, *Der Geist*) and passages from Schubart's *Ideen zu einer Aesthetik der Tonkunst*: 'Blätter und Blümchen aus der goldenen Aue. Gesammelt und zusammengebunden von Robert Schumann, genannt Skülander. 1823 (November und Dezember)'. A second book, begun in 1825 and continued until 1828, *Allerley aus der Feder Roberts an der Mulde* – Zwickau stands on the river Mulde – consists mainly of his own verses, which reflect the school exercise of making metrical German versions of Latin poems (he collected into a still extant volume his translations of selected odes of Horace). Schumann's love of 'societies' began to show itself in 1825, when he took a leading part in the founding of two schoolboy bodies, a secret 'Schülerverbindung' (founded 19 May) which cultivated fencing, but apparently not politics, and a literary 'Schülerverein' (first meeting 12 December) for the study of German literature. Probably to

101

this period of the Schülerverein belong the beginnings of some further dramatic essays: a *Coriolan*; a comedy, *Leonhard und Mantellier*; *Die beiden Montalti* and *Die Brüder Lanzendörfer* (said to be 'horror dramas').

Schumann's father now thought of sending him to study composition with Weber, but that master's death put an end to the plan and, a month or so after Weber, August Schumann himself died (10 August 1826). His 19-year-old daughter Emilie, Robert's only sister, a physical and mental invalid, had died a short time before. Calf-love for two girls whose names have been recorded, Nanni Petsch and Liddy Hempel, provided further emotional experiences which were characteristically worked into an autobiographical story, *Juniusabende und Julitage*, that he described two years later as 'my first work, my truest and my finest; how I wept as I wrote it and yet how happy I was'. Nanni and Liddy were soon partly but not entirely displaced in his affections by Agnes Carus, the young, intelligent and musical wife of a doctor at Colditz. At the end of July 1827, after visiting the Caruses at Colditz, he made a further excursion to Leipzig (where his closest school friends were now at the university), Dresden, Prague and Teplice (where he again met, and parted with, Liddy Hempel). His letters during this holiday refer frequently to champagne drinking, a habit he shook off only years later. Another habit formed in 1827 was that of keeping a diary or commonplace-book; his earliest diary was entitled 'Tage des Jünglinglebens'. From the summer of 1827, too, date his literary enthusiasm for Jean Paul, which had an overwhelming effect on his already flowery prose style, and his musical enthusiasm for Schubert. His notes record 'daily improvisation at the

instrument. Attempts to compose without instrument. . . .
Beginnings of a piano concerto in F minor'. Four
songs dating from 1827 have survived: *Verwandlung*
(words by E. Schulze), *Lied für xxx* (to his own poem
Leicht wie gaukelnde Sylphiden), *Sehnsucht* (also to his
own words), written in February and revised in June,
and a translation of Byron's *I saw thee weep* (in July);
no doubt they were written for Frau Carus, who was a
singer.

CHAPTER TWO

Leipzig and Heidelberg, 1828–31

On 15 March 1828 Schumann passed his school-leaving examination, and in doing so he won high praise. In obedience to the wishes of his mother and his guardian he then unwillingly matriculated as a law student in the University of Leipzig (29 March). Some autobiographical notes from a later period record: 'Easter 1828. Night raptures. Constant improvisation daily. Also literary fantasies in Jean Paul's manner. Special enthusiasm for Schubert, Beethoven too, Bach less. Letter to Franz Schubert (not sent)'.

Before settling at Leipzig he and a new acquaintance made there, Gisbert Rosen, set out on an expedition in the latter part of April to Bayreuth, Nuremberg, Augsburg and Munich, where on 8 May they introduced themselves to Heine and then parted – Rosen to study at Heidelberg, Schumann to return by way of Zwickau to Leipzig, where he was to share lodgings with his old school friend Emil Flechsig. He reached Leipzig again on 15 May; his friends the Caruses had now settled there, and at their house he met musicians, including Marschner, and took part in chamber music; otherwise his circle of friends was limited to Flechsig and a few other students. Despite his promises to his mother and guardian that he would devote himself to legal study, he did not attend a single lecture (according to Flechsig), but spent hours daily on his imitations of Jean Paul and

in improvisation at the piano. He was already haunted from time to time by fears of insanity. The literary productions of this period, recorded in a new notebook, 'Hottentottiana' (May 1828–June 1830), include notes for a novel, *Selene*, in which his idealized self figures as Gustav, fragments of shorter stories, *Die Harmonika* and *Weltteil*, and the beginning of a book on the aesthetics of music. Admiration for Schubert's *Erlkönig* led directly to the composition of another Goethe ballad, *Der Fischer* (June); it was followed by three settings of poems by Justinus Kerner (29 June–10 July), and these, perhaps with some earlier songs, were on 15 July submitted to the criticism of Gottlob Wiedebein (1779–1854), who then enjoyed some reputation as a song composer. In a second letter to Wiedebein (5 August), Schumann confessed that he was 'neither a connoisseur of harmony and thoroughbass nor a contrapuntist, but purely and simply guided by nature'; he would now set about the proper study of composition. Nevertheless, of four other songs written at this period he was afterwards able to use the musical substance of three (*An Anna II*, *Im Herbste* and *Hirtenknabe*) with little change in mature compositions: the Piano Sonatas op.11 and op.22 and the Intermezzo op.4 no.4. In August he began a course of piano study with the celebrated teacher Friedrich Wieck and consequently made the acquaintance of Wieck's nine-year-old daughter Clara (Hummel's A minor Concerto was one of the works studied; he told Hummel himself that he worked at it for a year). During August and September Schumann composed a set of *VIII polonaises pour le piano à quatre mains* in imitation of Schubert's duet polonaises; they were numbered op.3, the number 2 having been

bestowed on a collection of 11 of the songs. Other compositions of the autumn and winter of 1828 include a set of four-hand variations on a theme by Prince Louis Ferdinand of Prussia, whose chamber music Schumann admired, and (during January–March 1829) a would-be Schubertian Piano Quartet in C minor op.5, which in January 1830 he began to 'cobble into a symphony'. When in November he learnt of Schubert's death, his sobbing was heard by Flechsig all through the night.

Schumann was constantly hankering for the society of Rosen. Another attraction at Heidelberg university was one of the law professors, Justus Thibaut, who a few years before had published a book on musical aesthetics (*Über die Reinheit der Tonkunst*, Heidelberg, 1825). He persuaded his mother and guardian to allow him to move to the other university, left Leipzig on 11 May 1829 and, after a detour in the Rhineland which left him temporarily penniless, reached Heidelberg ten days later. He enjoyed life at Heidelberg and, under the influence of Thibaut's lively personality, even attended lectures for a short time. As a musician Thibaut was more enthusiastic than knowledgeable, but he had formed a musical society of students and friends, and in his private music room Schumann made the acquaintance of a considerable range of choral music, mainly Italian, from Palestrina and Victoria to Handel and Bach. The summer term ended on 20 August, and Schumann spent the holiday alone, touring Switzerland and northern Italy, where he got as far as Venice; at Milan he heard Pasta sing in Rossini's *La gazza ladra* at La Scala and for the first time fully appreciated Italian music. He returned to Heidelberg (on 20 October) by way of Augsburg and Stuttgart. Despite assurances to

his mother to the contrary, he completely neglected his law studies during the autumn term, and Thibaut seems to have irritated him even as a musician by his narrow conservatism and dogmatism. He told Wieck (letter of 6 November) that he was studying the last movement of Hummel's F♯ minor Sonata, 'a truly great, epic titan of a work', and asked him to send 'all Schubert's waltzes – only the two-hand ones (I believe there are 10–12 sets), Moscheles's G minor Concerto, Hummel's B minor Concerto' and 'anything *new* by Herz and Czerny'. He excused this last request on the ground that he was invited into family circles, but a later notebook entry suggests that he succumbed for a time to their 'shallow virtuosity' until it was eclipsed by Paganini's; from 1832, when his enthusiasm for Paganini was at its height, dates an unfinished *Phantasie satyrique* (*nach Henri Herz*). In the same letter to Wieck, Schumann spoke of having begun a number of symphonies, none of which he had completed; but it is clear from the context that he had done little more than improvise symphonic daydreams at the piano. At this period he was more concerned with piano playing than with composition, sometimes practising for seven hours a day, and in February 1830 he made a public appearance – his only one at Heidelberg – playing his old favourite, Moscheles's variations on *La marche d'Alexandre*, with a brilliant success that led to invitations to play at Mannheim and Mainz; these he declined. That winter he also led a full social life, attending a number of balls and masquerades.

On Easter Sunday 1830 Schumann heard Paganini play at Frankfurt; he was deeply impressed despite doubts about Paganini's artistic ideals. His year at Heidelberg

had now expired and he was due to return to Leipzig to complete his legal studies; but his guardian allowed him a term's respite. The summer was spent largely in composition; in April Schumann wrote (or perhaps only completed) a set of waltzes for piano, under the obvious influence of Schubert's waltzes, and in May an *Etude fantastique en double-sons* in D, which was afterwards styled 'Toccata' and (after revision and transposition to C in July 1832) published as op.7. In June he wrote a piano piece entitled *Papillote*, based on his song *Im Herbste*, which was later used as the third movement of the Sonata op.22. Later on he resumed work on a Concerto in F for piano and orchestra, of which a part – or perhaps an independent set of variations in the same key, on the lines of Moscheles's *Alexandre* variations for piano and orchestra – was completed in August for piano without orchestra and published in November 1831, as the definitive op.1: *Thème sur le nom 'Abegg' varié pour le pianoforte*, its name being that of a girl whose acquaintance he had made (the theme, 'Abegg-Walzer', dates from February 1830).

At the very time he was writing the 'Abegg' Variations, Schumann was trying desperately to persuade his mother to allow him to abandon law for music as a profession. At his request his mother appealed on 7 August to Friedrich Wieck, who replied two days later that Robert, with his talent and imagination, could in three years be made into one of the foremost living pianists provided he would work hard and steadily at the acquisition of a technique. Wieck made no secret of his doubt of the steadfastness of Schumann's character, of his possessing the resolution not only to work at the mechanics of piano playing, but to study 'dry, cold

theory' as well for two years with the Kantor of the Thomasschule, C. T. Weinlig; he ended by advising Frau Schumann to allow her son a trial period of six months. Robert eagerly accepted the condition (22 August). On 24 September he bade farewell to Heidelberg, going on a Rhine trip nearly as far as the Dutch border, and in October settled at Leipzig once more.

On 20 October he went to live in Wieck's own house (Grimmaische Gasse 36). His promises to reform his way of life – particularly in the matters of cigar smoking and heavy drinking – still seem to have been unfulfilled, for his letters to Zwickau continue to record his penniless condition due to overspending of his allowance. He dreamed of composing an opera on *Hamlet* and resumed his piano study with Wieck, but the latter seems to have been more interested in the formation of his prodigy daughter Clara; he was absent with her on a concert tour from Christmas 1830 until the end of January 1831, and the prospect of a still longer absence from September 1831 onwards obliged Schumann to look elsewhere. Accordingly on 20 August 1831 he wrote to Hummel asking to be accepted as a pupil; nothing came of the idea, but he was still clinging to it in May of the following year. The letter to Hummel voices his dissatisfaction with Wieck's teaching and his diary entries for the summer often record deep despondency concerning his own playing as well as disagreement with Wieck's views of music. Nothing had come of the proposed study of theory with Weinlig; only in June 1831 did he at last approach Heinrich Dorn, then conductor at the Leipzig theatre, who on 12 July began his instruction in thoroughbass (according to Dorn's account to

Wasielewski 25 years later, Schumann's first exercise, a four-part chorale, was 'a model of part-writing in defiance of the rules'). In the meantime Schumann had found a new musical idol: Chopin. He was completely captivated by the latter's op.2, the variations on 'Là ci darem', which he unsuccessfully tried to master, and as early as 2 May he made up his mind to write about them.

On 8 June he wrote some poems headed 'Schmetterlinge' and recorded in his diary his decision to give his friends 'more beautiful, more suitable names': Wieck became 'Meister Raro', Clara 'Zilia', Christel (a girl who had just begun to play an important part in his sex life) 'Charitas', Dorn 'The Music Director', and so on. Five days later he had the idea of using them as characters in a novel, *Die Wunderkinder*, in which 'Florestan the *improvisatore*' appeared for the first time and Paganini, under a pseudonym, was to have a leading role. On 1 July, 'entirely new persons enter my diary today – two of my best friends.... Florestan and Eusebius', but it was not until 13 October that Florestan became 'my bosom friend; actually in the story he is to be my real self', while Eusebius, Meister Raro and the rest 'have changed their roles and from real persons have become fantasy characters'. Both Florestan and Eusebius appeared in the Chopin article, 'Ein Opus II', which was at last sent to the *Allgemeine musikalische Zeitung* on 27 September (two days after the departure of Friedrich and Clara Wieck on their seven-month tour), though it did not appear till 7 December.

Schumann's most ambitious composition of 1831 was a Sonata in B minor, of which the first movement (embodying material from some variations on

Paganini's *Rondo à la clochette*) was afterwards published as op.8; on 4 January 1832 he completed a set of variations on an original theme for Clara, and at the same time a set of piano pieces made up partly of the waltzes of the previous year, partly of revised versions of the four-hand polonaises of 1828, partly of new compositions, which he styled *Papillons* and related with incidents at the masked ball in Jean Paul's novel *Flegeljahre*. These were published by Kistner in April as op.2.

CHAPTER THREE

1832–40

I 1832–4

In April 1832 Heinrich Dorn refused to continue with
the lessons in thoroughbass and counterpoint, and
Schumann was left to continue his studies with the help
of Marpurg's *Abhandlung von der Fuge* in theory and
Bach's *Das wohltemperirte Clavier* as practical models.
The Wiecks returned at the beginning of May, but
Schumann did not go back to his old quarters in their
house, which he had left when they went away, nor
apparently was there any resumption of piano lessons.
About this time he began to have serious trouble with
his fingers. This has often been attributed to his use of a
mechanical device as an aid to piano playing. The device
has been blamed for maiming one of his fingers, putting
an end to his hopes of a career as a virtuoso; but the
weakness in the index and middle fingers of the right
hand preceded the use of the device, and may well have
been an effect of mercury poisoning induced by treatment for syphilis.

Compositions of the spring and summer of 1832
include an (apparently lost) *Exercice fantastique* op.5,
dedicated to Kuntzsch, sometimes confused with the
earlier *Etude fantastique* op.6 (i.e. the Toccata op.7, of
which the second version was completed at the same
time, July); piano transcriptions of six of Paganini's
Caprices for unaccompanied violin; a set of *Pièces*

phantastiques (or Intermezzos) for piano, mostly new but embodying portions of earlier compositions; a *Fandango: fantaisie rhapsodique pour le pianoforte, Oeuv. 4*, which was later expanded into the first movement of the F♯ minor Piano Sonata; and a set of 'XII Burlesken (Burle) in the style of the *Papillons*', which Schumann offered, in vain, to Breitkopf & Härtel later in the year, and of which one may survive as op.124 no.12 and possibly others as op.124 nos.1, 3 and 15, and the Intermezzo in the third movement of the F♯ minor Sonata. The Paganini Studies, the Intermezzos and the Toccata were published by Hofmeister of Leipzig. In October Schumann embarked on a much more ambitious work; in July he had confided to his old teacher, Kuntzsch, his intention to study 'score-reading and instrumentation', and on 2 November he approached Gottlieb Müller, Gewandhaus violinist and conductor of the Euterpe concerts, with a request for 'instruction in instrumentation' and 'to go through with you a symphony movement of my own composition' which he had 'worked at almost entirely according to my own ideas and without guidance'. A fortnight or so later he took the score of this movement, in G minor, with him to Zwickau where (and at Schneeberg near by) he spent four months of the winter. Clara Wieck and her father gave a concert there on 18 November, and the movement was performed on the same occasion; as a result Schumann completely rewrote it, and the revised version was played at Schneeberg in February 1833, when it shared the programme with Beethoven's Seventh Symphony, which exercised a marked influence on the second movement of his own work.

In March 1833 Schumann returned to Leipzig, oc-

10. Programme of the concert given by Clara Wieck and her father at the Zwickau Gewandhaus on 18 November 1832

114

cupying a new apartment in Riedels Garten, and on 29 April the first movement of the G minor Symphony was played at Clara Wieck's 'grand concert' in the Gewandhaus. By his own account it was a success, yet in May he abandoned the symphony in a not quite completed state. Some material from the fugal finale was embodied in the last number of a new composition, originally orchestrated though it survives only as a piano work, a set of so-called 'Impromptus' (really free variations) on the theme of Clara Wieck's *Romance variée* op.3; this was offered to Kistner as 'a second set of *Papillons*' but declined by him and issued at the composer's expense by his brothers' firm at Schneeberg in August. In June he wrote the first and third movements of the Piano Sonata in G minor op.22 and in July completed a second set of Paganini transcriptions, published as op.10. In June, too, Schumann, Wieck and a number of their friends began seriously to consider the desirability of founding a new musical periodical. A draft prospectus was prepared and (ultimately abortive) negotiations were begun with Hofmeister for its publication at the end of October. From July to the autumn (when in September he moved again, to Burgstrasse 21) Schumann suffered from the consequences of a feverish chill; in October he was thrown into a deeply melancholic state by the death of his sister-in-law Rosalie, and on the 'fearful night of the 17th' attempted or contemplated suicide by throwing himself from his fourth-floor window; he was left with a lifelong dread of living in upper storeys and promptly moved to a first-floor flat in the same house. A further blow was given by the death of his brother Julius on 18 November. Perhaps as a result of his mental condition – his diary

records his obsession with the fear that he would go mad – he was unable to finish the two works on which he was engaged 'from Michaelmas to Christmas': a set of *Scènes mignonnes* or *Scènes musicales sur un thème connu de Fr. Schubert* (variations on Schubert's so-called *Sehnsuchtswalzer* op.9 no.2) and *Etüden in Form freier Variationen über ein Beethovensches Thema* (the Allegretto of the Seventh Symphony), of which one variation was published years later as op.124 no.2. A new friendship formed in December with the 23-year-old composer Ludwig Schunke, who shared his new flat, did much to rescue Schumann from mental depression. Schunke in turn introduced him in January to a music-loving couple, Karl and Henriette Voigt, to the latter of whom Schumann dedicated his Schubert variations.

In his letters of March 1834 Schumann several times mentioned 'three sonatas'; these were the F♯ minor, the G minor and one in F minor which is quite distinct from that afterwards published as op.14. He continued to work intermittently at this never completed F minor Sonata until February 1837; nor were its companions finished in 1834. From March onwards Schumann's time was almost wholly taken up with the affairs of the new periodical. In February Breitkopf & Härtel had, like Hofmeister, declined to handle it, but a publisher was found at last in C. H. F. Hartmann, and the first number of the *Neue Leipziger Zeitschrift für Musik*, published twice weekly, appeared on 3 April. This at once became the chief field of Schumann's literary activity – articles had previously appeared in *Der Komet* and the *Leipziger Tageblatt*, besides his one contribution to the *Allegemeine musikalische Zeitung* – though in 1840–41 he also contributed to the *Deutsche allgemeine Zeitung*.

The first nominal editor of the new periodical was Julius Knorr, but Knorr's illness threw all the work on Schumann who, at the very same time, undertook to write the musical articles for Herlosssohn's *Damenkonversationslexikon*.

II Ernestine and Clara

On 21 April 1834 the 17½-year-old Ernestine von Fricken began boarding with the Wiecks as a piano pupil. She was the illegitimate child of a Captain Baron von Fricken in the little town of Asch. Schumann, who took her to be 'the daughter of a rich Bohemian baron', quickly fell in love with her and by 2 July was telling his mother he would like to marry her. Frau Voigt was their confidante, and the lovers met at the Voigts' house. On 1 September Baron von Fricken, disquieted by reports of the affair, came to Leipzig to take Ernestine away; a day or two before his arrival the lovers became betrothed, though only the Voigts were in the secret. On 5 September the Frickens went to Zwickau to interview Schumann's mother; Schumann himself rushed after them and there was some sort of discussion, but the Frickens left for Asch next day with the engagement still undisclosed. Just a week later Schumann conceived the idea of the set of piano pieces based on the musical letters – *S–C–H–A* – common to his name and to the town of Asch, which he called *Fasching: Schwänke auf vier Noten*, but afterwards renamed *Carnaval: scènes mignonnes sur quatre notes*; by about 23 September he had written some 'pathetic' variations on a 'Thema quasi marcia funebre' in C♯ minor by the baron, who was an amateur flautist, variations which were later expanded into the work generally known as the *Etudes sym-*

117

phoniques (op.13). Another variation-work of 1834, which cannot be dated more precisely, was a never completed set on Chopin's Nocturne op.15 no.3.

The new friend Schunke was dying of consumption; the Voigts were caring for him, but Schumann, unable to bear the sight, fled to Zwickau (25 October), where a month later he was struggling with a finale to the variations, beginning like the definitive Variation I, in which the funeral march was to be worked up 'gradually into a proud triumphal procession' (Schunke died on 7 December). On 4 December Schumann paid a visit to Asch; even then he did not know the truth about Ernestine's parentage. On 13 December Fricken formally adopted her as his daughter.

On 15 December Schumann was summoned back from Zwickau to Leipzig to negotiate a change of publisher for the *Neue Zeitschrift für Musik*. He and his group were dissatisfied with Hartmann, and from the first number of 1835 the paper was issued by J. A. Barth, with Schumann as the named editor and principal if not sole proprietor. In April 1835 Clara Wieck, now aged $15\frac{1}{2}$, returned to Leipzig after a concert tour that had lasted since the previous November, and until the end of July, when she left for a concert tour, Schumann was with her daily; a marked cooling of his attitude to Ernestine followed and in August, when he had discovered the circumstances of her birth, he began to try to withdraw from his engagement. On 20 October the *Neue Zeitschrift* published the first of 'Eusebius's' *Schwärmbriefe an Chiara*; on 25 November he and Clara exchanged their first kiss; we hear of more kisses at Zwickau in December; on New Year's Day 1836 Ernestine was formally jilted.

Little is known of Schumann's compositions, or his life generally, during 1835; but the F♯ minor Sonata had been completed by the end of August and in October most of the original finale of the G minor Sonata was written (the very end was still unwritten in March 1837). In August Mendelssohn paid a visit to Leipzig preparatory to his permanent settlement at the end of September as director of the Gewandhaus concerts, an event which revolutionized the musical life of the city and gave Schumann a new friend. At the beginning of October he also met two other of his idols at the Wiecks' house, Chopin and Moscheles. From this year, too, dates Schumann's long essay on Berlioz and his *Symphonie fantastique* which appeared in the *Neue Zeitschrift*, nos.33–49 of 1835.

In the middle of January 1836 Wieck took Clara to Dresden, probably to remove her from proximity to Schumann. But the latter was called to Zwickau by his mother's death on 4 February and, taking advantage of Wieck's temporary absence, he was able to see Clara between 7 and 11 February. Wieck, learning of this on his return, was furious with Clara and wrote to Schumann breaking off all relations with him: he may well have suspected that Schumann had contracted syphilis. When at last father and daughter returned to Leipzig on 8 April the lovers were obliged to avoid each other; Clara completely obeyed her father and even showed some leaning towards Carl Banck, whom Wieck engaged to give her singing lessons and who replaced Schumann as her mentor in composition. For his part, Schumann 'sought out Charitas' again and again, and afterwards noted that this had 'consequences' in January 1837.

When on 8 June the F♯ minor Sonata op.11 ('Dedicated to Clara by Florestan and Eusebius') was published, Schumann sent her a copy; she replied, doubtless under compulsion, by returning all his letters and asking him for hers. On 5 June he completed another sonata, the F minor op.14, originally in five movements, which was published later in the year by Haslinger of Vienna without its two scherzos and under the title 'Concert sans orchestre' (one of the scherzos was restored in 1853, when the second edition was issued as *Troisième grande sonate*; the other was published separately, on Brahms's initiative, in 1866). In the same month he sketched out yet another piano sonata, in C, into which he poured the expression of his desperate resignation, though the work was also intended as a contribution to the proposed Beethoven memorial at Bonn; it was practically completed by the beginning of December and offered on 19 December to Kistner as *Ruinen, Trophäen, Palmen: grosse Sonate für das Pianoforte, für Beethovens Monument, von Florestan und Eusebius*, with the suggestion that 100 copies should be given to the Bonn committee for them to sell; nothing came of this, and the work, particularly the last movement with its concealed reference to Beethoven's Seventh Symphony, underwent considerable revision before it was published as the Phantasie op.17 in April 1839.

Early in August 1836 Schumann was 'thinking of a quintet for strings and piano duet', but nothing came of the project. Then another visit from Chopin on 12 September seems to have sent him back to his *Etudes symphoniques* during the following week; he spent 'the whole day of 18 September at the piano', composing

'études with great gusto and excitement'. The present finale must have been written later still, for the young Mendelssohn disciple, Sterndale Bennett, who was musically saluted with a Marschner quotation in that finale, did not arrive at Leipzig until 29 October. Sterndale Bennett's first visit to Leipzig, during which he formed a warm friendship with Schumann, lasted until June of the following year; he paid a second visit from October 1838 to March 1839.

III 1837–8

In the first couple of months of 1837 Schumann seems to have been resigned to the loss of Clara; he was living quietly, reading *Ivanhoe*, *King John* and *Macbeth*, copying out Bach's *Art of Fugue*, studying 'older music', working at the finale of his other, never completed F minor Sonata, thinking of a Symphony in E♭. But in March this mood gave way to one of passionate despair, which he seems to have tried to drown in drink and by such riotous behaviour that his landlady, tolerant as she was, threatened to turn him out of his lodgings. On 3 May Clara Wieck returned to Leipzig after a concert tour of two or three months; a few weeks later Banck, too, fell into disgrace with her father – to her own consternation – and on 19 May Schumann published in his paper an article 'On the Last Art-historical Ball at Editor * *'s' evidently written before Banck's dismissal, in which Banck himself was ridiculed as 'de Knapp' and Clara as 'Ambrosia', the still-loved aspect of Clara being personified as 'Beda'. At this period Schumann's love for Clara was strongly mingled with hatred and, whether or not with the idea of 'avenging himself on her', as he afterwards said, he contemplated marriage to

another Clara whose surname remains unknown. From mid-June to early July he was much in the company of the beautiful 18-year-old Scottish pianist Robena Laidlaw, to whom he dedicated the *Phantasiestücke* op.12, most of which were composed in the period 22 May–4 July.

Then a step towards reconciliation was taken by Clara. Through their common friend E. A. Becker she asked Schumann to return to her the letters she had sent back more than a year before. He replied on 13 August in a letter which Clara justly described as 'cold, serious and yet so beautiful', assuring her she was still 'the dearest in the world' to him; the same day she played in public, in the Börsensaal, three of the *Etudes symphoniques*, and the day after – her letter is misdated 15 August – she formally pledged herself to him. Not until 8 or 9 September were the lovers able to meet, and on 13 September, Clara's birthday, Schumann wrote to her father asking for her hand. Both then and during the next few months Wieck seems to have been coldly and mockingly evasive rather than frankly hostile, and on 15 October he was able to take Clara away on a seven-month concert tour. During that separation the lovers corresponded secretly, but Clara's letters did not always give Schumann unalloyed consolation; for one thing, she was determined not to marry until Robert was in a financial position to assure her comfort. Even before the parting, Schumann's diary for early October shows that he was again contemplating suicide.

Schumann's first compositions after the reconciliation were the *Davidsbündlertänze* (by October), which were published the following year at his own expense through Friese, who in July 1837 had taken

11. Robert Schumann: lithograph by Feckert after a portrait by Adolf Menzel (1815–1905)

over the publication of the *Neue Zeitschrift* (its cir-
culation at the time was between 450 and 500).
Throughout October and November he was working
'furiously' at fugue, using Marpurg's textbook, but the
first compositions of 1838 show little trace of this
preoccupation; they were the *Novelletten* – the eight of
op.21 and some of the pieces published later in opp.99
and 124 – and the *Kinderszenen*, the latter nearly all
written in February; in March he completed the definitive form of the C major Phantasie op.17, and towards
the end of April the *Kreisleriana* 'in four days'. Then on
14 May the Wiecks returned to Leipzig, and before long
Clara's proximity – they were able to meet fairly often –
had an unfortunate effect on his work. On the other
hand, when Clara went to Dresden for a month in July,
he had days and nights of fearful depression, and his
diary records that one night he was 'within a moment of
bearing it no longer'. They had for some time thought of
settling together in Vienna, where Clara already enjoyed
a high reputation, and persuading Haslinger there to
publish the *Neue Zeitschrift*; in order to reconnoitre the
position Schumann spent the winter in the Austrian
capital, leaving Leipzig on 27 September and travelling
by way of Dresden and Prague. The editorial direction
of the *Neue Zeitschrift* was left in the hands of Oswald
Lorenz.

In Vienna Schumann occupied a first-floor room at
Schön Laternengasse 679. There in December he wrote
the definitive finale of the G minor Piano Sonata, the
Scherzo, Gigue and Romanze op.32, the little piece for
Clara (*An C——*, *Gruss zum Heiligen Abend* and
Wunsch, as it is headed in various manuscripts) after-
wards published without title as op.99 no.1, and the

Arabeske op.18; in January 1839 the *Blumenstück* op.19, and the first movement of a Piano Concerto in D minor; by about the end of February the *Humoreske* op.20; by the middle of March the beginning of 'a big *romantic sonata*' (possibly the work we now know as the *Faschingsschwank aus Wien*, of which four movements date from this period, though a lost Allegro in C minor may really be the first movement of this sonata), and soon afterwards the earliest of the *Nachtstücke* op.23 (as late as 10 December 1839 the Intermezzo from the *Faschingsschwank* was published as a supplement to the *Neue Zeitschrift* as a 'fragment from the *Nachtstücke* which are to appear shortly').

IV Towards marriage

Apart from creative work, 1839 opened auspiciously with a visit to Schubert's brother: Schumann unearthed a number of the master's manuscripts, including that of the 'Great' C major Symphony. But Schumann's hopes of Vienna were dashed one by one; he saw it would be impossible to publish the *Neue Zeitschrift* there; and on 30 March he received news of his brother Eduard's serious illness. He returned hurriedly to Leipzig, and after his brother's death on 6 April found himself confronted with a financial crisis in the family affairs; for a time he even contemplated the temporary abandonment of his career as a composer in order to take over the management of the family publishing and bookselling business. He had no sooner learnt that this would not be necessary than he was alarmed by the attitude of Clara, who had been sent to Paris in January and now began to insist again on financial security as a preliminary to marriage. Harmony between the lovers was

restored by the middle of May, and after another attempt by Wieck to break the engagement, Clara signed, on 15 June, the formal statement leading to legal proceedings for the setting aside of her father's consent. On 30 June Schumann placed the case in a lawyer's hands and on 15 or 16 July the plea was submitted to the courts, which on the 19th ordered an attempt at arbitration. At this period Schumann contemplated a marriage in Paris, and at the end of July he visited Berlin to enlist Clara's mother, now married to Adolf Bargiel, as an ally.

The course of the legal proceedings obliged Clara to return from Paris. Schumann met her at Altenburg on 19 August, and they spent a few happy days at Schneeberg with his relations, she a few more alone at Zwickau; on the 30th she followed him to Leipzig, staying with Friese. Her mother arrived next day and, in accordance with the court order, Archdeacon Fischer made two attempts to effect an arbitration; but the first time Wieck did not appear at all, the second too late. On 3 September Clara's mother took her away to Berlin. Ten days later Schumann followed her there and on the 17th, as the result of a fresh advance by Wieck, brought her back to Leipzig, where she stayed with maternal relatives. Clara's interviews with her father were abortive, as he imposed impossible financial conditions as the price of his consent to the marriage: on the one hand the sacrifice to him of all Clara's earnings during the last seven years, on the other a settlement on her of two-thirds of Schmann's capital. On 2 October the case came before the court of appeal, but Wieck did not appear, pleading that the previous order for arbitration had not yet been carried out! The result of

this manoeuvre was a further postponement until mid-December. The next day Clara returned to Berlin.

This was the most painful period of all for the lovers; Wieck's slanders and annoyances – now directed against Clara herself as well as against Schumann – reached a new height. Schumann's mental health had begun to suffer in September and things became worse during the next two months; the death of his old friend Henriette Voigt on 15 October was a further cause of depression. It is true that in one letter to Clara he speaks of 'about 50 new compositions begun', but there are many more references to 'complete lack of ideas' and 'inability to compose any more'. Indeed the second half of 1839 was almost completely blank as regards creative work. In June Schumann had begun two string quartets, and on 23 July a third, but nothing came of these; the little Fughette in G minor op.32 no.4 dates from the same period. In October he reported to Clara that he had been 'chewing for eight days at a stupid prelude and fugue' which he later condensed into the Präludium op.99 no.10. Only the imminence of Clara's return for the next court hearing seems to have released a fresh burst of activity, of which the principal fruit was the set of three Romanzen op.28.

On 18 December all parties appeared before the court of appeal and Wieck so lost control of himself that he had to be silenced by the president of the court. Judgment was reserved until 4 January 1840, and during the period of waiting the lovers spent Christmas together in Berlin. The decision, when it came, was not completely favourable; the court dismissed all Wieck's objections except one – the charge that Schumann was a heavy drinker. Wieck returned to the legal attack in a

Deductionsschrift, handed to the court on 26 January, and privately distributed lithographed copies of his original charges to Schumann's and Clara's friends; Schumann replied to the renewed legal accusation in a *Refutationsschrift* on 13 February, but was advised to take no action against the defamatory lithographs. On 28 March the parties were informed that the higher court of appeal had confirmed the judgment of 4 January; the onus of proving the charge of drunkenness rested with Wieck. In the meantime Schumann had had the idea of strengthening his position by acquiring a doctorate; on 31 January he inquired of Dr G. A. Keferstein the conditions for the granting of a Jena degree, and his friend was soon able to assure him that the university was prepared to make him a Doctor of Philosophy, without thesis or examination, in recognition of his achievements as composer, writer and editor. His curriculum vitae was sent in on 17 February, and he received his diploma 11 days later.

CHAPTER FOUR

1840-44

I The year of song

Schumann's wave of great creative energy gained considerable impetus·during January; a 'little Sonatina in B♭' was begun, and towards the end of the month the *Faschingsschwank* was taken up again and completed. Far more important: after an interval of 12 years, Schumann returned to the composition of songs. The earliest of the 1840 songs actually dated was the Fool's Song from *Twelfth Night*, op.127 no.5, written on 1 February. In a letter of 16 February, Schumann told Clara of the composition of 'six books of songs, ballads, big and little, and four-part'. Four of the 'sechs Hefte' consist of the songs published in October of the same year as *Myrthen* op.25, though not all of them had been composed when Schumann wrote to Clara. The only 'big ballad' was the setting of Heine's *Belsatzar* (on 7 February), published in 1846 as op.57. The Heine *Liederkreis* op.24 was completed by 24 February. The flow of songs was slowed down in March by preliminary work on an opera, *Doge und Dogaressa*, based on a story in the second part of Hoffmann's *Serapions-Brüder*; Schumann himself sketched out a prose libretto which Julius Becker tried, not very successfully, to versify, and the project was not finally abandoned until May. During March Liszt visited Leipzig and Schumann made his acquaintance.

12. Autograph
MS of 'Im
wunderschönen
Monat Mai' from
Schumann's
'Dichterliebe',
1840

In the meantime Clara had visited Leipzig again and the pair had spent a blissful fortnight together in Berlin (17–30 April), to which the Eichendorff *Liederkreis* op.39, composed during May, was the immediate sequel. The Eichendorff cycle was immediately followed by a Heine cycle, 20 songs composed between 24 May and 1 June: the *Dichterliebe* op.48 and four other songs (op.127 nos.2 and 3; op.142 nos.2 and 4) originally intended as part of the same set. On 5 June Clara came to Leipzig and songwriting was suspended.

On 7 July the lovers learnt that Wieck had failed to produce his proof of habitual drunkenness. They at once began a search for a dwelling and on the 16th found 'a little apartment in the Inselstrasse'. Legal consent to the marriage was granted on 1 August, and the banns were published on 16 August, Clara seeking distraction during this last nerve-racking period by a short concert tour in Thuringia while Schumann returned to his song-writing. During July he had composed the three Chamisso songs op.31, the five lieder op.40 and the Chamisso cycle *Frauenliebe und -leben* op.42 (on 11–12 July); now in August he completed the Geibel songs op.30, and the Reinick songs op.36. On 12 September Clara and Robert were married at the village church of Schönefeld near Leipzig.

The first composition after marriage was the vocal duet *Wenn ich ein Vöglein wär* op.43 no.1 (afterwards incorporated in the opera *Genoveva*). Ten days later (13 October) Schumann noted in his *Haushaltbuch*: 'Afternoon symphonic attempts'. A year before Clara had confided to her diary her belief that

it would be best if he composed for orchestra; his imagination cannot find sufficient scope on the piano. . . . His compositions are all orchestral

in feeling. . . . My highest wish is that he should compose for orchestra –
that is his field! May I succeed in bringing him to it.

But the impulse to songwriting was as yet still
predominant. After a patriotic potboiler, *Der deutsche
Rhein*, for solo voice, chorus and piano, of which 1500
copies were sold in a month or so, came an outpouring
of Kerner songs in November and December, most of
them published as op.35, others posthumously in
opp.127 and 142. Other scattered songs of this prolific
year, 1840, were collected in the first volume of *Lieder
und Gesänge* op.27 and the first three volumes of
Romanzen und Balladen opp.45, 49 and 53.
Schumann's 'great' song period ended with op.37, the
Gedichte aus 'Liebesfrühling' by Rückert, to which
Clara contributed three numbers (2, 4 and 11); they
were begun in January 1841, but not completed until
August. Schumann had in the intervening months turned
to a very different field.

II Orchestral music

From 23 to 26 January 1841, to Clara's joy, Schumann
sketched his symphony in B♭, op.38, suggested by a poem
of Adolph Böttger's and originally entitled 'Spring Symphony'. The orchestration took from 27 January to 20
February; the symphony was rehearsed by Mendelssohn on 28 March and performed under his direction three days later in the Leipzig Gewandhaus at
a concert given by Clara on behalf of the orchestra's
pension fund. It was well received, though in reality not
quite so enthusiastically as the Schumanns imagined.
But Schumann felt encouraged to proceed with 'all sorts
of other orchestral plans', of which the next was an
Overture in E, begun on 12 April and completed in

score five days later. Then came a 'scherzo to the over-ture' and a finale in E, the whole 'Suite' (as it was first called) being finished on 8 May; Schumann a little later spoke of it as 'die Symphonette', but when it was first performed (with the D minor Symphony) on 6 December of the same year, the work was described as 'Overture, Scherzo and Finale'. Next, in little more than a week, came a Fantasie in A minor for piano and orchestra, completed on 20 May, which Clara tried out at a Gewandhaus rehearsal on 13 August; this was the piece we now know as the first movement of the Piano Concerto. (Just over a fortnight later, on 1 September, she gave birth to the first of their eight children, Marie.) Ten days after the completion of the Fantasie, Schumann began a second symphony, in D minor, and at about the same time made some revision of no.1. This Symphony in D minor, which we now know in a different guise as 'no.4', was completed in its original form on 9 September, and within a fortnight Schumann had begun a successor to it; the first movement and scherzo of a Symphony in C minor ('Sinfonie III') were roughed out on 23 September, the Adagio and Rondo the following day, and by the 26th the sketch was 'practically finished'. But nothing more came of it; only the scherzo was later published as a piano piece, op.99 no.13. The wave of symphonic activity had spent itself; the *Neue Zeitschrift* was still taking up much of Schumann's time, and his thoughts were turning in a fresh direction – to opera. He considered subjects from Calderón and came under the spell of Moore's *Paradise and the Peri*. On 22 August he was sorking on a 'text to the *Peri*', presumably in the first place an opera libretto; later in the year he called in Böttger's help, and on 6

133

January 1842, the 'text' was finished, though possibly not that composed two years later. The only other major work of 1841 was a setting of Heine's *Tragödie* for chorus and orchestra, practically finished on 8 November, but abandoned in this form and finally published for one and two voices with piano as op.64 no.3.

III Chamber music

In November 1841 the Schumanns had been invited to Weimar, Clara to play, Robert to be present at the performance of the B♭ Symphony and songs. In February 1842 they set out on a similar tour to Bremen, Oldenburg and Hamburg, in the course of which Schumann became unpleasantly conscious of his rather passive part as Clara's shadow. (He had declined an invitation to conduct the symphony, pleading short sight.) Partly because of this, partly because it was difficult to absent himself longer from the *Neue Zeitschrift*, he returned alone to Leipzig on 12 March while Clara went on to Copenhagen, where she stayed for a month (20 March–18 April). Schumann spent this period of separation in deep melancholy, which he tried to drown in 'beer and champagne', unable to compose, working at counterpoint and fugue, brooding over the possibility of taking Clara to America, while Wieck spread a rumour that the pair had parted. Before setting out in February Schumann had been visited by 'quartet-ish thoughts'; now in his loneliness he returned to the study of Mozart's and Haydn's quartets, then of Beethoven's.

Clara's return on 26 April brought a happier mood. On 2 June he made 'quartet essays'; two days later the A minor Quartet was begun; on the 11th he began a

second quartet before the first was finished; and the third quartet of op.41 was written between 8 and 22 July. In the interval between the second and third quartets Schumann turned again on his old enemy Banck, attacking his *Wallfahrt zur heiligen Madonna* in a savage article that he did not reprint when he came to collect his writings in book form years later. At about the same time a libellous onslaught on another foe of some years' standing, Gustav Schilling, earned Schumann a sentence of six days' imprisonment – commuted to a five-thaler fine. After a short holiday at Karlsbad and Marienbad in August, and a successful rehearsal of the three quartets on 8 September, Schumann began a Piano Quintet on 23 September, completing the fair copy on 12 October; and despite the 'constant fearful sleepless nights' a Piano Quartet was begun on 24 October and finished a month later. The last products of this period of preoccupation with chamber music were a Piano Trio in A minor, completed in December 1842, and an Andante and Variations for two pianos, two cellos and horn, completed by the end of January 1843. Neither satisfied him in its original form, but the trio, seven years later, yielded the material for the *Phantasiestücke* op.88, and the variations were recast for pianos only in August and published as op.46 in February 1844; the first variation was now suppressed and the ending extended.

IV Choral music

The month of February 1843 was principally marked by a good deal of intercourse with Berlioz, who visited Leipzig twice in that month, and by Clara's visit of reconciliation to her father at Dresden. (In the following

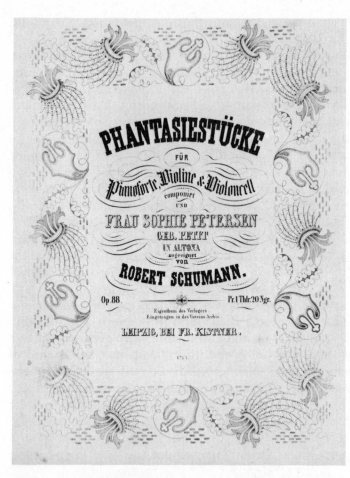

13. *Title-page of the first edition of Schumann's 'Phantasiestücke'*
op.88 (*Leipzig: Kistner, 1850*)

December Wieck made an embarrassed approach to Schumann himself and an uneasy peace was concluded.) And on the 20th or 23rd (Schumann's own dates are at variance) he also began a composition which had occupied his thoughts for 18 months, a setting of *Das Paradies und die Peri*, not now as an opera but as 'an oratorio not for the oratory'. The text was arranged by himself from a manuscript translation of Moore by his friend Flechsig and the published translation by Theodor Oelckers. The score, which Schumann considered his most important up to that date, was completed on 16 June. During the composition of the *Peri* the Leipzig Conservatory was opened (3 April) with Mendelssohn as its director and Schumann as one of the professors, responsible for 'piano playing, composition and playing from score'. And on 25 April a second daughter, Elise, was born. After the completion of the *Peri*, for which Peters paid 550 thalers, Schumann lay fallow for the rest of the year or struggled with projects that came to nothing. Opera in particular occupied his thoughts. On 1 December he conducted the first orchestral rehearsal of the *Peri* – his début as a conductor – and on the 4th and 11th the actual performances; he himself was satisfied and the Leipzig public were enthusiastic, but it is clear that he was passive and ineffectual as a conductor – as, indeed, he was also as a teacher at the conservatory.

The first five months of 1844 were spent in a concert tour of Russia, long desired by Clara, long dreaded by her husband, who at last allowed himself to be persuaded to go by Mendelssohn. Leaving the children with the relations at Schneeberg, the Schumanns set off on 25 January. They travelled by way of Königsberg, Riga,

14. Schumann's sketch of the Kremlin (made during the 1844 concert tour of Russia)

Mitau and Dorpat (where Schumann had a week's illness), Clara giving concerts in each town, and reached St Petersburg on 4 March. There Clara gave four successful public concerts and played to Nicholas I and the tsaritsa; they were warmly greeted both by foreign musicians, including their old friend Henselt, and by wealthy Russian dilettantes such as A. F. L'vov and Count Michal Wielhorski, whose orchestra played the B♭ Symphony in private under Schumann's direction on 21 March; but the visit did little to introduce Schumann's music to the wider musical public. He met neither Glinka nor Dargomïzhsky, though he and Clara attended a party at which the former was present. On 2 April they left St Petersburg, spent Easter at Tver with Schumann's maternal uncle, who had settled in Russia, and arrived in Moscow on 10 April. Clara's audiences were small, but the aristocracy were amiable and Schumann's Quintet was given at a matinée. They heard *A Life for the Tsar*, of which only the first act won Schumann's praise, and were greatly impressed by the Kremlin, which inspired him to write five poems. On 8 May they left again for St Petersburg and ten days later sailed from Kronstadt for Swinemünde; on 30 May they were back at Leipzig.

Throughout the Russian tour Schumann had been tortured by fits of melancholy, partly physical in origin, partly psychological – the result of consciousness that his part was markedly secondary to Clara's. He was also irritated by the fact that he was wasting time, unable to work at the opera he had been 'burning' to write since November. He took the second part of Goethe's *Faust* with him, selected certain scenes during his illness at Dorpat and even sketched music for the closing scene.

Still in ill-health, on returning to Leipzig his first care was to disembarrass himself of another obstacle to creative work, the editorship of the *Neue Zeitschrift für Musik* (in the previous November Dr Härtel had offered him the editorship of the much older *Allgemeine musikalische Zeitung*, but he had declined). He was succeeded at the end of June by Oswald Lorenz, who made way for Franz Brendel at the beginning of the next year. *Faust* was now superseded as an opera subject by Byron's *Corsair*; on 2 July he approached a librettist, and he actually composed a chorus of corsairs and an air for Conrad. But a visit from Hans Andersen on 22 July turned his attention to the poet's *Lykkens Kalosker* which he thought would make 'a fine *Zauberoper*', an idea to which he clung at least until the following April. In August he returned to *Faust* and completed the first three numbers of what we now know as part 3 of *Szenen aus Goethes Faust*, possibly a first draft of the Chorus Mysticus as well; in December he began to think of treating *Faust* as an oratorio instead of an opera, but laid the subject aside for several years.

The rather forced labour on *Faust* had completely exhausted Schumann's nervous energy, and at the end of August he had a very serious breakdown; it became intolerable to listen to music, 'which cut into my nerves as if with knives': in addition he felt slighted by the appointment of Gade to succeed Mendelssohn, who had left Leipzig, in the direction of the Gewandhaus concerts. A visit to the Harz in mid-September and treatment with Karlsbad salts both left Schumann in a worse state than before.

CHAPTER FIVE

Dresden, 1844–50

On 3 October 1844 the Schumanns went to Dresden; the first eight days they spent there were terrible – Schumann was sleepless, tortured by fearful imaginings, Clara found him 'swimming in tears' each morning, and even walking was difficult; but then a slight improvement set in, and they came to a decision to move to Dresden altogether. On 17 October they took a flat in the Waisenhausstrasse, no.35, and on 13 December they left Leipzig for their new home.

In the musically rather dull and conservative atmosphere of Dresden a mental convalescence began. In January 1845 Schumann started to teach Clara counterpoint, and on 28 February he himself wrote a fugue in D minor (either no.1 or no.2 of op.72). In April his *Haushaltbuch* begins to record happier moods and 'spring feelings'. On the 7th he began to guide Clara through Cherubini's *Cours de contrepoint* and started writing an organ fugue on B–A–C–H op.60 no.1; a second fugue was completed on the 18th (but the sixth not until 22 November). Later in April he had the idea of composing for the pedal piano – they had just hired a pedal attachment in order to practise organ playing – and the Studies and Sketches opp.56 and 58 were composed in the period 29 April–7 June. They were followed by a more important work, a Rondo for piano and orchestra, which with a middle movement (com-

pleted on 16 July) was appended to the Fantasie of 1841 to form the Piano Concerto op.54, which Clara played for the first time at the Leipzig Gewandhaus on 1 January 1846. Schumann's health was by no means fully restored in 1845, and he had to give up the idea of attending the unveiling of the Beethoven memorial at Bonn on 10 August. In October he rewrote the finale of op.52, finishing the score on the 20th. The previous day *Tannhäuser* had been given its first performance, but Schumann, who had formed a poor opinion of the work from the score given him by Wagner, was not present; when he did hear the opera, on 22 November, he radically changed his view. At this period he and Wagner belonged to the same circle, which with Hiller was concerned in founding regular subscription concerts at Dresden on the lines of the Leipzig Gewandhaus series.

A passage in a letter to Mendelssohn dating from the end of September 1845 has been generally interpreted as an indication that the Symphony in C was conceived then, but the earliest hint in the *Haushaltbuch* is dated 12 December; the sketch of the first movement was finished by the 17th and the whole work practically completed in draft by the 28th. But Schumann hesitated long over the orchestration; it was not begun until 12 February 1846 and, owing to aural nerve trouble (continual singing in the ears), even the first movement was not finished until 8 May, the whole work only on 19 October, less than three weeks before the first performance (Gewandhaus, 5 November). Except for a few partsongs, opp.55 and 59, the year 1846 was otherwise completely unproductive. Opera plans were considered, but rejected one after another; so too (in March) was an

autobiography (*Biographie eines Davidsbündlers*). The other events of the year included the birth of a fourth child, their first son, Emil (8 February), visits to Maxen (May), to Norderney for sea-bathing (July–August) and a concert-giving expedition to Vienna on which the Schumanns embarked on 24 November. Even Clara's playing failed to arouse much enthusiasm in Vienna, and at the third concert, on 1 January 1847, the Piano Concerto and B♭ Symphony conducted by the composer were very coolly received. The Schumanns left on 21 January, gave a concert at Brno and two in Prague, which were more successful, and returned to Dresden on 4 February. From 11 February to 24 March they were in Berlin, where Clara introduced the Piano Quintet at her concerts on 1 and 17 March and Schumann conducted a rather unsuccessful performance of *Das Paradies und die Peri* by the Singakademie on 17 February; here they were strongly tempted to make their home, for Clara had few close friends at Dresden, and in Berlin she had her mother and Mendelssohn's sister Fanny; but Fanny Hensel's death on 14 May put an end to the plan.

Directly after they returned from Berlin, which was at the end of March 1847, Schumann again began serious consideration of opera projects. For a few days he toyed with Słowacki's *Mazeppa*, but on 1 April decided definitely on Hebbel's *Genoveva*, which he had just read, and asked Robert Reinick to work out a libretto in accordance with his ideas. On 5 April he actually completed the first draft of the overture. But he soon became dissatisfied with Reinick's text and on 14 May approached Hebbel himself with a request for help; nothing came of this, though a personal visit from

Hebbel in the course of the summer made a deep impression on Schumann. Ultimately the libretto of *Genoveva* was compiled by the composer from Reinick's attempt, Hebbel's play and the play on the same subject by Tieck. While waiting for the *Genoveva* text Schumann took up *Faust* again on 17 April and composed and orchestrated the final chorus by 23 April; however, this failed to satisfy him, and on 22 May he began another version, which he completed by the end of July. The D minor Trio op.63 was written at the same time (3–16 June), and in May two Mörike songs were composed which, with the solo version of the Heine *Tragödie* of 1841, made up the fourth volume of *Romanzen und Balladen* op.64. Notwithstanding the death of their youngest child, Emil, on 22 June, the Schumanns went on 2 July to Zwickau, where for nearly a fortnight the little Saxon town fêted her now famous son with a serenade, a concert of his works and a popular concert at the Burgkeller; on 10 July Schumann conducted the C major Symphony and the choral *Beim Abschied zu singen* op.84, specially written for the occasion. On 14 July he completed a drastic revision of this symphony and in August began yet another trio – the F major op.80 – of which the sketch was completed on 25 October.

On 5 November Schumann was shocked by the news of Mendelssohn's death the previous day. He attended the memorial service at Leipzig on the 7th and on his return, with a view to a long article or even short book, began to note down some 'Reminiscences of Felix Mendelssohn-Bartholdy', published only in 1947. Two days after Mendelssohn's funeral came the farewell dinner for Ferdinand Hiller, who was leaving Dresden for

Düsseldorf. Schumann succeeded Hiller as master of the Dresden *Liedertafel* and composed for it the choruses opp.65 (November) and 62 (December); this activity interested him so much that at the end of November he conceived the idea of a parallel society for mixed voices, and accordingly the Verein für Chorgesang met for its first practice on 5 January 1848. He composed *solfeggi*, which have remained unpublished, for both the male-voice *Liedertafel* and the choral society.

On 26 December 1847 Schumann finished the orchestration of his *Genoveva* overture, sketched eight months earlier, and immediately started the composition of the first act; the excitement at once told heavily on his nerves, and he had to wrestle with the remoulding of the libretto in the intervals of composition, each act being completed (in the order of text, composition sketch, score) before the next was begun. Nevertheless the first act was finished in sketch by 3 January 1848, though not played to Clara until 16 days later (on the next day, the 20th, she gave birth to a son, Ludwig). Act 2 was sketched between 21 January and 4 February, and completed in full score on 30 March. In the meantime Schumann had been thrown into great excitement, and Clara into deep alarm, by the revolutionary outbreaks. On 18 March his *Haushaltbuch* hailed a 'springtime of the peoples' (*Völkerfrühling*), the next day he noted 'the *great* news from Berlin' and during 3–19 April he composed three patriotic and revolutionary songs for men's chorus and wind band, originally numbered op.65 but never published in his lifetime. Act 3 of *Genoveva* was sketched between 24 April and 3 May, and then on 6 May the B♭ major chorus 'Gerettet ist das edle Glied' for *Faust* (part 3 no.4, last section). On 25 June he was

145

able, semi-privately, to try out the whole of the *Faust* music so far written (part 3 of the *Szenen*) with chorus and orchestra – to his great satisfaction, for the last chorus had given him very great trouble. Two days later he finished the composition sketch of the fourth act of *Genoveva* and on 4 August the full score of the whole opera. The very next day he began to sketch some music for Byron's *Manfred*; this work was interrupted by the making of the four-hand arrangement of the C major Symphony and the composition during the fortnight 30 August–14 September of the *Album für die Jugend* op.68, of which the first seven pieces were given to Marie on her birthday (1 September); but the *Manfred* overture was 'practically finished' on 19 October, the whole first act was sketched in a single day (6 November) and the entire score was completed on 23 November.

Schumann was now in full flood of composition. On 25 November 1848 he started working on two very different works, both inspired by the writings of Rückert, the *Adventlied* for chorus and orchestra (completed in sketch on 30 November; orchestrated on 3–19 December) and the *Bilder aus Osten* for piano duet, suggested by Rückert's version of the Arabic *Makamen* of Hariri (completed 26 December). Another piano work of literary inspiration was the set of solo *Waldscenen* op.82, suggested by H. Laube's *Jagdbrevier* (composed 29 December–6 January 1849). The first three and a half months of 1849 were equally productive; works of the most diverse kinds followed in rapid succession – the touching-up of *Genoveva* (January), the *Phantasiestücke* for clarinet and piano (op.73; 11–12 February), the Adagio and Allegro for horn and piano

(op.70; 14–17 February), the *Conzertstück* for four horns and orchestra (op.86; sketched 18–20 February, orchestrated by 11 March), a number of *Romanzen und Balladen* for mixed chorus (opp.67, 75 and some at least of those posthumously published as opp.145–6; 6–16 March), two sets of *Romanzen* for women's voices (opp.69, 91; 17–22 March – op.91 no.6 added in August), the *Spanisches Liederspiel*, a cycle of vocal solos, duets and quartets (op.74; sketched 24–8 March). Then the trios in D minor and F were revised for publication, a labour that was completed on 9 April, on which day Schumann lost his last surviving brother, Karl. The 'choral ballads', in which Schumann considered he had discovered a new species, were at once tried out with the Chorverein and met with success. (He had given up the *Liedertafel* the previous year, finding it more trouble than it was worth.) On 13–15 April were written the five *Stücke im Volkston* for cello and piano, and on the 21st Schumann began the *Liederalbum für die Jugend* op.79.

Then came a dramatic interruption. On 3 May the Schumanns returned from a day in the country to find Dresden in revolutionary uproar; alarm bells rang and shots were fired. Next day the democrats formed a provisional government, and barricades were thrown up in the streets. Unwilling to take an active part like Wagner, Schumann on 5 May evaded forcible enrolment in the street guard by flight through the garden door with Clara and the seven-year-old Marie, abandoning the younger children; they took a train to Mügeln, then walked to Dohna and finally took refuge with their friend Major Serre at Maxen, where Schumann that evening wrote the *Frühlingslied* op.79 no.18. Clara was

anxious to return to fetch the younger children; accompanied by two other women, she set off at three o'clock in the morning, found the children asleep despite the firing, and fetched them safely to Maxen where 'my poor Robert had also spent anxious hours' (at this period she was advanced in another pregnancy, for she gave birth to a third son, Ferdinand, on 16 July). On the afternoon of the 10th they ventured back, the insurrection having been suppressed with Prussian help; Schumann at first waited at Strehla while Clara went into the city and collected things for a longer stay in the country; presently he joined her and they walked through the damaged streets 'swarming with Prussians'. They returned to Maxen in the evening and next morning moved the whole family to Kreischa near by, where they remained until 12 June, devouring the newspapers and gradually recovering their equilibrium. Even this period was not barren; on 13 May Schumann completed the *Liederalbum für die Jugend* op.79; during 18–21 May he returned to Laube's *Jagdbrevier* and set five numbers – perhaps begun in April before the insurrection – for male voices and four horns; during 23–6 May he made a setting of Rückert's *Verzweifle nicht im Schmerzenstal* op.93, for double men's chorus with an ad libitum organ part, which he orchestrated in May 1852; during 1–5 June (the autograph was, by a slip, misdated 1–5 May) came the *Minnespiel* from Rückert's *Liebesfrühling*, op.101, another cycle for several voices. A period of depression, which darkened Schumann's 39th birthday, then brought a pause and a rather sudden decision to return to Dresden. On the way back, on 12 June, Schumann's democratic emotion welled up to inspire the first of the Four Marches op.76; the other three, with a

fifth which, in slightly revised form, was published later as op.99 no.14, were written during the next four days. Schumann at once sent them to the publisher Whistling with an intimation that they were 'republican' and that he wished their content to be suggested by the date '1849' printed large; this was suppressed on more prudent reflection. But they were known in the Schumanns' intimate circle as 'the barricade marches'.

Schumann had included as the last song of the *Liederalbum* op.79 a setting of 'Kennst du das Land' from Goethe's *Wilhelm Meister*. This was written at Kreischa; now on his return to Dresden, perhaps with the approaching Goethe centenary in mind, he composed Mignon's other songs (also intended at first for op.79), the Harper's ballad and Philine's *Singet nicht in Trauertönen* (18–22 June), sketched the *Requiem für Mignon* which he later worked out for soloists, chorus and orchestra (2–3 July) and finally the Harper's three other songs (6–7 July). The solo songs, including *Kennst du das Land*, were later published as op.98*a*, the *Requiem für Mignon* as op.98*b*. This preoccupation with Goethe not unnaturally led Schumann back to *Faust*; during 13–18 July he composed the three numbers which now constitute the first part of the *Szenen aus Goethes Faust* – and during 24–6 July the sunrise scene with Ariel and the spirits and Faust's awakening; these were all orchestrated in August. On 29 August – the day after the actual centenary – the closing scenes (i.e. the third part of the *Szenen*) were performed in the Grosser Garten at Dresden; they also formed part of the Goethe celebrations at Weimar and Leipzig.

By the end of August Schumann had completed a little set of duets for soprano and tenor, op.78, begun on

25 July. Next, with his little daughter Marie again in mind, he set about the composition of piano duets 'for small and large children', one of which, the Birthday March op.85 no.1, he and Marie played as a birthday surprise for Clara on 13 September; the whole set, op.85, was composed in the periods 10–15 September and 27 September–1 October, the intervening period having been occupied with a more imposing work, the Introduction and Allegro for piano and orchestra op.92 (sketched 18–20 September, score completed on the 26th). During 11–16 October were composed three songs for double chorus which, with a similar setting of Goethe's *Talismane*, were published posthumously as op.141. A setting of Hebbel's *Nachtlied* op.108, was sketched on 4 November and orchestrated on 8–11 November. Later in the month Schumann completed a 'second *Spanisches Liederspiel*', this time with piano duet accompaniment and including at least one number (*Flutenreicher Ebro*) composed in April, presumably for the first set; this second *Liederspiel* was ultimately published, posthumously, as *Spanische Liebeslieder* op.138. December was equally fruitful, especially of experimental works; on 4–5 December three of Byron's *Hebrew Melodies* were set with harp accompaniment, op.95; on the 7th Schumann wrote the first of the three Romances for oboe and piano op.94; on the 22nd, after a week's inaction due to eye trouble, he composed a piano accompaniment for the declamation of Hebbel's *Schön Hedwig* op.106, which the poet himself considered 'extraordinarily beautiful'; at the turn of the year he was appropriately sketching a setting of Rückert's *Neujahrslied* for chorus and orchestra (27 December–3 January 1850).

The mood of Rückert's *Neujahrslied* reflected Schumann's own, faced as the composer was at that time by a serious decision. After five years at Dresden Schumann still had few real friends there and no place in or recognition by the official world; he had even been snubbed – for instance by Lüttichau, the intendant of the Opera, who had refused him the courtesy of complimentary tickets. In July 1849 he had put out feelers for the post of director of the Leipzig Gewandhaus, which he wrongly believed was about to fall vacant (two years earlier he had made similar private inquiries about the directorship of the Vienna Conservatory); and on 17 November he had received from Hiller a proposal that he should succeed the latter as municipal music director at Düsseldorf with a salary of 750 thalers; Schumann had remembered Mendelssohn's disparagement of the Düsseldorf musicians and, though tempted by the suggestion, he had replied by asking Hiller a number of questions about conditions. He was perturbed to learn that Düsseldorf possessed a lunatic asylum, for he disliked everything that reminded him of insanity. The year 1850 began more promisingly with two fairly successful performances of the *Peri* (5 and 12 January) which enhanced his reputation at Dresden; 'some influential people' misguidedly began trying to get him the post of second conductor at the Opera, vacant since Wagner's flight after the May insurrection; and he hoped that a successful production of *Genoveva* at Leipzig, promised for February after a postponement from the previous summer, might improve his position and make a move to Düsseldorf unnecessary. He told Hiller and the Düsseldorf authorities that he could give no definite answer before 1 April.

15. *Robert and Clara Schumann: daguerreotype, 1850*

The Schumanns went to Leipzig in February for the rehearsals of *Genoveva*, only to meet with a sharp disappointment: the production was postponed again, to make way for Meyerbeer's *Le prophète*. Nor was the first performance of the Introduction and Allegro by Clara at the Gewandhaus on the 14th very successful; on the other hand the *Conzertstück* for horns and the *Genoveva* overture, conducted by the composer at the orchestral pension fund concert on the 25th, aroused general enthusiasm. On 3 March the Schumanns left Leipzig for Bremen, where they gave only one concert (7 March), and Hamburg and Altona, where they stayed for more than a fortnight giving a number of concerts – two of them (21 and 23 March) with Jenny Lind – which brought them a clear profit of 800 thalers. After a short pause in Berlin they returned to Dresden on 29 March. Two days later Schumann sent his acceptance of the Düsseldorf post, though still secretly hoping that one less distant would offer itself.

These disturbances made composition impossible; January, February and March were completely unproductive. At the beginning of April Schumann busied himself with 'putting in order many compositions' – op.88 in its published form probably dates from this period – and, a second edition of the *Album für die Jugend* being contemplated, wrote on the 11th (in the blank spaces of the original sketchbook for the *Album*) the *Haus- und Lebensregeln* as an 'instructive appendix' to the *Album für die Jugend*; they first appeared as a supplement to the *Neue Zeitschrift* (no.36 of 1850) and were also published separately. Then the three songs of op.83 were written, and *Aufträge* op.77 no.5, after which Schumann returned yet again to *Faust*,

sketching the scenes of the four grey women and Faust's death during 25–8 April and orchestrating them by 10 May. The six settings of poems by 'Wielfried von der Neun' (F. W. T. Schöpff) op.89 were begun on 10 May and completed before the Schumanns' departure for Leipzig on 18 May for the long-delayed production of *Genoveva*. The first rehearsal with the soloists took place on 22 May, with Clara at the piano, the first rehearsal of the orchestra alone on the 29th, the first full rehearsal on 7 June. On 23 June there was an orchestral rehearsal attended by Spohr, Gade, Hiller and Moscheles, among others; next day was the dress rehearsal, and on the 25th the first performance, attended by a great concourse of friends including Liszt. Schumann himself conducted and, owing partly to a mishap on the stage, the success was only moderate. The second performance, on the 29th, went better; the third, on the 30th, conducted by Julius Rietz, went best of all. But the opera was then withdrawn and not heard again until Liszt produced it at Weimar on 9 April 1855. On 10 July the Schumanns returned to Dresden. During this stay at Leipzig Schumann took a leading part, with Otto Jahn, Dr Härtel and others, in founding a 'Bach Society' – Schumann's own use of the English word suggests that the inspiration came from the English Bach Society, founded the previous October – to commemorate the centenary of Bach's death by issuing a complete edition of his works; the first meeting was apparently held on 1 June.

July was marked by songwriting: the whole of op.96, op.77 nos.2 and 3, op.125 nos.1, 2, 3 and 5, and op.127 no.4. The last Dresden composition was op.90, a set of songs by Lenau, begun on 2 August, rounded off by a

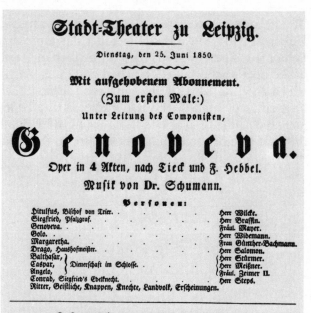

16. *Playbill for Schumann's opera 'Genoveva', first performed at the Stadttheater, Leipzig, on 25 June 1850*

'Requiem' translated from a Latin poem attributed to Héloïse. Schumann added this under the impression that Lenau was already dead; by a strange chance the news of the poet's actual death reached him on 25 August, on which day the songs were sung for the first time in a little circle of friends who had gathered to take leave of him and Clara. Early on the morning of 1 September they left Dresden, reaching Düsseldorf on the evening of the following day.

CHAPTER SIX

Düsseldorf, 1850–52

The Rhineland town of Düsseldorf, which had been paying Schumann's salary since 7 May, now welcomed them warmly. They were met by Hiller and the concert directors, greeted with a serenade from the *Liedertafel* on the evening of arrival and another from the orchestra two days later, and on the 7th they were treated to a concert of Schumann's works, a dinner with official speeches and a ball which (from sheer weariness) they did not attend. Both were exhausted and worried; Schumann was unwell and increasingly irritated by the street noises of their temporary uncomfortable apartments at the corner of the Allee- and Grabenstrassen, which made it almost impossible for him to work. However, he was able to orchestrate the Rückert *Neujahrslied*, and on 29 September they enjoyed a visit to Cologne and were profoundly impressed by the cathedral, where the Cardinal Archbishop Geissel was enthroned the next day. On 10 October Schumann's will to compose returned; he began the Cello Concerto, completing the sketch on the 16th and the full score on the 24th. On the day the concerto was completed Schumann also conducted the first of the ten subscription concerts of the season. He had brought from Leipzig a new leader, W. J. von Wasielewski, afterwards his biographer; he was satisfied with the orchestra; and Clara won her usual success as soloist. Altogether

Schumann conducted eight subscription concerts during this season of 1850–51, introducing new works of his own at four of them, the *Requiem für Mignon* (21 November), the *Neujahrslied* (11 January), the Symphony in E♭ (6 February) and the *Nachtlied* and overture to Schiller's *Die Braut von Messina* (both 13 March).

The new symphony had been begun on 2 November 1850, the first movement finished in sketch on the 9th (despite the interruption of another visit to Cologne), the scherzo on 25 November, the whole work in score on 9 December. The overture to *Die Braut von Messina* was sketched on 29–31 December and orchestrated 1–12 January 1851. Richard Pohl had sent Schumann an opera libretto on that subject; he rejected the libretto, after some consideration, but a re-reading of Schiller's tragedy suggested the overture. He then had the idea of writing a series of tragedy overtures, and sketched and orchestrated one to Shakespeare's *Julius Caesar* during 23 January–2 February. Between the two overtures he composed the songs op.107 nos.1, 2, 3 and 6 and op.125 no.4. In March came the four *Märchenbilder* op.113, for viola and piano, some Lenau songs op.117 and the long-unpublished *Frühlingsgrüsse*.

By this time the happy relationships of the first few months at Düsseldorf had begun to cloud over. There were temperamental misunderstandings between the silent, introspective composer and the sociable Rhinelanders, whom he found too talkative and, musically, not serious enough. At the same time the well-drilled chorus and orchestra he had inherited from Hiller soon felt the consequences of his relaxed discipline and shortcomings as a conductor. At the concert

on 13 March the choir sang badly, the new *Braut* overture was coldly received and an article in the Düsseldorf paper was frankly critical of Schumann's direction of the concerts. Four days later the *Haushaltbuch* records 'doubts about staying longer at Düsseldorf'. But the performance of Bach's *St John Passion* on 13 April appears to have been successful.

Schumann's productivity was unabated. A long correspondence with Pohl concerning an oratorio libretto on Luther led to no result, but in April Moritz Horn sent him a poem, *Der Rose Pilgerfahrt*, which attracted him, though he asked Horn to make a number of alterations before setting it; and on 3 May he enlisted Horn's help in altering the end of Uhland's ballad *Der Königssohn*, of which he had nearly completed a setting for soloists, chorus and orchestra. The *Rose* was completed in its original form for solo voices, chorus and piano by 11 May; Schumann's own catalogue of his compositions gives '12 May–1 June' for the bulk of *Der Königssohn*, but the dates cannot be reconciled with those in the *Haushaltbuch* and statements in dated letters, and must have been entered from (faulty) memory. From the end of May date the duets and solo songs to verses by the child poet Elisabeth Kulmann, opp.103 and 104, and two Uhland settings for unaccompanied chorus, op.145 no.3 and op.146 no.1; *Der Königssohn* was finished, with Horn's ending, in June; and on 12 June Schumann returned to a set of piano duets (originally called *Kinderball* but renamed *Ballszenen* op.109), of which four numbers had been written at some earlier date, and completed it during the next few days. At this period Schumann began negotiating with the Elberfeld publisher F. W. Arnold for the publication

159

of numerous oddments for piano, originally 30 or more, to be collectively entitled *Spreu* ('Chaff '); ultimately only 14 were published, as *Bunte Blätter*, the original idea being to issue the pieces separately with wrappers of various colours.

On 6 July *Der Rose Pilgerfahrt* was given a successful private performance in the Schumanns' own house by a little body of 24 picked singers from which in the following autumn Schumann formed a *Singekränzchen* which met fortnightly in private houses to sing Bach motets, Palestrina, Lassus (*Miserere*) and Lotti (*Crucifixus*), though it collapsed in a few months. *Der Königssohn* suggested the idea of a whole series of 'ballads' for chorus and orchestra, and Schumann asked Pohl to interrupt work on *Luther* to adapt Uhland's *Des Sängers Fluch* for musical setting. Then came a relaxation: on 18 or 19 July the Schumanns made a trip up the Rhine to Heidelberg and thence to Baden-Baden, Basle, Geneva, Chamonix and Vevey, returning to Düsseldorf on 5 August. After 11 days, in which Schumann wrote the song op.107 no.5 and the three piano pieces op.111, he set off again, to Antwerp to judge a competition for men's choirs, visiting Brussels at the same time. He returned home on 22 August to find himself quickly confronted with unpleasantness. On 25 August there was a meeting of the Gesangsverein at which Schumann made some sort of outburst, and on 6 September there was a 'storm with Wortmann', the assistant burgomaster who was also secretary of the Musikverein, at the first meeting to discuss the winter concerts; and Schumann was again beset with 'doubts about the future'. According to his own statement to Wasielewski he composed the Violin Sonata in A minor

op.105 (12–16 September) at a time when he was 'very angry with certain people'. It was followed, after the group of Pfarrius songs op.119, by two more chamber works, the G minor Trio op.110 (2–9 October) and the D minor Violin Sonata op.121 (26 October–2 November). Next he busied himself with orchestration: first he scored the piano part of *Der Rose Pilgerfahrt* (7–27 November), a labour, pressed on him by friends, which he considered both unnecessary and uninteresting; then he orchestrated the scherzo of Burgmüller's Second Symphony, of which he had found the manuscript score at Düsseldorf (1 and 2 December), made a piano score of his D minor Symphony, altering many details (3–11 December), and entirely reorchestrated it (12–19 December); the original title-page shows that he thought of calling this revised version – not that of 1841, as is commonly believed – a 'Symphonistische Fantasie', but it ultimately appeared as 'Symphony no.4'. The last composition of 1851 was an overture to an unwritten Singspiel on Goethe's *Hermann und Dorothea* (sketched 19–20 December; orchestration finished 23 December). The idea of such a play with music, with piano accompaniment, had been in Schumann's mind since 1846; now in 1851 Moritz Horn had recalled it, and there was some discussion about the preparation of a libretto; even a year after the composition of the overture Schumann was still debating with Horn the possibility of a 'concert oratorio' on the subject.

The year 1852 opened with the composition of the Uhland–Pohl ballad, *Des Sängers Fluch* op.139, sketched 1–6 January, scored 10–19 January. *Der Rose Pilgerfahrt* was given in its new form on 5 February,

(a)

17. Autograph score of the opening of Schumann's Symphony in D minor: (a) 1841 version; (b) 1851 version

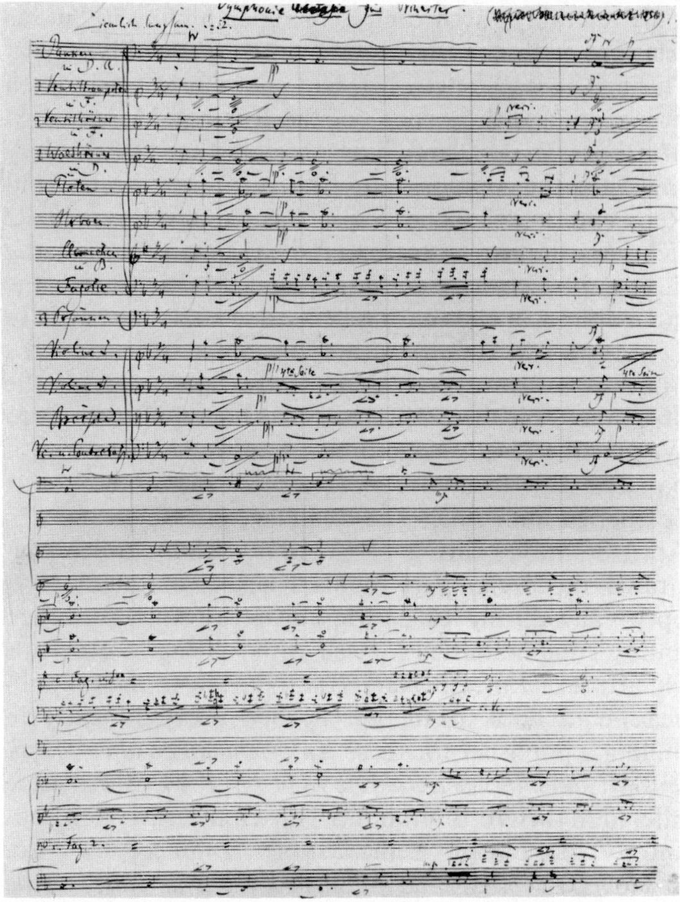

(*b*)

'only tolerably', according to Schumann himself; but his relations with the choir were growing steadily worse. During that winter he had continued his Bach propaganda with portions of the B minor Mass and *St Matthew Passion*, and it is possible that this activity suggested the composition of his own Mass op.147 and Requiem op.148. The Mass was sketched on 13–22 February and orchestrated (and the piano score made) during 24 February–5 March and 24–30 March, the interruption of the scoring being due to a short visit to Leipzig (5–22 March) with Clara when *Rose* and the *Manfred* overture were performed – the latter for the first time – on the 14th. They had hoped also to go to Weimar for Liszt's stage production of *Manfred* and, if possible, a performance of *Lohengrin*; but *Manfred* was postponed until June. The Requiem was sketched on 26 April–8 May; during 9–15 May Schumann orchestrated the organ part of his Rückert motet of 1849 op.93, and during 16–23 May his Requiem. From then until 4 June he busied himself with 'setting in order his old articles' (*Davidsbündlerei*) with a view to their republication in book form; but he had great difficulty in finding a publisher for these *Aufzeichnungen über Musik und Musiker aus den Jahren 1834 bis 1844*, which were offered in turn to Breitkopf & Härtel, to Senff, to Kahnt and to Georg Wigand, the last of whom accepted them in November 1853 and published them as *Gesammelte Schriften über Musik und Musiker* in four volumes in spring 1854.

CHAPTER SEVEN

Last years, 1852–6

It was at the beginning of April 1852 that Schumann experienced some sort of 'rheumatic attack' (according to Clara) with sleeplessness and depression, which grew worse in June and prevented his attending the Weimar production of *Manfred* on 13 June. (At this period he was putting out confidential inquiries about the post of court Kapellmeister at Sondershausen.) On recovering he sketched his setting of Geibel's four ballads *Vom Pagen und der Königstochter* op.140 for soloists, chorus and orchestra, and wrote a piano accompaniment for the recitation of Shelley's *The Fugitives* op.122 no.2. His malady was now beginning to show itself in greater hesitancy of speech and slowness of movement (reflected in his feeling for musical tempos), and in general apathy. From 26 June to 6 July he went to Godesberg for a cure, which only made him worse. On 28 July he began the scoring of the Geibel ballad cycle and on 1–4 August took a small part, smaller than had been intended, in the conducting of a big male-choir festival at Düsseldorf (it was on this occasion that he brought out the *Julius Cäsar* overture). In the middle of August he and Clara went to Scheveningen to try the effect of sea-bathing and remained there until 17 September; this seems to have been beneficial, and the orchestration of *Vom Pagen und der Königstochter* was actually finished in Holland; but

Schumann was ordered to avoid all exertion, such as conducting, on his return to Düsseldorf. He occupied himself with the mechanical work of writing vocal scores (*Sängers Fluch*, Requiem, and Part 2 of the *Szenen aus Goethes Faust*), and at his request the first two concerts of the 1852–3 season were directed by Julius Tausch, the deputy. In the middle of October he had a serious attack of giddiness and on 21 November he noted 'remarkable aural symptoms'. He reappeared at the concert on 3 December, when *Vom Pagen und der Königstochter* had its first performance, but he was coldly received. Moreover, the choir disliked working under him again, after the more efficient Tausch, and the concert directors of the Gesangsverein actually invited him to resign; after a tremendous storm (11–23 December) the three directors resigned instead and the matter was smoothed over, but it was agreed that Tausch should take over all choral rehearsals – Schumann had already asked him to take the pre-liminary rehearsals of new works – leaving to Schumann only the orchestral rehearsals and public performances. The five *Gedichte der Königin Maria Stuart* op.135 were written during this unhappy period (9–16 December) and a piano duet arrangement of the D minor Symphony was made; the symphony was also played in its revised form at the subscription concert of 30 December.

Another more or less 'mechanical' work began the year 1853. A performance of the Bach violin Chaconne with Mendelssohn's piano accompaniment suggested to Schumann that other of Bach's works for violin solo would reach a wider public if provided with accompani-ments; he suggested the idea to Dr Härtel on 4 January

and completed the accompaniments to the sonatas on 5 February. After an interval in which he set an adaptation of Uhland's *Das Glück von Edenhall* for solo voices, men's chorus and orchestra (27 February–12 March), he wrote similar accompaniments for Bach's cello sonatas (19 March–10 April), and (15–19 April) a Festival Overture, with final chorus, on the *Rheinweinlied* – already begun the previous summer – specially for the Lower Rhine Music Festival next month. Schumann scored a notable success with the D minor Symphony on the first day of the festival (15 May), which ended two days later with the new overture, but friends who came to Düsseldorf on this occasion were disquieted to find him obsessed with 'magnetic experiments' in the form of table-turning (his first experiment was on 24 April, and on the 27th he wrote an article on the phenomena – unfortunately lost). Later in the summer there were further disturbing physical symptoms, including, on 30 July, an apparent stroke during a visit to Bonn.

Schumann completed a number of works during the summer of 1853: the Overture, Scherzo and Finale arranged for piano solo (20–24 April), the piano score of *Faust* 'put in order' (by 24 May), the seven Fughettas for piano op.126 composed (28 May–9 June), the *Drei Clavier-Sonaten für die Jugend* op.118 (11–24 June), piano solo arrangements of the string quartets op.41 nos.1 and 2 (4–11 August), an overture to *Faust* (13–17 August), an Allegro with Introduction for piano and orchestra op.134 (24–30 August), a Fantasie for violin and orchestra op.131 (2–7 September), a piano accompaniment to the recitation of Hebbel's *Ballade vom Heideknaben* op.122 no.1 (15 September), five pieces

167

for piano duet which, with an earlier *Menuett*, make up the *Kinderball* op.130 (18–20 September) and a Violin Concerto (21 September–3 October).

The violin Fantasie and concerto were both written for the 22-year-old Joachim, whose performance of Beethoven's concerto had deeply impressed Schumann at the Lower Rhine Festival in May and who had revisited Düsseldorf during 28–31 August, and again profoundly impressed him by his 'wonderful' playing (the excitement caused a sudden 'affection of the speech' on 30 August). On 30 September the Schumanns were visited by Joachim's new friend, the 20-year-old Brahms, who at once made a profound impression on them as composer and pianist; Schumann promptly expressed his enthusiasm in an article (written 9–13 October) entitled 'Neue Bahnen' and published in the *Neue Zeitschrift für Musik* on 28 October. Brahms stayed at Düsseldorf until 3 November, and his visit coincided almost exactly with Schumann's last period of creative activity: the *Märchenerzählungen* op.132 for clarinet, viola and piano were composed 9–11 October and the *Gesänge der Frühe* for piano (also called *An Diotima* after Hölderlin's heroine), which Schumann described to the publisher Arnold as 'characteristic pieces which depict the emotions on the approach and advance of morning, but more as expression of feeling than painting', 15–18 October. On the 21st he returned to the harmonization of unaccompanied string works, this time writing an accompaniment to Paganini's Caprice no.24 (on the theme of which Brahms later composed his variations). Joachim was to come to give the first performance of the violin Fantasie at the first subscription concert of the season on 27 October, and

also to hear his own *Hamlet* overture, and Schumann, Brahms and Schumann's disciple Albert Dietrich agreed to collaborate in the composition of a Violin Sonata in A minor on *F–A–E*, the initials of Joachim's motto 'Frei aber einsam'; Schumann's contributions were the second and fourth movements, an intermezzo in F and finale in A minor and major (composed 22–3 October). The collective sonata was duly presented to Joachim the day after the concert and played at sight by him and Clara; next morning (the 29th) Schumann set about the replacement of Dietrich's first movement and Brahms's scherzo with two movements of his own, and thus on the 31st completed his third violin sonata. During 2–4 November he wrote five Romanzen for cello and piano.

The concert of 27 October was the last conducted by Schumann at Düsseldorf. As the result of a disastrous performance of a mass by Hauptmann at the Maximilian Church on 16 October the choir refused to sing Mendelssohn's *Erste Walpurgisnacht* under Schumann on the 27th, and he had to allow Tausch to conduct it. The rehearsal of Joachim's *Hamlet* on the afternoon of the concert was chaotic. The Musikverein committee was obliged to take some action, and on 7 November the chairman and another member called on Schumann with the committee's unanimous proposal that he should in future conduct only his own compositions, while Tausch should deputize for him on all other occasions; they were seen by Clara, who had completely blinded herself to her husband's condition and ability, and who saw in this suggestion nothing but an 'infamous intrigue' on Tausch's part and an 'insult for Robert'. On the 9th Schumann replied with the assertion that such a proposal – although it was only a proposal – was a

18. *Robert Schumann: chalk drawing (1853) by J. J .B. Laurens; the artist commented on the abnormal enlargement of the pupils of Schumann's eyes*

breach of contract; he himself actually broke the contract next day by failing to appear at either rehearsal or concert. He thought of leaving Düsseldorf for either Vienna or Berlin. On the 14th the committee replied courteously to Schumann, but at the same time implemented its proposal by a decision to that effect. On 6 December the burgomaster wrote to Schumann offering arbitration, or at least investigation, by a sub-committee of the municipal council (the town continued to pay his salary as municipal director of music until the middle of 1855).

Apparently Schumann never replied to the burgomaster; he and Clara had already left on 24 November for a concert tour in Holland in the course of which Clara played the new Introduction and Allegro op.134 for the first time, at Utrecht on the 26th. They were enthusiastically received there, at The Hague, Rotterdam (where they were honoured with a torchlight serenade) and Amsterdam, Schumann conducting the well-drilled Dutch orchestras in his Second and Third Symphonies, and returned to Düsseldorf on 22 December.

During 19–30 January 1854 the Schumanns visited Hanover, where Joachim played the violin Fantasie and conducted the D minor Symphony, Clara played twice at court, and they again enjoyed the company of Brahms. Before, during and after this expedition to Hanover Schumann was occupied in the compilation of a 'Dichtergarten', an anthology of sayings on music by great writers, and at the beginning of February he wrote a preface to it (now lost). During 6–8 February he was searching through Plato and Homer in the Düsseldorf

Municipal Library, to the alarm of Clara, who feared the results of the mental exertion.

On 10 February Schumann recorded that he had had 'very strong and painful aural symptoms'; this was repeated the next night and grew worse the following day; he now had the illusion of 'wonderfully beautiful music' constantly sounding in his head, an illusion which, especially in conjunction with his other sufferings, supports the diagnosis of syphilis. There was little respite, and in the night of the 17th he rose and wrote down a theme in E♭ which he said the angels had sung to him (actually an echo of the slow movement of the Violin Concerto); on the 18th and 19th the angels were replaced by devils in the form of tigers and hyenas who threatened him with Hell, though sometimes the angel voices brought comfort. This state lasted for a week, though in lucid intervals he was able to write two business letters and to compose five variations on the E♭ theme. On the evening of the 26th he asked to be taken to a lunatic asylum, but was persuaded by Clara and the doctor to go to bed. Next morning he was making a fair copy of the variations when, being left alone for a few moments, he ran out of the house to the Rhine bridge and threw himself into the river. He was rescued by fishermen and brought home. After being kept at home for several days, during which Clara was not allowed to see him, he was taken on 4 March to Dr Richarz's private asylum at Endenich near Bonn. After a gradual improvement, with many setbacks, Schumann suddenly wished for a letter from Clara and was able to reply rationally on 14 September; then for seven months he was able to correspond with her and with Brahms, Joachim and the publisher Simrock. On 24 December

he was visited by Joachim and on 11 January 1855 by Brahms. In March 1855 he asked for the Paganini capriccios and resumed his accompaniment writing, 'not', he wrote, 'in canonically complicated style as with the A minor Variations [i.e. no.24], but simple harmonizations'. Brahms saw him again on 2 April, but these visits agitated him, and he was never allowed to see Clara. There was a speedy relapse. On 5 May he wrote Clara the last letter she received from him, and on 10 September Dr Richarz told her there was no longer hope of a complete recovery. On 8 June 1856, his birthday, Brahms found him making alphabetical lists of towns and countries. On 23 July Clara was summoned to Endenich by telegram, as he was not expected to live; but the crisis passed, and she returned to Düsseldorf still without having seen him, though Brahms did so. The suspense was unbearable; she returned to Endenich with Brahms on the 27th and saw her husband for the first time after nearly two and a half years. He appeared to recognize her but could not speak intelligibly. She and Brahms were constantly with him or near him on the 28th, and at 4 p.m. on the 29th he died. He was buried two days later in the cemetery by the Sternentor at Bonn.

CHAPTER EIGHT

Influences and piano music

Schumann's earliest works seem to have been attempts to imitate Schubert and Weber, Hummel and Moscheles, Spohr, Prince Louis Ferdinand of Prussia and other minor figures. Despite his unbounded admiration for Beethoven, there are comparatively few traces of Beethoven's influence in his melodic invention or harmonic procedure, and he hardly ever attempted to rival Beethoven's sustained, continuous flights of thought. Chopin, another idol of his younger days, affected his style even less. Bach on the other hand, particularly Bach's fugal themes, did so profoundly in his later years; the sketches for Schumann's later works, and his fugal studies, show that a number of themes – ultimately not treated fugally – were originally conceived as Bachian fugue subjects. Angular themes derived from conventional Bachian shapes, even Bachian passage-work, are fused with 19th-century Romantic harmony and treated on Romantic lines with varying success in such very different works as the *Manfred* and *Faust* overtures, the cathedral scene and the tenor solo, 'Ewiger Wonnebrand', in *Faust*, the Adagio of the E♭ Symphony and 'Verrufene Stelle' in the *Waldszenen*. But in his early days, and to some extent throughout his life, Schumann's musical thoughts were engendered by dance rhythms, particularly the waltz and the polonaise, by the metres of lyrical verse, and above all by dreamy

keyboard improvisation; he was for years – and the years in which he created his most individual work – unable to compose except at the piano and, with all his efforts, he never completely emancipated himself; he believed in the advice he gave to young musicians in the *Haus- und Lebensregeln*, 'to make everything in the head', but the things he made in his own head were inferior to the things he found at the keyboard.

Schumann's earliest surviving works are songs, and it is significant that, although he published none of them himself in that form, he used the material of three of them with little alteration in piano works which he did publish. Many of the lyrical melodies of the years between 1828 and 1840, when he devoted himself almost entirely to piano composition, are so square-cut, 'rhyming' and stanzaic as to suggest very strongly that they too were inspired by verse, even if they never existed in an intermediate stage as actual songs. (In 1833 he contemplated writing *Musikalische Gedichte, mit unterlegten Liedern von H. Heine*.) Equally characteristic of the younger Schumann of the piano pieces are the aphoristic themes and figures discovered by his fingers and fitted together, in the manner of a mosaic, into short pieces of the type of Schubert's short lyrical piano pieces, a type that in the 1820s and 1830s became very popular. (cf Tomášek's *Eclogues*, Ludwig Berger's *Etudes*, Wilhelm Taubert's *Minnelieder*, Mendelssohn's *Lieder ohne Worte*). Sometimes a Schumann piece as we have it (e.g. the Intermezzo op.4 no.4) consists of fragments of as many as three different abandoned compositions.

Schumann's short pieces are distinguished from those of his predecessors and contemporaries not only by their intrinsic charm and fantasy but also by their liter-

ary and musical allusiveness and by their autobiographical nature. While some of Schumann's early music
has its origin (afterwards concealed) in lyrical verse,
other of it – purely musical in origin – was given
literary or pictorial titles or brought into relationship
with some literary idea later. A notable case is the cycle
Papillons: constructed partly from earlier waltzes and
four-hand polonaises written in imitation of Schubert;
provided with a programmatic finale suggesting the end
of a ball, and related number by number to paragraphs
in a chapter of Jean Paul's *Flegeljahre* (though the
relationship was never made public); and finally published with an enigmatic title bearing no relation to the
work's origin or its acquired connection with Jean Paul
and fully significant only to the composer himself ('lar-
vae' and 'butterflies' played important parts in his
private world of thought). Such literary connections
persisted to the end in Schumann's piano music, though
the public was not allowed to know about, for instance,
the relationship between the *Waldszenen* and Laube's
Jagdbrevier or between the *Gesänge der Frühe* and
Hölderlin's *Diotima* poems. Sometimes, instead, they
were teased with hints and suggestions, with quotations
from Goethe (op.4 no.2) or Shakespeare (the Intermezzo
of op.21 no.3), with the pretence that Schumann's compositions were the work of his fictional Florestan and
Eusebius, with themes that spell out proper names
('Abegg', 'Asch', 'Gade'). His music is full of musical
quotations and allusions; he alluded to the *Marseillaise*,
to the traditional German *Grossvatertanz*, to Beethoven,
to Marschner, to Clara's compositions, to other music
of his own sometimes openly but more often under
subtle disguises and with a significance often difficult

and sometimes impossible to guess (e.g. the reference to the 'Abegg' theme in op.4 no.6). He obviously took pleasure in hiding behind masks, in burying these secrets in his music – doubtless there are a number that have never been discovered – and it is clear that his music meant more to him than it can ever mean to anyone else. All his most individual music is completely introvert, pages from a secret autobiography or, rather, diary; it has been argued that a number of his themes or motifs derive from a private method of musical cryptography, with correspondences between letters and notes (the most pervasive of them involving an encipherment of the name 'Clara'). His sketchbooks make it clear that he often cherished a theme or a harmonic progression not only for its intrinsic musical sake, but because it recalled to him the precise moment and mood in which it was conceived; he dated a theme used in the finale of the Phantasie op.17 '30.11.36. and wallowed blissfully in it when I was sick', or another '29 April 38, since no letter came from you'. He took pleasure less in communicating a mood or emotion than in hugging the secret circumstance of the mood. In many of these respects – love of extra-musical associations, fundamental lyricism, emphasis on self-expression – Schumann is the typical musical Romantic; he is equally so in his earlier technique.

The individual theme or melody being specially valuable for its own sake, its function as structural material tends to be neglected. Indeed with the earlier Schumann structure is merely a framework on which to spread the themes; the parts matter much more than the whole. Consequently the forms are simple, and the simpler the more satisfactory. The material is seldom 'developed' in

177

the Classical sense, but continually remoulded, as if under an improviser's fingers. Schumann's preference for variation writing when he wished to create a larger work is characteristic, as is also the fact that his variations are less often ornamental or Beethovenian in method than plastic remodellings of the theme. The same principle of thematic remodelling is employed in the three piano sonatas for the purpose of giving unity to these larger compositions, but Schumann's static, mosaic-like conception of form and the lack of germinal quality in the themes are here more serious defects than in the short pieces. They are overcome more successfully in the C major Phantasie than in the sonatas; the Phantasie, however, was conceived as a whole, not assembled from heterogeneous earlier compositions. Schumann also experimented with the naive concatenation of a number of simple formal units, contrasted or related, to form long pieces such as the *Humoreske* and *Blumenstück*, but such pieces depend solely on the charm of the separate sections and cannot be said to exist as wholes.

The texture of Schumann's piano music is much more individual and contributes much more to the effect of the separate melodies and aphorisms than the structure. In a criticism dating from 1835 of a piano sonata by Loewe, Schumann asserted his growing conviction that 'the piano expresses itself essentially and peculiarly in three things above all – through richness of part-writing [*Stimmenfülle*] and harmonic change (as in Beethoven and Schubert), through use of the pedal (as in Field), or through volubility (as in Czerny and Herz)'. All three are fully exploited in his own, though the element of volubility becomes less noticeable in the music written

after the finger trouble of 1832 (from the same time begins his neglect of the brilliant upper register of the instrument which he had used freely enough in the 'Abegg' Variations, *Papillons* and the Allegro op.8). Rapidly changing, often boldly chromatic harmony; pedal effects novel in the 1830s and passages impossible without pedal; cross-rhythms and syncopation; endless variety of accompaniment figures – chordal, arpeggiated, broken-chordal, counter-melodic, broken-chord figures suggesting counter-melodies that are never explicit – all help to envelop essentially clear and simple melodic ideas in a rich, diffused, Romantic light.

CHAPTER NINE

Songs

Schumann's later compositions are really seen in true perspective only when considered in relation to, or contrast with, the piano music of 1831–9. The great outpouring of songs in 1840 is, as Brendel said, a 'continuation of his character-pieces for piano'; but the songs are not only piano pieces with another dimension, an additional tone-colour: they are explicit, whereas the piano pieces are reserved. The lyrical element is set free and its emotional content made precise. Schumann himself acknowledged to Zuccalmaglio (letter of 31 December 1840) that 'the *Myrthen* certainly allow a deeper insight into my inner musical workings'. Moreover the poem (which Schumann nearly always chose with care because it answered to something in himself, and generally chose with good literary taste) acted as a lens for his musical thought, sharpening, concentrating, shaping it. In holding the balance between poem and music and between voice and piano, Schumann generally stands midway between Schubert and Wolf; there are a number of purely lyrical songs, such as *Widmung* op.25 no.1, in which the voice sings and the piano 'accompanies', and there are declamatory songs such as *Auf einer Burg* op.39 no.7 (though in both these particular cases the bulk of the piano parts is self-contained, embodying the melodic line). But Schumann's

most typical songs are those in which the melody is shared by voice and piano either simply as in *Der Nussbaum* op.25 no.3, or more subtly as in *Kommen und Scheiden* op.90 no.3, and those in which some other, non-melodic element in the piano part provides a perfect complement to the vocal part and is equally important to the total effect (e.g. *Im Rhein im heiligen Strome* op.48 no.6; *Der Gärtner* op.107 no.3). Another common feature of Schumann's songs is the piano epilogue, often extensive; this is particularly noticeable in the songs of the *Dichterliebe* cycle and those originally intended to form part of it, such as *Mein Wagen rollet langsam* op.142 no.4, indeed of the Heine songs generally.

It is noteworthy that Schumann was at his happiest with Heine, a poet of double or veiled meanings. He was specially happy, too, with Eichendorff and in his few Mörike songs; Chamisso and Kerner led him on to ground where he was weaker, though he rose superior to Chamisso in the *Frauenliebe und -leben* cycle. Goethe's lyrics seldom drew his best from him, and Schiller he neglected almost entirely. His settings of Burns (in translation) – more successful when solo songs than when he composed them for unaccompanied chorus – and his quasi-*Volkslieder* in German vein include some attractive things (e.g. the *Volksliedchen* op.51 no.2 and *Marienwürmchen* op.79 no.13), but are not specially characteristic. The narrative or quasi-narrative ballad attracted him again and again to experiment not only in the forms of solo song and unaccompanied chorus, but as 'melodrama' (accompanied recitation) and, towards the end of his life, in pieces for soloists, chorus and

orchestra. As always with Schumann, the lessening of the personal, subjective element was accompanied by a weakening of inspiration.

Parallel with Schumann's love of 'cycles' of piano pieces, connected by threads of varying tenuity, is his cultivation of the *Liederkreis*, sometimes united only by the circumstance that all the verses come from a single poet, as in the cases of the Heine and Eichendorff cycles actually so called, op.24 and op.39, sometimes adumbrating a story, as in *Frauenliebe und -leben*. Schumann sought to make the latter type a little more dramatic by distributing the songs among four voices, which sometimes unite in duets and quartets, in the two *Liederspiele* on Geibel's translations from the Spanish, opp.74 and 138, and the Rückert *Minnespiel* op.101.

CHAPTER TEN

Orchestral and chamber music

The turning-point in Schumann's creative career was in 1841, after his marriage. Encouraged by his wife, he felt the need to strike out in the larger forms and in less limited media; he did not cease to be a Romantic, but his Romantic conception of music first as a medium of self-expression was now modified by the older Classical view of musical composition as a craft to be practised. His first essay in orchestral composition, the never completed G minor Symphony of 1832–3, had been discouraging; and the first completed symphony, the B♭, suffers (though less seriously) from the same principal defect: that it is inflated piano music with mainly routine orchestration. The basic substance of this symphony is similar to that of the short piano pieces; the opening motto theme was probably even verbally inspired (by the line 'Im Tale blüht der Frühling auf ' in the poem by Böttger which suggested the composition of the symphony in the first place); the four movements originally bore 'characteristic' titles connected with springtime. But Schumann's inability to cover a large canvas with his playful aphorisms and lyrical melodies is as painfully apparent here as in the piano sonatas, and the inventiveness which seldom failed to produce new and interesting piano texture almost dried up when he had at his disposal a medium capable of figuration far richer but not shaped under his fingers. Only in his last

symphony, the E♭, did Schumann hit upon at least an opening Allegro theme that was genuinely symphonic in character, capable of expansion; elsewhere in avoiding the lyrical he adopted mere patterns, more like passage-work than true themes, capable of endless manipulation but lifeless and infertile.

The charm of Schumann's symphonies lies in their never long-repressed lyricism, their interest in the devices by which Schumann sought to unify them thematically, to overcome his natural tendency to loose, suite-like structure. In the B♭ Symphony the brass theme of the slow introduction generates the main theme of the following Allegro; slow movement and scherzo are not only played without a break but are thematically related through a trombone passage at the end of the former. Thematic interrelationship is carried much further in the D minor Symphony, especially in its original form, and in the definitive version the movements are directed to be played without a break; indeed the whole work is so closely knit and so novel in structure that, despite its weakness of invention and (in the definitive version) muddy orchestration, it constitutes a landmark in the history of the symphony. The C major Symphony, too, has a slow introduction presenting not only a motto theme on the brass but much of the thematic material of the following Allegro; the finale refers back to this and also to the slow movement. The finale of the C major Symphony is marked by another favourite symphonic device of Schumann's: the introduction of a new, lyrical theme towards the end of a movement (cf the first movement of the B♭ Symphony and the first and last movements of the D minor). In the E♭ Symphony Schumann gave freer rein to his tendency

towards the suite; there are five 'picturesque' movements, of which only the finale looks back at moments to the fourth, though here Schumann dispensed with the double trios of the symphonies in B♭ and C.

Schumann's concertos are even farther from the Viennese Classical models than his symphonies. Like Chopin's, his models were Hummel and Moscheles rather than Mozart and Beethoven. It is significant that his first completed essay in this field was the lyrical, essentially monothematic Fantasie for piano and orchestra which he later converted into a full-length concerto by the addition of an intermezzo and finale. All the tonal and thematic subtleties of the Classical concerto are jettisoned: the A minor Concerto is essentially a piano work with a lightly, transparently scored orchestral accompaniment which here and there takes over cantabile melodies from the soloist: as Schumann himself said, 'something between symphony, concerto and grand sonata'. The two later pieces for piano and orchestra adopt the same formula, but with less success. The Cello Concerto, in three connected movements, is (as the composer put it in his own catalogue) really a 'Concert Piece for cello with orchestral accompaniment' (in the slow movement a decidedly pianistic accompaniment). Only in the late Violin Concerto, the second of the two works for violin and orchestra inspired by Joachim's playing, did Schumann return to the Classical concerto model with 'double exposition' in the first movement; even so, he failed to grasp the point of the Classical ritornello, for the tutti simply anticipates the solo exposition.

Of Schumann's other orchestral works, the *Manfred* overture is outstanding, a 'character study' in which

185

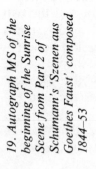

19. Autograph MS of the beginning of the Sunrise Scene from Part 2 of Schumann's 'Szenen aus Goethes Faust', composed 1844–53

Schumann could be as subjective as he wished, identifying himself with Byron's hero, and where neither crabbed thematic invention nor sombre scoring was out of place. It has affinities, besides identity of key, with that other sombre masterpiece, the fourth movement of the E♭ major Symphony.

Schumann seems hardly ever to have been able to think directly in terms of the orchestra, and his scoring is very often a matter of rather conservative routine; but his earlier orchestral works (e.g. the Symphony in B♭, the original version of the D minor, the Piano Concerto) are free from the thickness, the unnecessary doublings, of (for instance) the second version of the D minor Symphony. The scherzo of the B♭ Symphony has some delightful 'open-work' scoring, and two passages in the finale of the D minor Symphony were actually so thin in the original version that the editor (Franz Wüllner) strengthened them in the published score of that version. Two of Schumann's most delightful passages of orchestration occur in works seldom heard, the Waterfall Scene in *Manfred* and the Sunrise Scene at the beginning of Part 2 of the *Szenen aus Goethes Faust*, where he used harp, divided strings and the other apparatus of the Romantic orchestra with great skill (see fig.19).

Schumann's chamber music may, even more than the orchestral, be regarded as an extension of his piano music. The piano is physically present in everything but the three string quartets and even there its spiritual presence is frequently felt in themes and figuration. In the works with piano the strings tend to double it or to be opposed to it as a block (particularly in the Quartet and Quintet). In other respects Schumann's chamber scoring is happier than his orchestral scoring; it

presented him with fewer problems, and he was more comfortable in the more intimate media with their opportunities for the complex figuration and harmonic subtlety so characteristic of his thought. Like the symphonies and piano sonatas, the chamber works are often linked internally by unobtrusive thematic references in one movement to another.

CHAPTER ELEVEN

Choral and dramatic music

It was in his choral works that Schumann tried to venture farthest from his fundamental piano style. They date from the period when he was trying to emancipate himself from the habit of composition at the keyboard; the texts generally exclude subjectivity, and the medium makes intimacy impossible. Schumann's choral writing is seldom very enterprising: a large proportion of it (as in his later music generally) is homophonic and rather plainly chordal, square-cut and rhythmically monotonous; and his contrapuntal studies make themselves felt in the shaping of his themes rather than in the flow of his general texture, despite the fairly numerous passages of deliberate (but always essentially harmonic) counterpoint. But his works for soloists, chorus and orchestra were courageously experimental. The first of them, the 'secular oratorio' *Das Paradies und die Peri*, and its later and weaker companion *Der Rose Pilgerfahrt* were new of their kind, 'through-composed' in the sense that the numbers are distinct but not separated, with the older type of recitative replaced by a more melodic type of declamation and symphonically accompanied. *Elijah* had not yet been written in 1843, though Mendelssohn's *Erste Walpurgisnacht* had just been given in the form in which we know it; but Schumann's models, particularly in the part of the Peri herself, were evidently found rather in the newest German Romantic operas,

Marschner's *Der Vampyr* and *Der fliegende Holländer*. Much more successful artistically, despite the beautiful lyrical passages of *Paradies und die Peri*, is the exquisite *Requiem für Mignon*, intimate and subjective, a work falling into no definite category.

Two other important works stand with one foot in the concert hall and one on the stage: the *Manfred* music and the *Szenen aus Goethes Faust*, the former incidental music to a drama not intended for the stage, the latter a collection of compositions spread over nine years and connected only by the circumstance that they are all settings of portions of a drama part of which can be staged and part can only be read. Both contain many fine lyrical pages, and *Manfred* shows Schumann attempting (as he also attempted with the piano in op.106 and op.122) the never satisfactory species of accompaniment to recitation, while the first two sections of the *Faust* music consist largely of music that might well be part of an opera and was probably conceived as such.

The lack of genuine dramatic talent revealed in *Manfred* and *Faust* is the mortal weakness of Schumann's one opera. The score of *Genoveva* has many beauties and is written in a convention little less advanced than Wagner's at the same period, with continuous texture and almost complete absence of bare recitative, but suffers from lyrical expansiveness and feeble characterization. *Genoveva* employs thematic reminiscence – particularly in association with the villain Golo – quite as much as *Tannhäuser* or *Lohengrin* (Wagner himself had not yet arrived at the true symphonic leitmotif), but it is characteristic of Schumann's lack of theatre sense that his points are made so unobtrusively that they have often passed unnoticed, whereas

Wagner's are driven home with the necessary emphasis.

Harmonically and in melodic contours the Schumann of *Genoveva* (and *Manfred* and other late works) speaks a language strikingly akin to that of Wagner in the 1850s; indeed both spring from many of the same roots. It has been generally agreed that Schumann's later music shows a falling off in inspiration which has been attributed to the deterioration of his mind, to over-production, to the influence of Mendelssohn and to various other causes. It is certainly true that a number of his late works, particularly those of the last two years, are failures and that even ten years earlier an element of heavy, solemn, sometimes bombastic banality begins to appear and then appears with increasing frequency; but such fairly late works as the E♭ Symphony and the D minor Violin Sonata or, among the miniatures, the delightful setting of Mörike's *Der Gärtner* op.107 no.3 show that the general decline must not be dated too early. Some of the manifestations that have been taken for symptoms of mental decay – melodic and thematic angularity, increased harmonic complication – really betoken nothing more than a normal development of style influenced partly by the *Zeitgeist*, partly by Bach.

CHAPTER TWELVE

Critical writings

During the period between the years 1831 and 1844 Schumann was active as a musical journalist. He had at his command an exuberant, florid prose style, modelled on that of Jean Paul Richter, and the best of his writing reveals the fantastic, lyrical, aphorism-scattering personality of the composer of the early piano pieces. Here, too, he loved to conceal himself behind the same fictional masks, Florestan and Eusebius, and from his projected novel of 1831 he also took other masks to accompany them, the members of the 'Davidsbund' – 'Meister Raro' and the rest – masks which now concealed real persons such as Friedrich Wieck, now himself, now mere puppets of his imagination. Even his reviews of new music were sometimes dressed in a fantastic fictional garb. As a critic Schumann uttered many acute and often quoted dicta on his contemporaries and on the older masters he adored. But, like that of most creative artists, his judgment was intensely subjective, conditioned almost entirely by affinities with his own creative nature. Thus he was apt to be blind to the merits of important composers whose art had nothing in common with his own, and over-kind to lesser but more congenial men. The fact that his first and his last published writings both hailed the advent of young and still unrecognized geniuses, Chopin and Brahms, has earned him a reputation for unusual perspicacity which he scarcely de-

serves. Nevertheless the whole corpus of his critical writing is valuable for the light it throws on his own creative personality and on the emotional and intellectual climate of musical Germany in the 1830s, the high noon of Romanticism.

WORKS

Edition: *R. Schumann: Werke*, ed. C. Schumann, J. Brahms and others (Leipzig, 1881–93) [SW]

Numbers in the right-hand column denote references in the text.

THEATRICAL

op.	Title	Libretto	Composed	Publication/MS	Production	Remarks	SW	
								139–40, 142, 190
—	Der Corsar, opera	O. Marbach, after Byron	1844	1983	—	unfinished; chorus of corsairs, interlude and a sketch for Conrad's air	—	140
81	Genoveva, opera	R. Reinick, after L. Tieck and C. F. Hebbel, with alterations by Schumann	1847–9	1851	Leipzig, 25 June 1850	in 4 acts; first pubd in piano reduction	ix/2	131, 143–4, 145, 146, 151, 153, 154, *155*, 190–91, 246
115	Manfred, incidental music	Byron, trans. K. A. Suckow	1848–9	1853	Leipzig, 13 June 1852	first pubd in piano reduction	ix/4, 1	146, 164, 165, 185, 190, 191, **246**

Opp.98b, 108, 112, 116, 139, 140 and 143 are also pubd in vocal score in SW ix/8, and opp.144, 147, 148 and the Szenen aus Goethes Faust in SW ix/9.   *135–40, 189–90*

CHORAL WITH ORCHESTRA

op.	Title, forces	Text	Composed	Publication	First performance	Remarks	SW	
—	Psalm cl, S, A, pf, orch	biblical	1822	—	—		—	100
—	Overture and chorus (Chor von Landleuten), chorus, orch	—	1822	—	—		—	100
—	Tragödie	Heine	1841	—	—	orch version of op.64 no.3	—	134, 144
50	Das Paradies und die Peri, solo vv, chorus, orch	T. Moore's Lalla Rookh, trans. and adapted	1843	1845	Leipzig, 4 Dec 1843	first contemplated Aug 1841	ix/1, 3	133–4, 137, 143, 151, 189, 190
—	Szenen aus Goethes Faust, solo vv, chorus, orch	Goethe	1844–53	1858	Cologne, 13 Jan 1862	first pubd in piano reduction	ix/7	140, 144, 145–6, 149, 153–4, **166**, 167, 174, 187, 190
71	Adventlied, S, chorus, orch	F. Rückert	1848	1849	—	first pubd in piano reduction	ix/1, 43	146
84	Beim Abschied zu singen, chorus, wind insts	E. von Feuchtersleben	1847	1850	—		ix/3, 1	144
93	Verzweifle nicht im Schmerzenstal, double chorus, orch	Rückert	1852	1893	—	orch version of motet op.93	ix/3, 6	164

op.	Title, key, forces		Composed	Publication/MS	First performance	Remarks	SW	
98b	Requiem für Mignon, solo vv, chorus, orch	from Goethe's Wilhelm Meister	1849	1851	Düsseldorf, 21 Nov 1850	see also Songs, op.98a	ix/3, 67	149, 158, 190
108	Nachtlied, chorus, orch	Hebbel	1849	1853	Düsseldorf, 13 March 1851		ix/3, 114	150, 158
112	Der Rose Pilgerfahrt, solo vv, chorus, orch	M. Horn	1851	1852	Düsseldorf, 5 Feb 1852	first pubd in piano reduction	ix/3, 138	159, 160, 161, 164, 189
116	Der Königssohn, solo vv, chorus, orch	L. Uhland	1851	1853	—		ix/4, 101	159, 160
123	Fest-Ouverture, T, chorus, orch	W. Müller, M. Claudius	1852–3	1857	Düsseldorf, 17 May 1853	on J. André's Rheinweinlied	ii, 145	167
139	Des Sängers Fluch, solo vv, chorus, orch	R. Pohl, after Uhland	1852	1858	—		ix/4, 184	160, 161, 166
140	Vom Pagen und der Königstochter, solo vv, chorus, orch	E. Geibel	1852	1857	Düsseldorf, 3 Dec 1852		ix/5, 1	165, 166
143	Das Glück von Edenhall, solo vv, chorus, orch	L. Hasenclever, after Uhland	1853	1860	—		ix/5, 99	167
144	Neujahrslied, chorus, orch	Rückert	1849–50	1861	Düsseldorf, 11 Jan 1851		ix/5, 148	150–51, 157, 158
147	Mass, chorus, orch	liturgical	1852–3	1862	—	first pubd in piano reduction	ix/9, 2	164
148	Requiem, chorus, orch	liturgical	1852	1864	—		ix/9, 3	164, 166

ORCHESTRAL

op.	Title, key, forces	Composed	Publication/MS	First performance	Remarks	SW	
—	Pf Concerto, E♭	1828	—	—	unfinished	—	107, 132–4, 183–8
—	Pf Concerto, F	1829–31	—	—	unfinished	—	
—	Introduction and Variations on a theme of Paganini	1831	—	—	introduction, theme and sketches for 4 variations; variations 3 and 4 used in opp.4 and 8	—	108
—	Symphony, E♭	1831–2	1981	—	sketches for a 'Sinfonia per il Hamlet'	—	

op.	Title, key, forces	Composed	Publication/MS	First performance	Remarks	SW	
—	Symphony, g	1832–3	1972	Zwickau, 18 Nov 1832 (1st movt only); Schneeberg, 12 Feb 1833 (rev. and completed)	3 movts completed; sketch for 4th movt	—	113, 115, 183
—	Pf Concerto, d	1839	—	—		—	125
—	Symphony, c	1840–41	—	—	1 movt only; sketches for 4 movts; scherzo used in Bunte Blätter, pts. of Adagio used in Adagio and finale of Symphony no.2	—	133
38	Symphony no.1, B♭, 'Spring'	1841	1841	Leipzig, 31 March 1841	movts orig. entitled 1 Frühlingsbeginn, 2 Abend, 3 Frohe Gespielen, 4 Voller Frühling	i, 1	132, 134, 139, 143, 183, 184, 185, 187
52	Overture, Scherzo and Finale, e–E	1841; last movt rev. 1845	1846	Leipzig, 6 Dec 1841	orig. title Suite, then Symphonette	ii, 1	132–3, 142, 167
54	Pf Concerto, a	1st movt 1841; 2nd and 3rd movts 1845	1846	Leipzig, 1 Jan 1846	first movt orig. Fantasie, pf, orch	iii, 146	133, 141–2, 143, 185, 187
61	Symphony no.2, C	1845–6	1847	Leipzig, 5 Nov 1846		i, 109	142, 144, 146, 171, 184, 185
81	Genoveva, ov. to opera, c	1847	1850	Leipzig, 25 June 1850		i, 47	147, 153
86	Concertstück, 4 hn, F	1849	1851	Leipzig, 25 Feb 1850		iii, 69	150, 153, 185
92	Introduction and Allegro appassionato (Concertstück)	1849	1852	Leipzig, 14 Feb 1850		iii, 239	158, 172, 174, 184–5, 187, 191
97	Symphony no.3, E♭, 'Rhenish'	1850	1851	Düsseldorf, 6 Feb 1851		i, 243	158, 159
100	Die Braut von Messina, ov., c	1850–51	1851	Düsseldorf, 13 March 1851	to Schiller's play	ii, 70	
115	Manfred, ov., e♭	1848–9	1852	Weimar, 14 March 1852	see Theatrical	ii, 104	164, 174, 185, 186
120	Symphony no.4, d	1841 as no.2; rev. 1851 as no.4	1853	Leipzig, 6 Dec 1841; Düsseldorf, 30 Dec 1852	first version pubd (1891)	i, 310	133, 161, 162–3, 166, 167, 171, 184, 187
128	Julius Cäsar, ov., f	1851	1854	Düsseldorf, 3 Aug 1852	to Shakespeare's play	ii, 175	158, 165

op.	Title, key, forces	Composed	Publication/MS	Remarks	SW	
129	Vc Concerto, a	1850	1854	Leipzig, 9 June 1860	iii, 29	157, 185
131	Fantasie, vn, C	1853	1854	Hanover, Jan 1854	iii, 1	167, 168, 171
134	Introduction and Allegro, pf, d–D	1853	1855	Utrecht, 26 Nov 1853	ii, 291	167, 171, 185
136	Hermann und Dorothea, ov., b	1851	1857	— to Goethe's epic poem	ii, 214	161
—	Vn Concerto, d	1853	1937	Berlin, 26 Nov 1937	—	168, 172, 185
—	Szenen aus Goethes Faust, ov., d	1853	1858	Cologne, 13 Jan 1862 see Choral with orchestra	i, 231	167, 174

CHAMBER

op.	Title, key, forces	Composed	Publication/MS	Remarks	SW	
—	Quartet, vn, va, vc, pf, c	1828–30	1979	orig. op.5	—	134–5, 187–8
—	Quartet, f	1829	—		—	
—	Quartet, vn, va, vc, pf, B	1831–2	—	unfinished	—	106
—	Quartet	1838	—	lost	—	
—	2 string quartets, D, E♭	1839	D-Bds	sketches	—	
41	3 string quartets, a, F, A	1842	1843		iv, 1, 22, 41	127
44	Quintet, 2 vn, va, vc, pf, E♭	1842	1843		v/1, 1	134–5, 167, 187
47	Quartet, vn, va, vc, pf, E♭	1842	1845		v/1, 2	135, 139, 143, 187
—	Andante and variations, 2 pf, 2 vc, hn	1843	1893	orig. version of op.46, see Keyboard	xiv/1, 1	135, 187
63	Trio no.1, vn, vc, pf, d	1847	1848		v/2, 2	135
70	Adagio and Allegro, hn (vn/vc ad lib), pf, A♭	1849	1849	orig. title Romanze und Allegro	v/3, 2	144, 147
73	Phantasiestücke, cl (vn/vc ad lib), pf	1849	1849	orig. title Soiréestücke	v/3, 12	146–7
80	Trio no.2, vn, vc, pf, F	1847	1849		v/2, 50	146
88	Phantasiestücke, vn, vc, pf: 1 Romanze, 2 Humoreske, 3 Duett, 4 Finale	1849	1850	based on Piano Trio, a, 1842	v/2, 124	135, 136, 153
94	3 Romanzen, ob (vn/cl ad lib), pf	1849	1851		v/3, 100	150
102	5 Stücke im Volkston, vc (vn ad lib), pf	1849	1851		v/3, 110	147
105	Sonata no.1, vn, pf, a	1851	1852		v/3, 26	160
110	Trio no.3, vn, vc, pf, g	1851	1852		v/2, 90	161
113	Märchenbilder, va (vn ad lib), pf	1851	1852		v/3, 82	158
121	Sonata no.2, vn, pf, d	1851	1853		v/3, 48	161, 191

op.	Title, key, forces	Composed	Publication/MS	Remarks	SW	
—	Pf acc. to 6 vn sonatas by Bach	1853	1853		—	166–7
—	Pf acc. to 6 vc sonatas by Bach	1853	c1870		—	167
132	Märchenerzählungen, cl (vn ad lib), va, pf	1853	1854		v/2, 148	168
—	Sonata, vn, pf, 'F.A.E.'	1853	1935	2nd and 4th movts only; 1st and 3rd by Dietrich and Brahms	—	169
—	Sonata no.3, vn, pf, a	1853	1956	in 4 movts, 2 being those which Schumann wrote for 'F.A.E.' sonata	—	
—	5 Romanzen, vc, pf	1853	—	lost	—	169
—	Pf acc. to Paganini's vn capriccios	1853–5	1941		—	168, 173

PARTSONGS FOR MIXED VOICES

(SATB, unaccompanied, unless otherwise stated; incipit given only if different from title)

op.	Title, forces	Incipit	Text	Composed	Publication	SW	
55	Fünf Lieder:		R. Burns, trans. Gerhard	1846	1847	xii, 1	142
	1 Das Hochlandmädchen	Nicht Damen tönt von hohem Rang					
	2 Zahnweh	Wie du mit gift'gem Stachel fast					
	3 Mich zieht es nach dem Dörfchen hin						
	4 Die alte, gute Zeit	Wer lenkt nicht gern den heitern Blick					
	5 Hochlandbursch	Schönster Bursch, den je ich traf					
59	Vier Gesänge [orig. pubd as 4, 1, 2, 3; 5 added later]			1846	1848	xii, 11	142
	1 Nord oder Süd!	Schwelle die Segel, günstiger Wind!	K. Lappe				
	2 Am Bodensee	Zierlich ist des Vogels Tritt im Schnee	A. Platen				
	3 Jägerlied	Die gute Nacht, die ich dir sage	Mörike				
	4 Gute Nacht	Heloe! Heloe! Komm du auf unsre Heide	Rückert				
	5 Hirtenknaben-Gesang, SSTT		A. von Droste-Hülshoff	1846	1930	—	
67	Romanzen und Balladen, i:			1849	1849	xii, 20	147
	1 Der König von Thule	Es war ein König in Thule	Goethe				
	2 Schön-Rohtraut	Wie heisst König Ringangs Töchterlein?	Mörike				
	3 Heidenröslein	Sah ein Knab' ein Röslein steh'n	Goethe				
	4 Ungewitter	Auf hohen Burgeszinnen	Chamisso				
	5 John Anderson	John Anderson, mein Lieb!	Burns, trans. Gerhard				

75	Romanzen und Balladen, ii:			1849		
	1 Schnitter Tod	Es ist ein Schnitter, der heisst Tod	Des Knaben Wunderhorn (Brentano)	1850	xii, 28	147
	2 Im Walde (2nd setting)	Es zog eine Hochzeit den Berg entlang	Eichendorff			
	3 Der traurige Jäger	Zur ew'gen Ruh' sie sangen die schöne Müllerin	Eichendorff			
	4 Der Rekrut	Sonst kam mein John mir zu	Burns, trans. Gerhard			
	5 Vom verwundeten Knaben	Es wollt' ein Mädchen früh aufsteh'n	Herder's Volkslieder			
141	Vier doppelchörige Gesänge:			1849		
	1 An die Sterne	Sterne, in des Himmels Ferne!	Rückert	1858	xii, 36	150
	2 Ungewisses Licht	Bahnlos und pfadlos	J. C. von Zedlitz			
	3 Zuversicht	Nach oben musst du blicken	Zedlitz			
	4 Talismane	Gottes ist der Orient!	Goethe			
145	Romanzen und Balladen, iii:			1849–51		147
	1 Der Schmidt	Ich hör' meinen Schatz	Uhland	1860	xii, 60	
	2 Die Nonne	Sie steht am Zellenfenster	anon.			159
	3 Der Sänger	Noch singt den Widerhallen	Uhland			
	4 John Anderson	John Anderson, mein Lieb!	Burns, trans. Gerhard			
	5 Romanze vom Gänsebuben	Helf' mir Gott	O. Malsburg			
146	Romanzen und Balladen, iv:			1849–51		147
	1 Brautgesang	Das Haus benedei ich und preis' es laut	Uhland	1860	xii, 68	159
	2 Der Bänkelsänger Willie	O Bänkelsänger Willie, du ziehst zum Jahrmarkt aus	Burns, trans. Gerhard			
	3 Der Traum	Im schönstem Garten wallten zwei Buhlen	Uhland			
	4 Sommerlied	Seinen Traum, lind wob	Rückert			
	5 Das Schifflein, fl, hn	Ein Schifflein ziehet leise	Uhland			
—	Des Glockentürmers Töchterlein	Mein hochgebornes Schätzelein	Rückert	1851	—	
—	Bei Schenkung eines Flügels, pf	Orange und Myrthe hier	Schumann	1853	—	

PARTSONGS FOR WOMEN'S VOICES
(SSAA: incipit given only if different from title)

op.	Title, accompaniment	Incipit	Text	Composed	Publication	SW	
69	Romanzen, i, pf ad lib:						
	1 Tamburinschlägerin	Schwirrend Tamburin	Alvaro de Ameida, trans. Eichendorff	1849	1849	x/2, 16	147
	2 Waldmädchen	Bin ein Feuer hell	Eichendorff				
	3 Klosterfräulein	Ich armes Klosterfräulein	Kerner				
	4 Soldatenbraut (2nd setting)	Ach, wenn's nur der König auch wüsst	Mörike				
	5 Meerfey	Still bei Nacht fährt manches Schiff	Eichendorff				
	6 Die Kapelle	Droben stehet die Kapelle	Uhland				
91	Romanzen, ii, pf ad lib:						
	1 Rosmarien	Es wollt die Jungfrau früh aufsteh'n	Des Knaben Wunderhorn	1849	1851	x/2, 32	147
	2 Jäger Wohlgemut	Es jagt' ein Jäger wohlgemut	Des Knaben Wunderhorn				
	3 Der Wassermann	Es war in des Maien mildem Glanz	Kerner				
	4 Das verlassene Mägdelein (2nd setting)	Früh wann die Hähne kräh'n	Mörike				
	5 Der Bleicherin Nachtlied	Bleiche, bleiche weisses Lein	Reinick				
	6 In Meeres Mitten		Rückert				147

PARTSONGS FOR MEN'S VOICES
(TTBB; unaccompanied unless otherwise stated; incipit given only if different from title)

op.	Title	Incipit	Text	Composed	Publication	SW	
33	Sechs Lieder:						
	1 Der träumende See	Der See ruht tief im blauen Traum	J. Mosen	1840	1842	xi, 1	
	2 Die Minnesänger	Zu dem Wettgesange schreiten	Heine				
	3 Der Lotosblume (2nd setting)	Die Lotosblume ängstigt	Heine				
	4 Der Zecher als Doktrinär	Was quälte dir dein banges Herz?	Mosen				
	5 Rastlose Liebe	Dem Schnee, dem Regen	Goethe				
	6 Frühlingsglocken	Schneeglöckchen tut läuten	Reinick				
62	Drei Gesänge:						
	1 Der Eidgenossen Nachtwache	In stiller Bucht	Eichendorff	1847	1848	xi, 12	145
	2 Freiheitslied	Zittr', o Erde dunkle Macht	Rückert				
	3 Schlachtgesang	Mit unserm Arm ist nichts getan	Klopstock				

op.	Title, forces	Incipit	Text	Composed	Publication/MS	SW	
65	Ritornelle in canonischen Weisen [orig. order 5, 4, 2, 1, 6, 7, 8, 3]: 1 Die Rose stand im Tau, 2 Lasst Lautenspiel und Becherklang, 3 Blüt' oder Schnee!, 4 Gebt mir zu trinken!, 5 Zürne nicht des Herbstes Wind, 6 In Sommertagen rüste den Schlitten, 7 In Meeres Mitten ist ein offener Laden, 8 Hätte zu einem Traubenkerne [pubd 1906]		Rückert	1847	1849	xi, 20	145
—	Zum Anfang	Mache deinem Meister Ehre	Rückert	1847	1928	—	145
—	Drei Freiheitsgesänge, wind insts ad lib: 1 Zu den Waffen 2 Schwarz-Rot-Gold 3 Deutscher Freiheitsgesang	Vom Angesicht die Mask' herab! In Kümmernis und Dunkelheit Der Sieg ist dein, mein Heldenvolk!	Ullrich F. Freiligrath J. Fürst	1848	1913	—	
93	Verzweifle nicht im Schmerzenstal, motet, double chorus, org ad lib [orchd 1852]		Rückert	1849	1851	—	148, 164
137	Fünf Gesänge aus H. Laubes Jagdbrevier, 4 hn ad lib [orig. order 1, 2, 3, 5, 4]: 1 Zur hohen Jagd 2 Habet acht! 3 Jagdmorgen 4 Frühe 5 Bei der Flasche	Frisch auf zum fröhlichen Jagen Habet Acht auf der Jagd O frischer Morgen, frischer Mut Früh steht der Jäger auf Wo gibt es wohl noch Jägerei	Laube	1849	1857	ix/4, 175	148, 176

105, 129–32, 134, 175, 180–82

SONGS

(duets, trios etc and works for vocal declamation with pf acc. and/or other insts ad lib: incipit given only if different from title)

op.	Title, forces	Incipit	Text	Composed	Publication/MS	SW	
—	Verwandlung	Wenn der Winter sonst entschwand	E. Schulze	1827	—	—	103
—	Lied für xxx	Leicht wie gaukelnde Sylphiden	Schumann	1827	—	—	
—	11 songs:						
	1 Sehnsucht	Sterne der blauen himmlischen Auen	Schumann	1827	1933	—	103
	2 Die Weinende	Ich sah dich weinen!	Byron, trans.	1827	1933	—	103
	3 Erinnerung	Glück der Engel!	J. G. Jacobi	1828	1933	—	
	4 Kurzes Erwachen	Ich bin im Mai gegangen	Kerner	1828	1933	—	
	5 Gesanges Erwachen	Könnt' ich einmal wieder singen	Kerner	1828	1933	—	
	6 An Anna I	Lange harrt ich	Kerner	1828	1933	—	
	7 An Anna II [used in op.11]	Nicht im Tale	Kerner	1828	1893	xiv/1, 34	105
	8 Im Herbste [used in op.22]	Zieh' nur, du Sonne	Kerner	1828	1893	xiv/1, 36	105, 108
	9 Hirtenknabe [used as op.4 no.4]	Bin nur ein armer Hirtenknab	Schumann	1828	1893	xiv/1, 37	105
	10 Der Fischer	Das Wasser rauscht, das Wasser schwoll	Goethe	1828	1933	—	105
	11 Klage [lost]		Jacobi	1828	—	—	

op.	Title, forces	Incipit	Text	Composed	Publication/MS	SW
—	Vom Reitersmann	Sie schlingt um meinen Nacken	Old Ger.	—	D-Zsch	—
—	Maultreiberlied [lost]	Sie sollen ihn nicht haben	—	1838	—	—
—	Ein Gedanke		E. Ferrand	1840	1942	
—	Patriotisches Lied (Der deutsche Rhein, 1v, chorus, pf)		N. Becker	1840	1840	x/2, 168
—	Der Reiter und der Bodensee [frag.]	Der Reiter reitet durchs helle Tal	G. Schwab	1840	1897	—
—	Die nächtliche Heerschau [frag.]	Nachts um die zwölfte Stunde	Zedlitz	1840	1897	—
24	Liederkreis:		Heine	1840	1840	xiii/1, 3 129, 182
		1 Morgens steh ich auf und frage, 2 Es treibt mich hin, 3 Ich wandelte unter den Bäumen, 4 Lieb Liebchen, 5 Schöne Wiege meiner Leiden, 6 Warte, warte, wilder Schiffmann, 7 Berg und Burgen schaun herunter, 8 Anfangs wollt ich fast verzagen, 9 Mit Myrten und Rosen				
25	Myrten			1840	1840	
1	Widmung	Du meine Seele, du mein Herz	Rückert			xiii/1. 24 129, 180
2	Freisinn	Lasst mich nur auf meinem Sattel gelten!	Goethe			180
3	Der Nussbaum	Es grünet ein Nussbaum vor dem Haus	Mosen			181
4	Jemand	Mein Herz ist betrübt	Burns, trans. Gerhard			
5	Lieder aus dem Schenkenbuch im Divan I	Sitz ich allein	Goethe			
6	Lieder aus dem Schenkenbuch im Divan II	Setze mir nicht	Goethe			
7	Die Lotosblume	Die Lotosblume ängstigt	Heine			
8	Talismane	Gottes ist der Orient	Goethe			
9	Lied der Suleika	Wie mit innigstem Behagen	Goethe, attrib. Marianne von Willemer			
10	Die Hochländer-Witwe	Ich bin gekommen ins Niederland	Burns, trans. Gerhard			
11	Lieder der Braut aus dem Liebesfrühling I	Mutter, Mutter! Glaube nicht	Rückert			
12	Lieder der Braut aus dem Liebesfrühling II	Lass mich ihn am Busen hangen	Rückert			
13	Hochländers Abschied	Mein Herz ist im Hochland	Burns, trans. Gerhard			
14	Hochländisches Wiegenlied	Schlafe, süsser kleiner Donald	Burns, trans. Gerhard			

op.	Title, forces	Incipit	Text	Composed	Publication/MS	SW	
31	Drei Gesänge:			1840	1841	xiii/1, 92	131
	1 Die Löwenbraut	Mit der Myrte geschmückt	Chamisso				
	2 Die Kartenlegerin	Schlief die Mutter endlich ein	Chamisso, after Béranger				
	3 Die rote Hanne, chorus ad lib	Den Säugling an der Brust	Chamisso, after Béranger				
34	Vier Duette, S, T:			1840	1841	x/1, 2	
	1 Liebesgarten	Die Liebe ist ein Rosenstrauch	Reinick				
	2 Liebhabers Ständchen	Wachst du noch, Liebchen, Gruss and Kuss!	Burns, trans. Gerhard				
	3 Unterm Fenster	Wer ist vor meiner Kammertür?	Burns, trans. Gerhard				
	4 Familien-Gemälde	Grossvater und Grossmutter	A. Grün				
35	Zwölf Gedichte:		Kerner	1840	1841	xiii/1, 108	132
	1 Lust der Sturmnacht	Wenn durch Berg und Tale					
	2 Stirb, Lieb und Freud!	Zu Augsburg steht ein hohes Haus					
	3 Wanderlied	Wohlauf! noch getrunken den funkelnden Wein!					
	4 Erstes Grün	Du junges Grün, du frisches Gras!					
	5 Sehnsucht nach der Waldgegend	Wär ich nie aus euch gegangen					
	6 Auf das Trinkglas eines verstorbenen Freundes	Du herrlich Glas					
	7 Wanderung	Wohlauf und frisch gewandert					
	8 Stille Liebe	Könnt ich dich in Liedern preisen					
	9 Frage	Wärst du nicht, heil'ger Abendschein!					
	10 Stille Tränen	Du bist vom Schlaf erstanden					
	11 Wer machte dich so krank?	Dass du so krank geworden					
	12 Alte Laute	Hörst du den Vogel singen?					
36	Sechs Gedichte:		Reinick	1840	1842	xiii/1, 132	131
	1 Sonntags am Rhein	Des Sonntags in der Morgenstund					
	2 Ständchen	Komm in die stille Nacht					
	3 Nichts schöneres	Als ich zuerst dich hab gesehn					
	4 An den Sonnenschein	O Sonnenschein!					
	5 Dichters Genesung	Und wieder hatt ich der Schönsten gedacht					
	6 Liebesbotschaft	Wolken, die ihr nach Osten eilt					

37	Zwölf Gedichte aus 'Liebesfrühling' [nos.2, 4, 11, by Clara Schumann]: 1 Der Himmel hat ein Träne geweint, 3 O ihr Herren, 5 Ich hab in mich gesogen, 6 Liebste, was kann denn uns scheiden?, S, T, 7 Schön ist das Fest des Lenzes, S, T, 8 Flügel! Flügel! um zu fliegen, 9 Rose, Meer und Sonne, 10 O Sonn, o Meer, o Rose, 12 So wahr die Sonne scheinet, S, T	Rückert	1840	1841	xiii/2, 2	132
39	Liederkreis (op.77/1 orig. incl. as 1st song, but omitted in 2/1850):	Eichendorff	1840	1842	xiii/2, 28	131, 182
	1 In der Fremde — Aus der Heimat hinter den Blitzen rot					
	2 Intermezzo — Dein Bildnis wunderselig					
	3 Waldesgespräch — Es ist schon spät					
	4 Die Stille — Es weiss und rät es doch keiner					
	5 Mondnacht — Es war, als hätt der Himmel die Erde still geküsst					
	6 Schöne Fremde — Es rauschen die Wipfel und schauern					180
	7 Auf einer Burg — Eingeschlafen auf der Lauer					
	8 In der Fremde — Ich hör die Bächlein rauschen					
	9 Wehmut — Ich kann wohl manchmal singen					
	10 Zwielicht — Dämm'rung will die Flügel spreiten					
	11 Im Walde — Es zog eine Hochzeit den Berg entlang					
	12 Frühlingsnacht — Überm Garten durch die Lüfte					
40	Fünf Lieder:	H. C. Andersen, trans. Chamisso	1840	1842	xiii/2, 50	131
	1 Märzveilchen — Der Himmel wölbt sich rein und blau	Andersen, trans. Chamisso				
	2 Muttertraum — Die Mutter betet herzig	Andersen, trans. Chamisso				
	3 Der Soldat — Es geht bei gedämpfter Trommel Klang	Andersen, trans. Chamisso				
	4 Der Spielmann — Im Städtchen gibt es des Jubels viel	Chamisso				
	5 Verratene Liebe — Da Nachts wir uns küssten	Chamisso				
42	Frauenliebe und -leben: 1 Seit ich ihn gesehen, 2 Er, der Herrlichste von allen, 3 Ich kann's nicht fassen, nicht glauben, 4 Du Ring an meinem Finger, 5 Helft mir, ihr Schwestern, 6 Süsser Freund, du blickest, 7 An meinem Herzen, an meiner Brust, 8 Nun hast du mir den ersten Schmerz getan	Chamisso	1840	1843	xiii/2, 62	131, 181, 182
43	Drei zweistimmige Lieder:		1840	1844	xi/1, 18	131
	1 Wenn ich ein Vöglein wär [later incorporated in op.81]	Des Knaben Wunderhorn				
	2 Herbstlied — Das Laub fällt von den Bäumen	S. A. Mahlmann				
	3 Schön Blümelein — Ich bin hinaus gegangen	Reinick				

op.	Title, forces	Incipit	Text	Composed	Publication/MS	SW	
45	Romanzen und Balladen, i:				1843	xiii/2, 78	132
	1 Der Schatzgräber	Wenn alle Wälder schliefen	Eichendorff	1840			
	2 Frühlingsfahrt	Es zogen zwei rüst'ge Gesellen	Eichendorff				
	3 Abends am Strand	Wir sassen am Fischerhause	Heine				
48	Dichterliebe:	1 Im wunderschönen Monat Mai, 2 Aus meinen Tränen spriessen, 3 Die Rose, die Lilie, die Taube, die Sonne, 4 Wenn ich in deine Augen seh, 5 Ich will meine Seele tauchen, 6 Im Rhein, im heiligen Strome, 7 Ich grolle nicht. 8 Und wüssten's die Blumen, die kleinen, 9 Das ist ein Flöten und Geigen, 10 Hör ich das Liedchen klingen, 11 Ein Jüngling liebt ein Mädchen, 12 Am leuchtenden Sommermorgen, 13 Ich hab im Traume geweinet, 14 Allnächtlich im Traume, 15 Aus alten Märchen, 16 Die alten, bösen Lieder	Heine	1840	1844	xiii/2, 88	*130, 131, 181*
49	Romanzen und Balladen, ii:			1840	1844	xiii/2, 122	132
	1 Die beiden Grenadiere	Nach Frankreich zogen zwei Grenadier'	Heine				
	2 Die feindlichen Brüder	Oben auf des Berges Spitze	Heine				
	3 Die Nonne	Im Garten steht die Nonne	A. Fröhlich				
51	Lieder und Gesänge, ii:				1850	xiii/2, 132	181
	1 Sehnsucht	Ich blick in mein Herz	Geibel	1840			
	2 Volksliedchen	Wenn ich früh in den Garten geh	Rückert	1840			
	3 Ich wandre nicht	Warum soll ich denn wandern	C. Christern	1840			
	4 Auf dem Rhein	Auf deinem Grunde haben sie an verborgnem Ort	K. L. Immermann	1846			
	5 Liebeslied	Dir zu eröffnen mein Herz	Goethe	1850			
53	Romanzen und Balladen, iii:			1840	1845	xiii/2, 142	132
	1 Blondels Lied	Spähend nach dem Eisengitter	J. G. Seidl				
	2 Loreley	Es flüstern und rauschen die Wogen	W. Lorenz				
	3 Der arme Peter	1 Der Hans und die Grete tanzen herum / 2 In meiner Brust / 3 Der arme Peter wankt vorbei	Heine				
57	Belsatzar	Die Mitternacht zog näher schon	Heine	1840	1846	xiii/3, 2	129

op.	Title, forces	Incipit	Text	Composed	Publication/MS	SW	
79	Lieder-Album für die Jugend:						
	1 Der Abendstern	Du lieblicher Stern	Hoffmann von Fallersleben	1849	1849	xiii/3, 30	147, 148, 149
	2 Schmetterling	O Schmetterling, sprich	Hoffmann von Fallersleben				
	3 Frühlingsbotschaft	Kuckuck, Kuckuck ruft aus dem Wald	Hoffmann von Fallersleben				
	4 Frühlingsgruss	So sei gegrüsst vieltausendmal	Hoffmann von Fallersleben				
	5 Vom Schlaraffenland	Kommt, wir wollen uns begeben	Hoffmann von Fallersleben				
	6 Sonntag	Der Sonntag ist gekommen	Hoffmann von Fallersleben				
	7 Zigeunerliedchen	1 Unter die Soldaten 2 Jeden Morgen, in der Frühe	Geibel Geibel				
	8 Des Knaben Berglied	Ich bin vom Berg der Hirtenknab	Uhland				
	9 Maiied, duet ad lib	Komm, lieber Mai	C. A. Overbeck				
	10 Das Käuzlein	Ich armes Käuzlein kleine	Des Knaben Wunderhorn				
	11 Hinaus ins Freie!	Wie blüht es im Tale	Hoffmann von Fallersleben				181
	12 Der Sandmann	Zwei feine Stieflein hab ich an	H. Kletke				
	13 Marienwürmchen	Marienwürmchen, setze dich	Des Knaben Wunderhorn				
	14 Die Waise	Der Frühling kehret wieder	Hoffmann von Fallersleben				
	15 Das Glück, duet	Vöglein vom Zweig	Hebbel				
	16 Weihnachtslied	Als das Christkind ward zur Welt gebracht	Andersen, trans.				
	17 Die wandelnde Glocke	Es war ein Kind	Goethe				147
	18 Frühlingslied, duet ad lib	Schneeglöckchen klingen wieder	Hoffmann von Fallersleben				
	19 Frühlings Ankunft	Nach diesen trüben Tagen	Hoffmann von Fallersleben				
	20 Die Schwalben, duet	Es fliegen zwei Schwalben	Des Knaben Wunderhorn				
	21 Kinderwacht	Wenn fromme Kindlein schlafen gehn	anon.				

op.	Title, forces	Incipit	Text	Composed	Publication/MS	SW	
96	Lieder und Gesänge, iv:						
	1 Nachtlied	Über allen Gipfeln ist Ruh	Goethe	1850	1851	xiii/3, 136	154
	2 Schneeglöckchen	Die Sonne sah die Erde an	anon.				
	3 Ihre Stimme	Lass tief in mir dich lesen	Platen				
	4 Gesungen!	Hört ihr im Laube des Regens	Neun [Schöpff]				
	5 Himmel und Erde	Wie der Bäume kühne Wipfel	Neun [Schöpff]				
98a	Lieder und Gesänge aus Wilhelm Meister:	1 Kennst du das Land, 2 Ballade des Harfners ('Was hör ich draussen vor dem Tor'), 3 Nur wer die Sehnsucht kennt, 4 Wer sein Brot mit Tränen ass, 5 Heiss mich nicht reden, 6 Wer sich der Einsamkeit ergibt, 7 Singet nicht in Trauertönen, 8 An die Türen will ich schleichen, 9 So lasst mich scheinen	Goethe	1849	1851	xiii/4, 2	149
101	Minnespiel:	1 Meine Töne still und heiter, T, 2 Liebster, deine Worte stehlen, S, 3 Ich bin dein Baum, A, B, 4 Mein schöner Stern!, T, 5 Schön ist das Fest des Lenzes, S, A, T, B, 6 O Freund, mein Schirm, mein Schutz!, A/S, 7 Die tausend Grüsse, S, T, 8 So wahr die Sonne scheinet, S, A, T, B	Rückert	1849	1852	x/2, 88	148, 182
103	Mädchenlieder, S, A/2S:		E. Kulmann	1851	1851	x/1, 42	159
	1 Mailied	Pflücket Rosen, um das Haar schön					
	2 Frühlingslied	Der Frühling kehret wieder					
	3 An die Nachtigall	Bleibe hier und singe, liebe Nachtigall!					
	4 An den Abendstern	Schweb empor am Himmel					
104	Sieben Lieder:	1 Mond, meiner Seele Liebling, 2 Viel Glück zur Reise, Schwalben!, 3 Du nennst mich armes Mädchen, 4 Der Zeisig ('Wir sind ja, Kind, im Mai'), 5 Reich mir die Hand, o Wolke, 6 Die letzten Blumen starben, 7 Gekämpft hat meine Barke	Kulmann	1851	1851	xiii/4, 27	159
106	Schön Hedwig, declamation	Im Kreise der Vasallen	Hebbel	1849	1853	xiii/4, 106	150, 190
107	Sechs Gesänge:			1851–2	1852	xiii/4, 40	
	1 Herzeleid	Die Weiden lassen matt die Zweige hangen	Ullrich				158
	2 Die Fensterscheibe	Die Fenster klär ich zum Feiertag	Ullrich				
	3 Der Gärtner	Auf ihrem Leibrösslein	Mörike				158, 181, 191
	4 Die Spinnerin	Auf dem Dorf in den Spinnstuben	P. Heyse				
	5 Im Wald	Ich zieh so allein in den Wald hinein!	W. Müller				160
	6 Abendlied	Es ist so still geworden	G. Kinkel				158

Op.	Title / movement	First line	Author	Date	Date	Ref	Page
114	Drei Lieder, 3 female vv:						
	1 Nänie	Unter den roten Blumen schlummere	L. Bechstein	1853	1853	x/2, 118	
	2 Triolett	Senkt die Nacht den sanften Fittig nieder	L'Egru				
	3 Spruch	O blicke, wenn den Sinn dir will die Welt	Rückert				158
117	Vier Husarenlieder, Bar:						
	1 Der Husar, trara!, 2 Der leidige Frieden, 3 Den grünen Zeigern, 4 Da liegt der Feinde gestreckte Schar		Lenau	1851	1852	xiii/4, 52	161
119	Drei Gedichte:						
	1 Die Hütte	Im Wald, in grüner Runde	G. Pfarrius	1851	1853	xiii/4, 60	
	2 Warnung	Es geht der Tag zur Neige					
	3 Der Bräutigam und die Birke	Birke, Birke, des Waldes Zier					190
122	Zwei Balladen, declamations:						
		Der Knabe träumt	Hebbel	1852–3	1853	xiii/4, 112	
		Der Hagel klirrt nieder	Shelley, trans.				
	1 Ballade vom Haideknaben						167
	2 Die Flüchtlinge						165
125	Fünf heitere Gesänge:						
	1 Die Meerfee	Helle Silberglöcklein klingen	Buddeus	1850–51	1853	xiii/4, 68	154
	2 Husarenabzug	Aus dem dunkeln Tor wallt	C. Candidus				154
	3 Jung Volkers Lied [orig. intended for op.107 no.4]	Und die mich trug im Mutterarm	Mörike				154
	4 Frühlingslied	Das Körnlein springt	F. Braun				158
	5 Frühlingslust	Nun stehen die Rosen in Blüte	Heyse	1854			154
127	Fünf Lieder und Gesänge:						
	1 Sängers Trost	Weint auch einst kein Liebchen	Kerner	1840	1854	xiii/4, 80	132
	2 Dein Angesicht [orig. intended for op.48]		Heine	1840			131
	3 Es leuchtet meine Liebe [orig. intended for op.48]		Heine	1840			131
	4 Mein altes Ross		Moritz, Graf von Strachwitz	1850			154
	5 Schlusslied des Narren	Und als ich ein winzig Bübchen war	Shakespeare, trans. Tieck and A. Schlegel	1840			129
—	Frühlingsgrüsse	Nach langem Frost	Lenau	1851	1942		158

op.	Title, forces	Incipit	Text	Composed	Publication/MS	SW	
135	Gedichte der Königin Maria Stuart:		trans. G. Vincke	1852	1855	xiii/4, 90	166
	1 Abschied von Frankreich	Ich zieh dahin					
	2 Nach der Geburt ihres Sohnes	Herr Jesu Christ					
	3 An die Königin Elisabeth	Nur ein Gedanke					
	4 Abschied von der Welt	Was nützt die mir noch zugemess'ne Zeit?					
	5 Gebet	O Gott, mein Gebieter					
138	Spanische Liebeslieder: 1 Vorspiel, pf 4 hands, 2 Tief im Herzen trag ich Pein, S, 3 O wie lieblich ist das Mädchen, T, 4 Bedeckt mich mit Blumen, S, A, 5 Flutenreicher Ebro, Bar, 6 Intermezzo, pf 4 hands, 7 Weh, wie zornig ist das Mädchen, T, 8 Hoch, hoch sind die Berge, A, 9 Blaue Augen hat das Mädchen, T, B, 10 Dunkler Lichtglanz, S, A, T, B		Geibel	1849	1857	x/2, 124	150, 182
139	From Des Sängers Fluch:		Pohl, after Uhland	1852	1858	—	160
	4 Provenzalisches Lied	In den Talen der Provence					
	7 Ballade	In der hohen Hall sass König Sifrid					
142	Vier Gesänge:			1840	1858	xiii/4, 98	
	1 Trost im Gesang	Der Wandrer, dem verschwunden	Kerner				132
	2 Lehn deine Wang [orig. intended for op.48]		Heine				131
	3 Mädchen-Schwermut	Kleine Tropfen, seid ihr Tränen	L. Bernhard				131
	4 Mein Wagen rollet langsam [orig. intended for op.48]		Heine				181
—	Mailied [duet]	Gern mach' ich dir	Schumann	1851	D-Zsch	—	
—	Liedchen von Marie und Papa, duet		Rückert	1852	1942	—	
—	Glockentürmers Töchterlein	Mein hochgebor'nes Schätzelein	Des Knaben Wunderhorn		Zsch	—	
—	Das Käuzlein [2nd setting]	Ich armes Käuzlein kleine	Rückert	—		—	
—	Deutscher Blumengarten [duet]			—		—	

INDEX TO THE SONGS

An Anna I, 1828; An Anna II, 1828; An den Abendstern, op.103 no.4; An den Mond, op.95 no.2; An den Sonnenschein, op.36 no.4; An die Königin Elisabeth, op.135 no.3; An die Nachtigall, op.103 no.3; An die Türen will ich schleichen, op.98a no.8; 'Anfangs wollt ich fast verzagen', op.24 no.8; 'An meinem Herzen, an meiner Brust', op.42 no.7; 'Auf das Trinkglas eines verstorbenen Freundes, op.35 no.6; 'Auf deinem Grunde haben sie an verborgnem Ort', op.51 no.4; 'Auf dem Dorf in den Spinnstuben', op.107 no.4

Auf dem Rhein, op.51 no.4; 'Auf einer Burg, op.39 no.7; 'Auf ihrem Grab', op.64 no.3; 'Auf ihrem Leibrösslein', op.107 no.3; 'Aufträge, op.77 no.5; 'Aus alten Märchen', op.48 no.15; 'Aus dem dunkeln Tor wallt', op.125 no.2; Aus den hebräischen Gesängen, op.25 no.15; 'Aus den östlichen Rosen, op.25 no.25; 'Aus der Heimat hinter den Blitzen rot', op.39 no.1; 'Aus meinen Tränen spriessen', op.48 no.2; Ballade des Harfners, op.98a no.2; Ballade vom Haideknaben, op.122 no.1

'Bedeckt mich mit Blumen', op.138 no.4; Belsatzar, op.57; 'Berg und schaun herunter', op.24 no.7; 'Bin nur ein armer Hirtenknab', 1828; 'Birke, Birke, des Waldes Zier', op.119 no.3; 'Blaue Augen hat das Mädchen', op.138 no.9; 'Bleibe hier und singe, liebe Nachtigall!', op.103 no.3; Blondels Lied, op.53 no.1; Botschaft, op.74 no.8; 'Da die Heimat', op.95 no.1; 'Da ich nun entsagen müssen', op.30 no.2; 'Da liegt Feinde gestreckte Schar', op.117 no.4; 'Dämm'rung will die Flügel spreiten', op.39 no.10; 'Da Nachts wir uns küssten', op.40 no.5; Das Glück, op.79 no.15

'Das ist ein Flöten und Geigen', op.48 no.9; Das Käuzlein, op.79 no.10; 'Das Käuzlein, after op.142; 'Das Körnlein springt', op.125 no.4; 'Das Laub fällt von den Bäumen', op.43 no.2; Das Schwert, 1848; 'Dass du so krank geworden', op.35 no.11; 'Dass ihr steht in Liebesglut', op.74 no.5; Das verlassne Mägdelein, op.64 no.2; 'Das Wasser rauscht, das Wasser schwoll', 1828; 'Dein Angesicht', op.127 no.2; 'Dein Bildnis op.39 no.2; 'Dein Tag ist aus, dein Ruhm fing an', op.95 no.3; Dem Helden, op.95 no.3

'Dem holden Lenzgeschmeide', op.90 no.2; 'Dem roten Röslein gleicht mein Lieb', op.27 no.2; 'Den grünen Zeigern', op.117 no.3; 'Den Säugling an der Brust', op.31 no.3; Der Abendstern, op.79 no.1; 'Der arme Peter, op.53 no.3; 'Der arme Peter wankt vorbei', op.53 no.3; Der Bräutigam und die Birke, op.119 no.3; Der deutsche Rhein, 1840; Der Einsiedler, op.83 no.3; 'Dereinst, dereinst, o Gedanke mein', op.74 no.3; Der Fischer, 1828; Der frohe Wandersmann, op.77 no.1; 'Der Frühling kehret wieder', op.79 no.14

'Der Frühling kehret wieder', op.103 no.2; Der Gärtner, op.107 no.3; 'Der Hagel klirrt nieder', op.122 no.2; Der Handschuh, op.87; 'Der Hans und die Grete tanzen herum', op.53 no.3; Der Hidalgo, op.30 no.3; 'Der Himmel hat eine Träne geweint', op.37 no.1; 'Der Himmel wölbt sich rein und blau', op.40 no.1; 'Der Husar, trara!', op.117 no.1; Der Knabe mit dem Wunderhorn, op.30 no.1; 'Der Knabe träumt', op.122 no.1; Der Kontrabandiste, op.74 no.10; 'Der leidige Frieden', op.117 no.2; 'Der Mond, der scheint', 1848; Der Nussbaum, op.25 no.3; Der Page, op.30 no.2; Der Reiter und der Bodensee, 1840; 'Der Reiter reitet durchs helle Tal', 1840; Der Sandmann, op.79 no.12

Der Schatzgräber, op.45 no.1; 'Der Schnee, der gestern noch in Flöckchen', op.79 no.26; Der schwere Abend, op.90 no.6; Der Soldat, op.40 no.3; 'Der Sonntag ist gekommen', op.79 no.6; Der Spielmann, op.40 no.4; 'Der Wandrer, dem verschwunden', op.142 no.1; Der weisse Hirsch, 1848; Der Zeisig, op.104 no.4; Des Buben Schützenlied, op.79 no.25; Des Knaben Berglied, op.79 no.8; Des Sennen Abschied, op.79 no.22; 'Des Sonntags in der Morgenstund', op.36 no.1; Deutscher Blumengarten, after op.142; Dichterliebe, op.48

Dichters Genesung, op.36 no.5; 'Die alten, bösen Lieder', op.48 no.16; Die Ammenuhr, 1848; Die beiden Grenadiere, op.49 no.1; Die Blume der Ergebung, op.83 no.2; 'Die dunklen Wolken hingen', op.90 no.6; Die feindlichen Brüder, op.49 no.2; 'Die Fenster klär ich zum Feiertag', op.107 no.2; Die Fensterscheibe, op.107 no.2; Die Flüchtlinge, op.122 no.2; Die Hochländer-Witwe, op.25 no.10; Die Hütte, op.119 no.1; Die Kartenlegerin, op.31 no.2; 'Die letzten Blumen starben', op.104 no.6

'Die Liebe ist ein Rosenstrauch', op.34 no.1; Die Lotosblume, op.25 no.7; 'Die Lotosblume ängstigt', op.25 no.7; Die Löwenbraut, op.31 no.1; Die Meerfee, op.125 no.1; 'Die Mitternacht zog näher schon', op.57; 'Die Mutter betet herzig', op.40 no.2; Die nächtliche Heerschau, 1840; Die Nonne, op.49 no.3; 'Die Rose, die Lilie, die Taube, die Sonne', op.48 no.3; Die rote Hanne, op.31 no.3; Die Schwalben, op.79 no.20; Die Sennin, op.90 no.4; Die Soldatenbraut, op.64 no.1; 'Die Sonne sah die Erdean', op.96 no.2; Die Spinnerin, op.107 no.4; Die Stille, op.39 no.4; 'Die tausend Grüsse', op.101 no.7

Die Tochter Jephthas, op.95 no.1; Die Waise, op.79 no.14; Die wandelnde Glocke, op.79 no.17; 'Die Weiden lassen matt die Zweige hangen', op.107 no.1; Die Weinende, 1827; 'Dir zu eröffnen mein Herz', op.51 no.5; 'Du bist vom Schlaf erstanden', op.35 no.10; 'Du bist wie eine Blume', op.25 no.24; 'Du herrlich Glas', op.35 no.6; 'Du

junges Grün, du frisches Gras', op.35 no.2; 'Du lieblicher Stern', op.79 no.1; 'Du meine Seele, du mein Herz', op.25 no.1; 'Du nennst mich armes Mädchen', op.104 no.3

'Dunkler Lichtglanz', op.138 no.10; 'Durch die Tannen und die Linden', op.89 no.3; 'Du Ring an meinem Finger', op.42 no.4; 'Eia, wie flattert der Kranz', op.78 no.1; 'Ein Gedanke, 1840; 'Einschlafen auf der Lauer', op.39 no.7; 'Ein Jüngling liebt ein Mädchen', op.48 no.11; Einsamkeit, op.90 no.5; 'Einscheckiges Pferd'?1845; 'Entflich mit mir und sei mein Weib', op.64 no.3; 'Er, der Herrlichste von allen', op.42 no.2; Erinnerung, 1828; Er ist's, op.79 no.23; Erste Begegnung, op.74 no.1; Erstes Grün, op.35 no.4

Er und Sie, op.78 no.2; Es fiel ein Reif in der Frühlingsnacht', op.64 no.3; 'Es fliegen zwei Schwalben', op.79 no.20; 'Es flüstern und rauschen die Wogen', op.53 no.2; 'Es flüster'sder Himmel', op.25 no.16; 'Es geht bei gedämpfter Trommel Klang', op.40 no.3; 'Es geht der Tag zur Neige', op.119 no.2; 'Es gingen drei Jäger', 1848; 'Es grünet ein Nussbaum vor dem Haus', op.25 no.3; 'Es ist schon spät', op.39 no.3; 'Es ist so still geworden', op.107 no.6; 'Es ist so süss zu scherzen', op.30 no.3; Es ist verraten, op.74 no.5; 'Es leuchtet meine Liebe', op.127 no.3

'Es rauschen die Wipfel und schauern', op.39 no.6; 'Es stürmet am Abendhimmel', op.89 no.1; 'Es treibt mich hin', op.24 no.2; 'Es war, als hätt der Himmel die Erde still geküsst', op.39 no.5; 'Es war ein Kind', op.79 no.17; 'Es weiss und rät es doch keiner', op.39 no.4; 'Es zog eine Hochzeit den Berg entlang', op.39 no.11; 'Es zogen zwei rüst'ge Gesellen', op.45 no.2; Familiengemälde, op.34 no.4; 'Fein Rösslein, ich beschlage dich', op.90 no.1; 'Flügel! Flügel! um zu fliegen', op.37 no.8; 'Flutenreicher Ebro', op.138 no.5

Frage, op.35 no.9; Frauenliebe und -leben, op.42; Freisinn, op.25 no.2; 'Frühling lässt sein blaues Band', op.79 no.23; Frühlings Ankunft, op.79 no.19; Frühlings-botschaft, op.79 no.3; Frühlingsfahrt, op.45 no.2; Frühlingsgruss, op.79 no.4; Frühlingsgrüsse, 1851; Frühlingslied, op.79 no.18; Frühlingsgrüsse, op.103 no.2; Frühlingslied, op.125 no.4; Frühlings-lust, op.125 no.5; Frühlingsnacht, op.39 no.12; 'Früh wann die Hähne krähn', op.64 no.2; Gebet, op.135 no.5; Gedichte die Königin Maria Stuart, op.135; Geisternähe, op.77 no.3; 'Gekämpft hat meine Barke', op.104 no.7; 'Gern mach' ich dir', 1852

Gesanges Erwachen, 1828; Geständnis, op.74 no.7; Gesungen, op.96 no.4; Glockentürmers Töchterlein, after op.142; 'Glück der Engel', 1828; 'Gottes ist der Orient', op.25 no.8; 'Grossvater und Grossmutter', op.34 no.4; 'Grün ist der Jasminenstrauch', op.27 no.4;

Hauptmanns Weib', op.25 no.19; Heimliches Verschwinden, op.89 no.2; 'Heiss mich nicht treden', op.98a no.5; 'Helft mir, ihr Schwestern', op.42 no.5; 'Helle Silberglöcklein klingen', op.125 no.1; 'Herbstlied', op.43 no.2; 'Herbstlied, op.89 no.3; 'Herr Jesu Christ', op.135 no.2 Herzeleid, op.107 no.1; 'Hier in diesen erdbeklommenen Lüften', op.25 no.26; Himmel und Erde, op.96 no.5; Hinaus ins Freie, op.79 no.11; Hirtenknabe, 1828; 'Hoch, hoch sind die Berge', op.138 no.8; Hochländers Abschied, op.25 no.13; Hochländisches Wiegenlied, op.25 no.14; 'Hoch zu Pferd', op.25 no.19; 'Hör ich das Liedchen klingen', op.48 no.10; 'Hörst du den Vogel singen?', op.35 no.12; 'Hört ihr im Laube des Regens', op.96 no.4; Husarenabzug, op.125 no.2; 'Ich armes Käuzlein kleine', op.79 no.10; 'Icharmes Käuzlein kleine', after op.142

'Ich bin dein Baum', op.101 no.3; 'Ich bin der Kontrabandiste', op.74 no.10; 'Ich bin die Blum' in Garten', op.83 no.2; 'Ich bin ein lust'ger Geselle', op.30 no.1; 'Ich bin gekommen ins Niederland', op.25 no.10; Ich bin geliebt, op.74 no.9; 'Ich bin hinaus gegangen', op.43 no.3; 'Ich bin im Mai gegangen', 1828; 'Ich bin vom Berg der Hirtenknab', op.79 no.8; 'Ich blick in mein Herz', op.51 no.1; 'Ich denke dein', op.78 no.3; 'Ich grolle nicht', op.48 no.7; 'Ich hab im Traum geweinet', op.48 no.13; 'Ich hab in mich gesogen', op.37 no.5; 'Ich hab mein Weib allein', op.25 no.22

'Ich hör die Bächlein rauschen', op.39 no.8; 'Ich kann's nicht fassen, nicht glauben', op.42 no.3; 'Ich kann wohl manchmal singen', op.39 no.9; 'Ich sah dich weinen', 1827; 'Ich schau über Forth hinüber', op.25 no.23; 'Ich sende einen Gruss', op.25 no.25; 'Ich wandelte unter den Bäumen', op.24 no.3; Ich wandre nicht, op.51 no.3; 'Ich will meine Seele tauchen', op.48 no.5; 'Ich zieh dahin', op.135 no.1; 'Ich zieh so allein in den Wald hinein', op.107 no.5; Ihre Stimme, op.96 no.3; 'Ihr Matten, lebt wohl, ihr sonnigen Weiden', op.79 no.22

'Im Garten steht die Nonne', op.49 no.3; Im Herbste, 1828; 'Im Kreise der Vasallen', op.106; 'Im Rhein, im heiligen Strome', op.48 no.6; 'Im Schatten des Waldes', op.29 no.3; 'Im Städtchen gibt es des Jubels viel', op.40 no.4; Im Wald, op.107 no.5; Im Walde, op.39 no.11; 'Im Wald, in grüner Runde', op.119 no.1; Im Westen, op.25 no.23; Im wunderschönen Monat Mai', op.48 no.1; 'In den Talen der Provence', op.139 no.4; In der Fremde, op.39 no.1; In der Fremde, op.39 no.8; 'In der hohen Hall sass König Sifrid', op.139 no.7; In der Nacht, op.74 no.4

'In einsamen Stunden drängt Wehmut sich auf', op.77 no.4; 'In meinem Garten die Nelken', op.29 no.2; 'In meiner Brust', op.53 no.3; Ins Freie, op.89 no.5; Intermezzo, op.39 no.2; Intermezzo, op.74 no.2;

Jasminenstrauch, op.27 no.4; 'Jeden Morgen, in der Frühe', op.79 no.7; Jemand, op.25 no.4; Jung Volkers Lied, op.125 no.3; 'Kennst du das Land', op.79 no.28; 'Kennst du das Land', op.98a no.1; Kinderwacht, op.79 no.21; Klage, 1828; 'Kleine Tropfen, seid ihr Tränen', op.142 no.3; Kommen und Scheiden, op.90 no.3

'Komm in die stille Nacht', op.36 no.2; 'Komm, lieber Mai', op.79 no.9; 'Komm, Trost der Welt', op.83 no.3; 'Kommt, wir wollen uns begeben', op.79 no.5; 'Könnt ich dich in Liedern preisen', op.35 no.8; 'Könnt'ich einmal wieder singen', 1828; 'Kuckuck, Kuckuck ruft aus dem Wald', op.79 no.3; Kurzes Erwachen, 1828; Ländlisches Lied, op.29 no.1; 'Lange harrt ich', 1828; 'Lass mich ihm am Busen hangen', op.25 no.12; 'Lass tief in dir mich lesen', op.96 no.3; 'Lasst mich nur auf meinem Sattel gelten', op.25 no.2; 'Lehn deine Wang', op.142 no.2; 'Leicht wie gaukelnde Sylphiden', 1827; 'Leis rudern hier', op.25 no.17

'Lieben, von ganzer Seele lieben', op.83 no.1; Liebesbotschaft, op.36 no.6; Liebesgarten, op.34 no.1; Liebesgram, op.74 no.3; Liebeslied, op.51 no.5; Liebhabers Ständchen, op.34 no.2; 'Lieb Liebchen', op.24 no.4; 'Liebster, deine Worte stehlen', op.101 no.2; 'Liebste, was kann denn uns scheiden?', op.37 no.6; Lied, op.29 no.2; Liedchen von Marie und Papa, 1852; Lied der Suleika, op.25 no.9; Lied eines Schmiedes, op.90 no.1; Lied-Album für die Jugend, op.79; Lieder aus dem Schenkenbuch im Divan I, op.25 no.5; Lieder aus dem Schenkenbuch im Divan II, op.25 no.6; Lieder der Braut aus dem Liebesfrühling I, op.25 no.12; Lieder der Braut aus dem Liebesfrühling II, op.25 no.12; Liederkreis, op.24; Liederkreis, op.39

Lied für xxx, 1827; Lied Lynceus des Türmers, op.79 no.27; Loreley, op.53 no.2; Lust der Sturmnacht, op.35 no.1; Mädchenlieder, op.103; Mädchen-Schwermut, op.143 no.3; Mailied, op.79 no.9; Mailied, op.103 no.1; Mailied, after op.142; Marienwürmchen, op.79 no.13; 'Marienwürmchen, setze dich', op.79 no.1; Märzveilchen, op.40 no.1; Maultreiberlied, 1838; 'Mein altes Ross', op.127 no.4; 'Mein Aug ist trüb', op.27 no.3; Meine Rose, op.90 no.2

'Meine Töne still und heiter', op.101 no.1; Mein Garten, op.77 no.2; 'Mein Herz ist betrübt', op.25 no.4; 'Mein Herz ist im Hochland', op.25 no.13; 'Mein Herz ist schwer', op.25 no.15; 'Mein hochgebornes Schätzelein', after op.142; 'Mein schöner Stern', op.101 no.4; 'Mein Wagen rollet langsam', op.142 no.4; Melancholie, op.74 no.6; Mignon, op.79 no.28; Minnespiel, op.101; 'Mir ist's so eng allüberall!', op.89 no.5; 'Mit dem Pfeil, dem Bogen', op.79 no.25; 'Mit der Myrte geschmückt', op.31 no.1; 'Mit Myrten und Rosen', op.24

no.9; 'Mögen alle bösen Zungen', op.74 no.9; 'Mond, meiner Seele Liebling', op.104 no.1

Mondnacht, op.39 no.5; 'Morgens steh ich auf und frage', op.24 no.1; 'Mutter, Mutter glaube nicht', op.25 no.11; Muttertraum, op.40 no.2; Myrthen, op.25; Nach der Geburt ihres Sohnes, op.135 no.2; 'Nach diesen trüben Tagen', op.79 no.19; 'Nach Frankreich zogen zwei Grenadier', op.49 no.1; 'Nach langem Frost', 1851; Nachtlied, op.96 no.1; 'Nachts um die zwölfte Stunde', 1840; 'Nachts zu unbekannter Stunde', op.89 no.2; Nänie, op.114 no.1; 'Nelken wind ich und Jasmin', op.74 no.8; 'Nicht im Tale', 1828; 'Nicht so schnelle', op.77 no.5

Nichts Schöneres, op.36 no.3; 'Niemand', op.25 no.22; 'Nun hast du mir den ersten Schmerz getan', op.42 no.8; 'Nun scheidet vom sterbenden Walde', op.89 no.4; 'Nun stehen die Rosen in Blüte', op.125 no.5; 'Nur ein Gedanke', op.135 no.3; 'Nur wer die Sehnsucht kennt', op.98a no.3; 'Oben auf des Berges Spitz', op.49 no.2; 'O blicke, wenn den Sinn dir will die Welt', op.114 no.3; 'O Freund, mein Schirm, mein Schutz, op.101 no.6; 'O Gott, mein Gebieter', op.135 no.5

'O ihr Herren', op.37 no.3; 'O Schmetterling, sprich', op.79 no.2; 'O Sonnenschein!', op.36 no.4; 'O Sonn, o Meer, o Rose', op.37 no.10; 'O wie lieblich ist das Mädchen', op.138 no.3; 'Pflücket Rosen, um das Haar schön', op.103 no.1; Provenzalisches Lied, op.139 no.4; Rätsel, op.25 no.16; 'Reich mir die Hand, o Wolke', op.104 no.5; Requiem, op.90 no.7; Resignation, op.83 no.1; 'Rose, Meer und Sonne', op.37 no.9; Röselein, Röselein', op.89 no.6; 'Ruh von schmerzensreichen Mühen aus', op.90 no.7

'Sag an, o lieber Vogel mein', op.27 no.1; Sängers Trost, op.127 no.1; 'Schlafe, süsser kleiner Donald' op.25 no.14; 'Schlaf, Kindlein, schlaf', op.78 no.4; 'Schlafflose Sonne, melanchol'scher Stern', op.95 no.2; 'Schlief die Mutter endlich ein', op.31 no.2; Schlusslied des Narren, op.127 no.5; Schmetterling, op.79 no.2; Schneeglöckchen, op.79 no.26; Schneeglöckchen, op.96 no.2; 'Schneeglöckchen klingen wieder', op.79 no.18; Schön Blümelein, op.43 no.3; Schöne Fremde, op.39 no.6; 'Schöne Sennin, noch einmal singe', op.90 no.4

'Schöne Wiege meiner Leiden', op.24 no.5; Schön Hedwig, op.106; 'Schön ist das Fest des Lenzes', op.37 no.7; 'Schön ist das Fest des Lenzes', op.101 no.5; 'Schweb empor am Himmel', op.103 no.4; 'Seh ich in das stille Tal', op.78 no.2; Sehnsucht, 1827; Sehnsucht, op.51 no.1; 'Sehnsucht nach der Waldgegend', op.35 no.5; 'Seit ich ihn gesehen', op.42 no.1; 'Senkt die Nacht den sanften Fittig nieder',

op.114 no.2; 'Setze mir nicht', op.25 no.6; 'Sie schlingt um meinen Nacken', 1840; 'Sie sollen ihn nicht haben', 1840; 'Singet nicht in Trauertönen', op.98a no.7; 'Sitz ich allein', op.25 no.5

'So lasst mich scheinen', op.98a no.9; Soldatenlied, ?1845; Sommerruh, 1849; 'Sommerruh, wie schön bist du', 1849; Sonntag, op.79 no.6; Sonntags am Rhein, op.36 no.1; 'So oft sie kam', op.90 no.3; 'So sei gegrüsst vieltausendmal', op.79, no.4; 'So wahr die Sonne scheinet', op.37 no.12; 'So wahr die Sonne scheinet', op.101 no.8; 'Spähend nach dem Eisengitter', op.53 no.1; Spanische Liebeslieder, op.138; Spanisches Liederspiel, op.74; Spinnlied, op.79 no.24; 'Spinn, spinn', op.79 no.24; Spruch, op.114 no.3; Ständchen, op.36 no.2; 'Sterne der blauen himmlischen Auen', 1827; Stille Liebe, op.35 no.8; Stille Tränen, op.35 no.10; Stiller Vorwurf, op.77 no.4

Stirb, Lieb und Freud!, op.35 no.2; 'Süsser Freund, du blickest', op.42 no.6; Talismane, op.25 no.8; Tanzlied, op.78 no.1; 'Tief im Herzen trag ich Pein', op.138 no.2; Tragödie, op.64 no.3; Triolett, op.114 no.2; Trost im Gesang, op.142 no.1; 'Über allen Gipfeln ist Ruh', op.96 no.1; 'Überm Garten durch die Lüfte', op.39 no.12; 'Und als ich ein winzig Bübchen war', op.127 no.5; 'Und die mich trug im Mutterarm', op.125 no.3; 'Und schläfst du, mein Mädchen, auf', op.74 no.2

'Und wenn die Primel schneeweiss blickt', op.29 no.1; 'Und wieder hatt ich der Schönsten gedacht', op.36 no.5; 'Und wüssten's die Blumen, die kleinen', op.48 no.8; 'Unter den roten Blumen schlummere', op.114 no.1; 'Unter die Soldaten', op.79 no.7; Unterm Fenster, op.34 no.3; 'Veilchen, Rosmarin, Mimosen', op.77 no.2; Verratene Liebe, op.40 no.5; Verwandlung, 1827; 'Viel Glück zur Reise, Schwalben', op.104 no.2; Vier Husarenlieder, op.117; 'Vöglein vom Zweig', op.79 no.15; Volksliedchen, op.51 no.2; Vom Reitersmann, after 1828; Vom Schlaraffenland, op.79 no.5

'Von dem Rosenbusch', op.74 no.1; 'Vor seinem Löwengarten', op.87; 'Wachst du noch, Liebchen, Gruss und Kuss!', op.34 no.2; Waldesgespräch, op.39 no.3; Wanderlust, op.35 no.3; Wanderung,

op.35 no.7; 'Wann, wann erscheint der Morgen', op.74 no.6; 'Wär ich nie aus euch gegangen', op.35 no.5; Warnung, op.119 no.2; 'Wärst du nicht, heil'ger Abendschein!', op.35 no.9; 'Warte, warte, wilder Schiffmann', op.24 no.6; 'Warum soll ich denn wandern', op.51 no.3; 'Was hör ich draussen vor dem Tor', op.98a no.2; 'Was nützt die mir noch zugemess'ne Zeit?', op.135 no.4; 'Was soll ich sagen?, op.27 no.3

'Was weht um meine Schläfe', op.77 no.3; 'Was will die einsame Träne?', op.25 no.21; Wehmut, op.39 no.9; 'Weh, wie zornig ist das Mädchen', op.138 no.7; Weihnachtslied, op.79 no.16; 'Weint auch einst kein Liebchen', op.127 no.1; Weit, weit, op.25 no.20; 'Wem Gott will rechte Gunst erweisen', op.77 no.1; 'Wenn alle Wälder schliefen', op.45 no.1; 'Wenn der Winter sonst entschwand', 1827; 'Wenn durch Berg und Tale', op.35 no.1; 'Wenn durch die Piazzetta', op.25 no.18

'Wenn fromme Kindlein schlafen gehn', op.79 no.21; 'Wenn ich ein Vögleinwar', op.43 no.1; 'Wenn ich frühlinden Gartengeh', op.51 no.2; 'Wenn ich in deine Augen seh', op.48 no.4; 'Wer ist vor meiner Kammertür?', op.34 no.3; 'Wer machte dich so krank?', op.35 no.11; 'Wer nie sein Brot mit Tränen ass', op.98a no.4; 'Wer sich der Einsamkeit ergibt', op.98a no.6; Widmung, op.25 no.1; 'Wie blüht es im Tale', op.79 no.11; 'Wie der Bäume kühne Wipfel', op.96 no.5; Wiegenlied, op.78 no.4

'Wie kann nich froh', op.25 no.20; 'Wie mit innigstem Behagen', op.25 no.9; 'Wild verwachs'ne dunkle Fichten', op.90 no.5; 'Wir sassen am Fischerhause', op.45 no.3; 'Wir sind ja, Kind, im Maie', op.104 no.4; 'Wohlauf! noch getrunken den funkelnden Wein', op.35 no.3; 'Wohlauf und frisch gewandert', op.35 no.7; 'Wolken, die ihr nach Osten eilt', no.6; 'Zieh' nur du Sonne', 1828; 'Zigeunerleben, op.29 no.3; Zigeunerliedchen, op.79 no.7; 'Zu Augsburg steht ein hohes Haus', no.2

Zum Schluss, op.25 no.26; 'Zum Sehen geboren', op.79 no.27; 'Zur Schmiede ging ein junger Held', 1848; 'Zwei feine Stieflein hab ich an', op.79 no.12; Zwei Venetianische Lieder I, op.25 no.17; 'Zwei Venetianische Lieder II, op.25 no.18; Zwielicht, op.39 no.10

175–9, 180, 188

105
106

KEYBOARD
(for solo pf unless otherwise stated)

op.	Title, key	Composed	Publication/MS	Remarks	SW
—	8 polonaises, pf 4 hands	1828	1933	some material used in Papillons, op.2	—
—	Variations on a theme of Prince Louis Ferdinand of Prussia, pf 4 hands	1828	—		—

	Title	Date		Notes	Pubn	Pages
—	Romanze, f	1829		unfinished	—	108
—	6 Walzer	1829–30		some material used in Papillons, op.2	—	108, 179
1	Thème sur le nom Abegg varié pour le pianoforte	1829–30	1831	also orch sketches	vii/1, 2	
—	Variations on a theme of Weber	1831		on a theme from Preziosa		
—	Valse, E♭	1831		unfinished		
—	Valse per F. Wieck	1831–2		unfinished		
—	Sonata, A♭	1831–2		1st movt and Adagio only		
—	Andante with variations on an orig. theme, G	1831–2		inscribed 'Mit Gott', some material used in op.124 no.2	—	
—	Prelude and fugue	1832			—	
2	Papillons	1829–31	1831	includes some material from the 4-hand polonaises (1828) and some used also in 6 Walzer (1829–30)	vii/1, 12	111, 113, 115, 176, 179
3	6 Studien nach Capricen von Paganini, i	1832		orig. op.2	vii/1, 22	112, 113
4	6 intermezzos	1832		orig. op.3 and entitled Pièces phantastiques	vii/1, 46	105, 113, 175, 176, 177
—	Phantasie satyrique	1832		on a theme of Henri Herz; fragments only		107
—	Fandango, F♯	1832		later used in op.11		113
—	Exercice fantastique	1832		orig. op.5; lost		112
—	Rondo, B♭	1832		unfinished		
—	12 Burlesken (Burle)	1832		later used in op.124		113
—	Fugue, d	?1832	D-Bds	sketch		
—	Movt in B♭	?1832	S-Skma	sketch		
—	Fugal piece, b♭ [one of many]	?1832	Skma	sketch		
—	Canonic piece, A	?1832	Skma			
—	Fugue no.3	1832–3	—	probably intended as finale of op.5		
—	5 short pieces:					
	1 Notturnino			unfinished		
	2 Ballo					
	3 Burla					
	4 Capriccio			unfinished		
	5 Ecossaise			unfinished		
—	Sehnsuchtswalzer Variationen: scènes musicales sur un thème connu	1832–3		also entitled Scènes mignonnes and Scènes musicales sur un thème connu de Fr. Schubert; opening used as opening of Carnaval, op.9	—	116
5	10 Impromptus über ein Thema von Clara Wieck	1833		last no. includes material from finale of Symphony, g (1832–3); 2nd version of 1850 omits 2 variations but introduces a new variation, no.3	vii/1, 68	115
—	Etüden in Form freier Variationen über ein Beethovensches Thema	1833	1976	Allegretto of Beethoven's Symphony no.7: one variation pubd as op.124 no.2	—	116

op.	Title, key	Composed	Publication/MS		Remarks	SW	
			D-Zsch				
—	Variations sur un nocturne de Chopin	1834	Chopin's op.15 no.3, g			—	118
—	Sonata movt, B♭	1836				—	
—	Sonata no.4, f	1836–7		—	unfinished	—	116, 121
6	Davidsbündlertänze: 18 character-pieces	1837		1837	title in 2nd edn. (1850–51) Die Davidsbündler	vii/1, 96	122
7	Toccata, C	1829–32		1834	orig. op.6; orig. title Etude fantastique en double-sons, D	vii/1, 146	108, 112, 113
8	Allegro, b	1831		1835	1st movt of projected sonata	vii/1, 156	110–11, 179
9	Carnaval: scènes mignonnes sur quatre notes:	1833–5		1837	orig. title Fasching: Schwänke auf vier Noten für Pianoforte von Florestan, op.12	vii/2, 2	117, 258
	1 Préambule, 2 Pierrot, 3 Arlequin, 4 Valse noble. 5 Eusebius, 6 Florestan, 7 Coquette, 8 Réplique, Sphinxes, 9 Papillons, 10 ASCH-SCHA (Lettres dansantes), 11 Chiarina, 12 Chopin, 13 Estrella, 14 Reconnaissance, 15 Pantalon et Colombine, 16 Valse allemande, 17 Intermezzo: Paganini, 18 Aveu, 19 Promenade, 20 Pause, 21 Marche des Davidsbündler contres les Philistins						
10	6 Konzert-Etüden nach Capricen von Paganini, ii	1833		1835	orig. title Capricen für das Pianoforte, auf dem Grund der Violinstimme von Paganini zu Studien frei bearbeitet	vii/2, 30	115
11	Sonata no.1, f♯	1832 5		1836	on title-page 'Pianoforte-Sonata. Clara zugeignet von Florestan und Eusebius'	vii/2, 52	105, 113, 116, 119, 120, 178, 258
12	Phantasiestücke:	1837		1838	orig. title Phantasien; no.7 composed not later than 1832	vii/2, 82	122
	1 Des Abends, 2 Aufschwung, 3 Warum?, 4 Grillen, 5 In der Nacht, 6 Fabel, 7 Traumes Wirren, 8 Ende vom Lied						
9 ***		1837		1935	omitted from op.12	—	
13	Symphonische Etüden (Etudes symphoniques)	1834–7		1837	orig. title Etüden im Orchestercharakter für Pianoforte von Florestan und Eusebius; 2nd version (1852) entitled Etudes en formes de variations; 5 extra variations pubd in 1873 and incl. in SW xiv, 40; variation unpubd	vii/2, 108	117, 120–21, 122
14	Concert sans orchestre, f	1835–6		1836	3 movts of orig. 5 pubd 1836; rev. and pubd 1853 as Sonata no.3 with 1 scherzo restored	vii/3	116, 120, 178
—	Scherzo	1836		1866	rejected movt of op.14	—	120
—	2 Variations	1836		1984	omitted from 3rd movt of op.14	—	—

op.	Title, key	Composed	Publication/MS	Remarks	SW
68	Album für die Jugend	1848	1848	Orig. title Weihnachtsalbum; facs. of autograph (Leipzig, 1956); facs. of sketchbook (London, 1924), with 4 other pf pieces by Schumann, ed. L. Windesperger: Kuckuck im Versteck, Lagune in Venedig, Haschemann, waltz in G; these and 5 other pf pieces from sketchbook, ed. J. Werner (London, c1958): Für ganz Kleine, Puppenschlaffliedchen, Linke Hand soll sich auch zeigen, Auf der Gondel, untitled piece; other unpubd material, ed. J. Demus (Milan, 1973), with 2 further pieces: Rebus, Bärentanz	146, 153

Pt. 1, Für Kleinere: 1 Melodie, 2 Soldatenmarsch, 3 Trällerliedchen [orig. Kinderstückchen], 4 Ein Choral, 5 Stückchen, 6 Armes Waisenkind [orig. Armes Bettlerkind], 7 Jägerliedchen, 8 Wilder Reiter, 9 Volksliedchen [orig. Volkslied], 10 Fröhlicher Landmann, von der Arbeit zurückkehrend, 11 Sizilianisch [orig. Zwei Sizilianische], 12 Knecht Ruprecht, 13 Mai, lieber Mai [orig. Mai, schöner Mai], 14 Kleine Studie, 15 Frühlingsgesang, 16 Erster Verlust [orig. Kinderunglück], 17 Kleiner Morgenwanderer, 18 Schnitterliedchen
Pt. 2, Für Erwachsenere: 19 Kleine Romanze, 20 Ländliches Lied, 21***, 22 Rundgesang, 23 Reiterstück, 24 Ernteliedchen, 25 Nachklänge aus dem Theater, 26***, 27 Canonisches Liedchen [orig. Canon], 28 Erinnerung [orig. Erinnerung an Mendelssohn], 29 Fremder Mann, 30***, 31 Kriegslied, 32 Sheherazade, 33 Weinlesezeit – fröhliche Zeit, 34 Thema, 35 Mignon, 36 Lied italienischer Marinari [orig. Schifferlied], 37 Matrosenlied, 38 Winterszeit I, 39 Winterszeit II, 40 Kleine Fuge, 41 Nordisches Lied (Gruss an G), 42 Figurierter Choral, 43 Sylversterlied [orig. Zum Schluss]

op.	Title, key	Composed	Publication/MS	Remarks	SW
72	4 fugues, d, d, f, F	1845	1850		vii/5 · 141
76	4 marches, E♭, g, B♭ (Lager-Scene), E♭	1849	1849		vii/5 · 148–9
82	Waldscenen:	1848–9	1850		vii/5 · 146, 174, 176

1 Eintritt, 2 Jäger auf der Lauer [orig. Jägersmann auf der Lauer], 3 Einsame Blumen, 4 Verrufene Stelle [orig. Verrufener Ort], 5 Freundliche Landschaft [orig. Freier Ausblick], 6 Herberge [orig. Jägerhaus], 7 Vogel als Prophet, 8 Jagdlied, 9 Abschied

op.	Title, key	Composed	Publication/MS	Remarks	SW
85	12 vierhändige Klavierstücke für kleine und grosse Kinder:	1849	1850	no.2 from sketches in op.68; no.3 orig. title Gartenlied	vi, 48 · 150

1 Geburtstagsmarsch, 2 Bärentanz, 3 Gartenmelodie, 4 Beim Kränzewinden, 5 Kroatenmarsch, 6 Trauer, 7 Turniermarsch, 8 Reigen, 9 Am Springbrunnen, 10 Versteckens, 11 Gespenstermärchen, 12 Abendlied

op.	Title, key	Composed	Publication/MS	Remarks	SW
99	Bunte Blätter:		1852		vii/6, 2 · 124, 160
	Drei Stücklein:				
	1	1838		MS inscribed 'An meine geliebte Braut zum heiligen Abend 1838'	124
	2	1839			
	3	1839		orig. title Jagdstück	

No.	Title / Contents	Date	Remarks		
	Fünf Albumblätter:				
	4	1841			
	5	1838	theme used by Brahms for Variations, op.9		
	6	1836	orig. title Fata Morgana		
	7	1838	rejected from Carnaval, op.9		
	8	1838	orig. title Jugendschmerz		
	9 Novellette	1838			127
	10 Präludium	1839			
	11 Marsch	1843			
	12 Abendmusik	1841			
	13 Scherzo	1841			133
	14 Geschwindmarsch	1849			148
109	Ballszenen, pf 4 hands:	1851	orig. title Kinderball	vi, 94	159
	1 Préambule, 2 Polonaise, 3 Walzer, 4 Ungarisch, 5 Française, 6 Mazurka, 7 Ecossaise, 8 Walzer, 9 Promenade				
111	Phantasiestücke: 3 pieces, c, Ab, c	1851		vii/6, 36	160
118	Drei Clavier-Sonaten für die Jugend, G, D, C	1853		vii/6, 44	167
124	Albumblätter:	1854	orig. title Cyclus für Pianoforte — nos. 1, 3, 12, 15 perhaps among the 12 Burlesken offered to Breitkopf & Härtel in 1832	vii/6, 78	113, 116, 124
	1 Impromptu (1832), 2 Leides Ahnung (1832), 3 Scherzino (1832), 4 Walzer (1835), 5 Phantasietanz (1836), 6 Wiegenliedchen (1843), 7 Ländler (1836), 8 Lied ohne Ende (1837), 9 Impromptu (1838), 10 Walzer (1838), 11 Romanze (1835), 12 Burla (1832), 13 Larghetto (1832), 14 Vision (1838), 15 Walzer (1832), 16 Schlummerlied (1841), 17 Elfe (1835) [orig. intended for op.9], 18 Botschaft (1838), 19 Phantasiestück (1839), 20 Canon (1845)				
126	Sieben Klavierstücke in Fughettenform	1853		vii, 102	167
130	Kinderball, pf 4 hands:	1853		vi, 142	167-8
	1 Polonaise, 2 Walzer, 3 Menuett, 4 Ecossaise, 5 Française, 6 Ringelreihe				
133	5 Gesänge der Frühe	1853		vii/6, 114	168, 176
—	Canon on F. H. Himmel's An Alexis send ich dich, Ab	1855 1859	MS inscribed 'An Diotima' in Julius Knorr's op.30		
—	Thema, Eb	1893		xiv/1, 67	
—	Variations on an original theme	1939	theme used by Brahms for Variations for pf duet op.23		

221

BIBLIOGRAPHY

CATALOGUES

A. Dörffel: *Thematisches Verzeichniss sämmtlicher in Druck erschienenen Werke Robert Schumanns* (Leipzig, 1860, 4/1868/*R*1966)
G. Eismann: 'Nachweis der internationalen Standorte von Notenautographen Robert Schumanns', *Sammelbände der Robert-Schumann-Gesellschaft*, ii (Leipzig, 1966), 7–37
K. Hofmann: *Die Erstdrucke der Werke von Robert Schumann* (Tutzing, 1979)

BIBLIOGRAPHIES

G. Abraham, ed.: *Schumann: a Symposium* (London, 1952), 301
P. Mies: 'Literatur um Robert Schumann: ein Beitrag zu seinem 100. Todestag', *Musikhandel*, vii (1956), 93
F. Munte: *Verzeichnis des deutschsprachigen Schrifttums über Robert Schumann 1856–1970* (Hamburg, 1972)
A. Walker, ed.: *Robert Schumann: the Man and his Music* (London, 1972, rev. 2/1976), 442
K. Hofmann: *Die Erstdrucke der Werke von Robert Schumann: Bibliographie* (Tutzing, 1979)

GENEALOGIES

E. Bienenfeld: 'Die Stammtafel der Familie Robert Schumanns: Ahnen und Nachkommen eines Genies', *Archiv für Rassen- und Gesellschaftsbiologie*, xxvi (1932), 57
C. Eismann: 'Bemerkenswertes zur Genealogie Robert Schumanns', *Mf*, xxii (1969), 61
A. Walker: 'Schumann's Family Tree', *Robert Schumann: the Man and his Music* (London, 1972, rev. 2/1976), 418

ICONOGRAPHY

W. Gertler: *Robert Schumann* (Leipzig, 1936)
G. Eismann: *Eine Biographie in Wort und Bild* (Leipzig, 1956, enlarged 2/1964; Eng. trans., 1964)
R. Petzold and E. Crass: *Robert Schumann: sein Leben in Bildern* (Leipzig, 1956)
G. Eismann: 'Das authentische Schumann-Bild', *Sammelbände der Robert-Schumann-Gesellschaft*, i (Leipzig, 1961), 86

DIARIES

G. Eismann, ed.: *Robert Schumann: Tagebücher*, i: *1827–38* (Leipzig, 1971)

Bibliography

G. Nauhaus, ed.: *Robert Schumann: Tagebücher*, iii: *Haushaltbücher 1837–56* (Leipzig, 1982)

LETTERS

C. Schumann, ed.: *Jugendbriefe von Robert Schumann* (Leipzig, 1885, 4/1910; Eng. trans., 1888)

F. Jansen, ed.: *Robert Schumanns Briefe: neue Folge* (Leipzig, 1886, 2/1904; Eng. trans., 1890)

H. Erler: *Robert Schumanns Leben: aus seinen Briefen geschildert* (Berlin, 1886–7, 3/1927)

J. Gensel: 'Robert Schumanns Briefwechsel mit Henriette Voigt', *Die Grenzboten*, li (1892), 269, 324, 368; enlarged offprint (Leipzig, 1892)

F. Jansen: 'Briefwechsel zwischen Robert Franz und Robert Schumann', *Die Musik*, viii (1908–9), 280, 346

F. Krautwurst: 'Briefe von Chirstian Heinrich Rinck, Felix Mendelssohn Bartholdy und Robert Schumann aus dem Nachlass J. G. Herzogs in der Erlangen Universitätsbibliothek', *Jb für Frankische Landesforschung*, xxi (1976), 149

D. A. Wells: 'Letters of Mendelssohn, Schumann and Berlioz in Belfast', *ML*, lx (1979), 180 [letter of 10 Dec 1852 from Schumann to a publisher]

WRITINGS

R. Schumann: *Gesammelte Schriften über Musik und Musiker* (Leipzig, 1854, 4/1891/R1968, 5/1914; Eng. trans., 1877; new Eng. trans. [selection], 1947)

H. Deiters: 'Schumann als Schriftsteller', *AMZ*, iii (1865), 761, 777, 793

F. Liszt: 'Ein Kapitel zur Reform der musikalischen Kritik', *Gesammelte Schriften*, iv (Leipzig, 1882), 115–55

G. Wustmann: 'Die Davidsbündler: ein verloren geglaubter Aufsatz Robert Schumanns', *Die Grenzboten*, xlviii/4 (1889), 23

P. Spitta: 'Ueber Robert Schumanns Schriften', *Musikgeschichtliche Aufsätze* (Berlin, 1894), 383

G. Noren-Herzberg: 'Robert Schumann als Musikschriftsteller', *Die Musik*, v/4 (1905–6), 100

G. Wustmann: 'Zur Entstehungsgeschichte der Schumannscher Zeitschrift für Musik', *ZIMG*, viii (1906), 396

P. Kehm: *Die 'Neue Zeitschrift für Musik' unter Schumanns Redaktion: 1834–44* (diss., U. of Munich, 1943)

G. Eismann, ed.: *R. Schumann: Erinnerungen an Felix Mendelssohn Bartholdy* (Zwickau, 1947, enlarged 2/1948; Eng. edn., 1951)

I. Forger: *Robert Schumann als Kritiker: ein Beitrag zur Geschichte der musikalischen Kritik und zum Schumann-Problem* (diss., U. of Münster, 1948)

J. Alf: 'Der Kritiker Robert Schumann', *110. Niederrheinisches Musikfest 1956*, 50

H. Homeyer: *Grundbegriffe der Musikanschauung Robert Schumanns: ihr Wesen, ihre Bedeutung und Funktion in seinem literarischen Gesamtwerk* (diss., U. of Münster, 1956)

H. Pleasants: *The Musical World of Robert Schumann* (New York and London, 1965)

K. Laux: 'Was ist ein Musikschriftsteller?: Carl Maria von Weber und Robert Schumann als Vorbild', *Sammelbände der Robert-Schumann-Gesellschaft*, ii (Leipzig, 1966), 38

L. Plantinga: *Schumann as Critic* (New Haven, 1967/*R*1977)

——: 'Schumann and the "Neue Zeitschrift für Musik" ', *Robert Schumann: the Man and his Music*, ed. A. Walker (London, 1972, rev. 2/1976), 162

H. Pleasants: 'Schumann the Critic', *Robert Schumann: the Man and his Music*, ed. A. Walker (London, 1972, rev. 2/1976), 179

DOCUMENTARY COMPILATIONS

F. Kerst: *Schumann-Brevier* (Berlin, 1905)

A. Schumann, ed.: *Der junge Schumann: Dichtungen und Briefe* (Leipzig, 1910)

W. Boetticher: *Robert Schumann in seinen Schriften und Briefen* (Berlin, 1942)

G. Eismann: *Robert Schumann: ein Quellenwerk über sein Leben und Schaffen* (Leipzig, 1956)

R. Münnich: *Aus Robert Schumanns Briefen und Schriften* (Weimar, 1956)

W. Schwarz: 'Robert Schumann und der deutsche Osten: aus unveröffentlichten Tagebuchaufzeichnungen, Briefen und Berichten', *Musik des Ostens*, ii (1963), 193

GENERAL STUDIES

E. Grieg: 'Robert Schumann', *Nyt tidskrift*, new ser., ii (1893–4), 217; also in *Century Magazine*, xlvii (1894), 440

R. Pugno: *Leçons écrites sur Schumann* (Paris, 1914)

W. Boetticher: *Robert Schumann: Einführung in Persönlichkeit und Werk* (Berlin, 1941)

H. Wolff: 'Robert Schumann: der Klassizist', *Musica*, ii (1948), 47

G. Abraham, ed.: *Schumann: a Symposium* (London, 1952)

H. Moser and E. Rebling, eds.: *Robert Schumann: aus Anlass seines 100. Todestages* (Leipzig, 1956)

Bibliography

E. Melkus: 'Schumanns letzte Werke', *ÖMz*, xv (1960), 565

——: 'Zur Revision unseres Schumann-Bildes', *ÖMz*, xv (1960), 182

A. Walker, ed.: *Robert Schumann: the Man and his Music* (London, 1972, rev. 2/1976)

J. W. Finson and R. L. Todd, eds.: *Mendelssohn and Schumann: Essays on their Music and its Context* (Durham, N. Carolina, 1984)

BIOGRAPHY, MEMOIRS

F. Brendel: 'Schumanns Biographie von Wasielewski', *NZM*, xlviii (1858), 113, 125, 137, 157, 169, 181, 193

W. von Wasielewski: *Robert Schumann* (Dresden, 1858, enlarged 2/1906; Eng. trans., 1871/*R*1975)

R. Pohl: 'Erinnerungen an Robert Schumann', *Deutsche Revue*, ii (1878), 169, 306

F. Jansen: *Die Davidsbündler: aus Robert Schumanns Sturm- und Drangperiode* (Leipzig, 1883)

W. von Wasielewski: *Schumanniana* (Bonn, 1883)

——: 'Robert Alexander Schumann', *ADB*

E. Hanslick: *Aus meinem Leben*, i (Berlin, 1894/*R*1971), 66, 105

W. von Wasielewski: 'Robert Schumanns Herzenserlebnisse: ein wichtiger Beitrag zur Schumann-Biographie', *Deutsche Revue*, xxii (1897), 40, 226

E. Hanslick: 'Robert Schumann in Endenich', *Am Ende des Jahrhunderts* (Berlin, 1899), 317

V. Joss: *Der Musikpädagoge Wieck und seine Familie: mit besonderer Berücksichtigung seines Schwiegersohnes R. Schumann* (Dresden, 1902)

B. Litzmann: *Clara Schumann: ein Künstlerleben nach Tagebüchern und Briefen* (Leipzig, 1902–8/*R*1971; Eng. trans., abridged, 1913/ *R*1972)

E. van der Straeten: 'Mendelssohns und Schumanns Beziehungen zu J. H. Lübeck und Johann J. H. Verhulst', *Die Musik*, iii (1903–4), 8, 94

——: 'Streiflichter auf Mendelssohns und Schumanns Beziehungen zu zeitgenössischen Musikern', *Die Musik*, iv (1904–5), 25, 105

C. Mauclair: *Schumann* (Paris, 1906)

F. May: *The Girlhood of Clara Schumann* (London, 1912)

M. Wieck: *Aus dem Kreise Wieck–Schumann* (Leipzig, 1912, rev. 2/1914)

F. Niecks: *Robert Schumann: a Supplementary and Corrective Biography* (London, 1925)

R. Pitrou: *La vie intérieure de Robert Schumann* (Paris, 1925)

E. Schumann: *Erinnerungen* (Stuttgart, 1925; Eng. trans., 1927)

V. Basch: *Schumann* (Paris, 1926)

——: *La vie douloureuse de Schumann* (Paris, 1928; Eng. trans., 1932)

K. Wagner: *Robert Schumann als Schüler und Abiturient* (Zwickau, 1928)

E. Schumann: *Robert Schumann: ein Lebensbild meines Vaters* (Leipzig, 1931)

P. Sutermeister: *Robert Schumann: sein Leben nach Briefen, Tagebüchern und Erinnerungen des Meisters und seiner Gattin* (Zurich, 1949)

E. Flechsig: 'Erinnerungen an Robert Schumann', *NZM*, Jg.117 (1956), 392

R. Stockhammer: 'Robert Schumann in Wien', *ÖMz*, xv (1960), 177

A. Walker: 'Schumann and his Background', *Robert Schumann: the Man and his Music*, (London, 1972, rev. 2/1976), 1–40

HEALTH

F. Richarz: 'Robert Schumanns Krankheit', *AMZ*, viii (1873), 597

H. Schaafhausen: 'Einige Reliquien berühmter Männer: Robert Schumanns Gehirn- und Gehörsorgane', *Correspondenz-Blatt der Deutschen Gesellschaft für Anthropologie*, xvi (1885), 147

P. Möbius: *Über Robert Schumanns Krankheit* (Halle, 1906)

H. Gruhle: 'Brief über Robert Schumanns Krankheit und P. Möbius', in P. Möbius: *Über Scheffels Krankheit* (Halle, 1907), 25

C. Pascal: 'Les maladies mentales de Robert Schumann', *Journal de psychologie normale et pathologique* (1908), March–April

E. Morselli: *La pazzia di Roberto Schumann e la psicologia supernormale* (Rome, 1909)

R. Bancour: 'La maladie de Schumann', *Chronique médicale* (1910), 481

F. Nussbaum: *Der Streit um Robert Schumanns Krankheit* (diss., U. of Cologne, 1923)

B. Springer: *Die genialen Syphilitiker* (Berlin, 1926), 143

H. MacMaster: *La folie de Robert Schumann* (Paris, 1928)

H. Kleinebreil: *Der kranke Schumann: Untersuchungen über Krankheit und Todesursache Robert Schumanns* (diss., U. of Jena, 1943)

E. Slater and A. Meyer: 'Contributions to a Pathography of the Musicians: I Robert Schumann', *Confinia psychiatrica*, ii (1959), 65

D. Kerner: 'Robert Schumann', *Krankheiten grosser Musiker* (Stuttgart, 1963), 123

H.-J. Rothe: 'Neue Dokumente zur Schumann-Forschung im Stadtarchiv Leipzig', *Arbeitsberichte zur Geschichte der Stadt Leipzig* (Leipzig, 1967), 1

E. Sams: 'Schumann's Hand Injury', *MT*, cxii (1971), 1156; cxiii (1972), 456

Bibliography

E. Slater: 'Schumann's Illness', *Robert Schumann: the Man and his Music*, ed. A. Walker (London, 1972, rev. 2/1976), 406

R. Henson and H. Urich: 'Schumann's Hand Injury', *British Medical Journal* (1978), no.1, p.900

L. Carerj: 'La mano invalida di Robert Schumann', *NRMI*, xiii (1979), 609

P. F. Ostwald: 'Florestan, Eusebius, Clara, and Schumann's Right Hand', *19th Century Music*, iv (1980–81), 17

LIFE AND WORKS

A. Ambros: 'Robert Schumanns Tage und Werke', *Culturhistorische Bilder aus der Gegenwart* (Leipzig, 1860), 51–96

A. Reissmann: *Robert Schumann: sein Leben und seine Werke* (Berlin, 1865, 2/1871; Eng. trans., 1886)

P. Spitta: 'Schumann, Robert', *Grove 1*

F. Liszt: 'Robert Schumann', *Gesammelte Schriften*, iv (Leipzig, 1882), 103, 156

H. Reimann: *Robert Schumanns Leben und Werke* (Leipzig, 1887)

H. Hadow: 'Robert Schumann and the Romantic Movement in Germany', *Studies in Modern Music*, i (London, 1892), 149–231

H. Abert: *Robert Schumann* (Berlin, 1903)

L. Schneider and M. Mareschal: *Schumann: sa vie et ses oeuvres* (Paris, 1905)

W. Dahms: *Schumann* (Berlin, 1916)

M. Ninck: *Schumann und die Romantik in der Musik* (Heidelberg, 1929)

M. Beaufils: *Schumann* (Paris, 1932)

C. Valabrega: *Schumann* (Modena, 1934)

W. Korte: *Robert Schumann* (Potsdam, 1937)

E. Bücken: *Robert Schumann* (Cologne, 1940)

R. Schauffler: *Florestan: the Life and Work of Robert Schumann* (New York, 1945)

J. Chissell: *Schumann* (London, 1948, rev. 2/1967)

K. Wörner: *Robert Schumann* (Zurich, 1949)

A. Coeuroy: *Robert Schumann* (Paris, 1950)

P. and W. Rehberg: *Robert Schumann: sein Leben und sein Werk* (Zurich, 1954)

M. Brion: *Schumann et l'âme romantique* (Paris, 1954; Eng. trans., 1956)

A. Boucourechliev: *Schumann* (Paris, 1957; Eng. trans., 1959/*R*1976)

E. Lippmann: 'Schumann, Robert', *MGG*

P. Young: *Tragic Muse: the Life and Works of Robert Schumann* (London, 1957, enlarged 2/1961)

I. Porena-Cappelli: 'Robert Schumann', *La MusicaE*

K. Laux: *Robert Schumann* (Leipzig, 1972)

R. Taylor: *Robert Schumann: his Life and Work* (London and New York, 1982)

PIANO MUSIC

F. Liszt: 'Robert Schumanns Klavierkompositionen op.5, 11 und 14', *Gesammelte Schriften*, ii (Leipzig, 1881), 99

R. Hohenemser: 'Formale Eigentümlichkeiten in R. Schumanns Klaviermusik', *Festschrift zum 50. Geburtstag Adolf Sandberger* (Munich, 1918), 21

J. Fuller Maitland: *Schumann's Pianoforte Works* (London, 1927)

M. Cohen: *Studien zur Sonataform bei Robert Schumann* (diss., U. of Vienna, 1928)

R. Goldenberg: *Der Klaviersatz bei Schumann* (diss., U. of Vienna, 1930)

W. Gertler: *Robert Schumann in seinen früheren Klavierwerken* (Wolfenbüttel, 1931)

M. Schweiger: *Die Harmonik in den Klavierwerken Robert Schumanns* (diss., U. of Vienna, 1931)

W. Schwarz: *Schumann und die Variation: mit besonderer Berücksichtigung der Klavierwerke* (Kassel, 1932)

K. Geiringer: 'Ein unbekanntes Klavierwerk aus Schumanns Jugendzeit' [Polonaises for piano duet], *Die Musik*, xxv (1932–3), 721

——: 'Ein unbekanntes Blatt aus Schumanns Endenicher Zeit', *Anbruch*, xvii (1935), 273

G. Kinsky: 'Ein unbekanntes Fantasiestück aus Schumanns Jugendzeit', *SMz*, lxxv (1935), 769

D. Tovey: 'Schumann: Carnaval', *Essays in Musical Analysis: Illustrative Music* (London, 1936), 109

W. Georgii: *Klaviermusik* (Zurich, 1941, 2/1950), 301

D. Tovey: 'Schumann: Novelette in F sharp minor op.21 no.8', *Essays in Musical Analysis: Chamber Music* (London, 1944)

G. Abraham: 'Schumann's Opp.II and III', *MMR*, lxxvi (1946), 123, 162, 222; repr. in *Slavonic and Romantic Music* (London, 1968), 261

R. Réti: 'Schumann's Kinderszenen: a Theme with Variations', *The Thematic Process in Music* (New York, 1951, rev. 2/1961), 31

K. Dale: 'The Piano Music', *Schumann: a Symposium*, ed. G. Abraham (London, 1952), 12–97

I. Parrott: 'A Plea for Schumann's Op.11', *ML*, xxxiii (1952), 55

R. Fiske: 'A Schumann Mystery', *MT*, cv (1964), 574 [on *Die Davidsbündlertänze* op.6]

Bibliography

K. Wörner: 'Schumanns "Kreisleriana"', *Sammelbände der Robert-Schumann-Gesellschaft*, ii (Leipzig, 1966), 58

W. Boetticher: 'Neue textkritische Forschungen an Robert Schumanns Klavierwerk', *AMw*, xxv (1968), 46–76

J. Chissell: *Schumann Piano Music* (London, 1972)

Y. Solomon: 'Solo Piano Music: (I) the Sonatas and Fantasie', *Robert Schumann: the Man and his Music*, ed. A. Walker (London, 1972, rev. 2/1976), 41

B. Vázsonyi: 'Solo Piano Music: (II) the Piano Cycles', *Robert Schumann: the Man and his Music*, ed. A. Walker (London, 1972, rev. 2/1976), 68

J. Weingarten: 'Interpreting Schumann's Piano Music', *Robert Schumann: the Man and his Music*, ed. A. Walker (London, 1972, rev. 2/1976), 93

J. L. Kollen: 'Robert Alexander Schumann (1810–1856): *Tema*, Opus 13', *Notations and Editions: a Book in Honor of Louise Cuyler* (Dubuque, Iowa, 1974), 163

L. C. Roesner: 'The Autograph of Schumann's Piano Sonata in F minor, opus 14', *MQ*, lxi (1975)), 98

W. Boetticher: *Robert Schumanns Klavierwerke: Entstehung, Urtext. Gestalt: Untersuchungen anhand unveröffentlicher Skizzen und biographischer Documente* (Wilhelmshaven, 1976–)

L. C. Roesner: 'Schumann's Revisions in the First Movement of the Piano Sonata in G Minor, Op. 22', *19th Century Music*, i (1977–8), 97

R. Polansky: 'The Rejected Kinderscenen of Robert Schumann's Opus 15', *JAMS*, xxxi (1978), 126

A. Walker: 'Schumann, Liszt and the C major Fantasie, Op.17: a Declining Relationship', *ML*, lx (1979), 156

C. S. Becker: 'A New Look at Schumann's Impromptus', *MQ*, lxvii (1981), 568

SONGS

M. Friedlaender: *Textrevision zu Robert Schumanns Liedern* (Leipzig, 1887)

V. Wolff: *Robert Schumann Lieder in ersten und späteren Fassungen* (Leipzig, 1914)

O. Bie: 'Robert Schumann', *Das deutsche Lied* (Berlin, 1926), 75–124

C. Spitz: 'Schumann's "Mary Stuart Songs"', *MMR*, lxvii (1937), 153

R. Felber: 'Schumann's Place in German Song', *MQ*, xxvi (1940), 340

R. Hernried: 'Four Unpublished Compositions by Robert Schumann', *MQ*, xxviii (1942), 50

G. Abraham: 'Schumann's Opp.II and III', *MMR*, lxxvi (1946), 123,

229

162, 222; repr. in *Slavonic and Romantic Music* (London, 1968), 261

W. Edelmann: *Über Text und Musik in Robert Schumanns Sololieder* (diss., U. of Münster, 1950)

M. Cooper: 'The Songs', *Schumann: a Symposium*, ed. G. Abraham (London, 1952), 98–137

E. Sams: 'Schumann's Year of Song', *MT*, cvi (1965), 105

——: *The Songs of Robert Schumann* (London, 1969, rev. 2/1975)

S. Walsh: *The Lieder of Schumann* (London, 1971)

A. Desmond: *Schumann Songs* (London, 1972)

E. Sams: 'The Songs', *Robert Schumann: the Man and his Music*, ed. A. Walker (London, 1972, rev. 2/1976), 120

A. Mayeda: 'Das Reich der Nacht in den Liedern Robert Schumanns', *De ratione in musica: Festschrift Erich Schenk* (Kassel, 1975), 202

K. Schlager: 'Erstarrte Idylle: Schumanns Eichendorff-Verständnis im Lied op.39/VII "Auf einer Burg"', *AMw*, xxxiii (1976), 119

R. Hallmark: 'The Sketches for "Dichterliebe" ', *19th Century Music*, i (1977–8), 110

——: *The Genesis of Schumann's 'Dichterliebe': a Source Study* (Ann Arbor, 1979)

G. Moore: *The Songs and Cycles of Schumann* (New York, 1981)

B. Turchin: 'Schumann's Conversion to Vocal Music: a Reconsideration', *MQ*, lxvii (1981), 392

OTHER VOCAL MUSIC

E. Hanslick: 'Szenen aus Goethes Faust von R. Schumann (3. Abteilung)'; 'Schumanns Musik zu Goethes Faust', *Aus dem Concert-Saal* (Vienna, 1870/*R*1971, 2/1896), 218; 304

——: 'Manfred von Robert Schumann', *Aus dem Concert-Saal* (Vienna, 1870/*R*1971, 2/1896), 190

——: 'Paradies und die Peri', *Aus dem Concert-Saal* (Vienna, 1870/*R*1971, 2/1896), 155

——: 'R. Schumann als Opernkomponist', *Die moderne Oper* (Berlin, 1875/*R*1971, 3/1911), 256

S. Bagge: 'Schumann und seine Faustszenen', *Sammlung musikalischer Vorträge* (Leipzig, 1879), no.4, p.121

Graf Waldersee: 'Über Schumanns *Manfred*', *Sammlung musikalischer Vorträge* (Leipzig, 1880), no.13, p.3

L. Torchi: 'R. Schumann e le sue "Scene tratte del Faust di Goethe"', *RMI*, ii (1895), 381–419, 629–65

R. Heuberger: *Robert Schumann: Scenen aus Goethes 'Faust'* [Musikführer no.62] (Frankfurt, n.d.)

——: *Robert Schumann: Das Paradies und die Peri* [Musikführer no.89] (Frankfurt, n.d.)

Bibliography

H. Abert: 'R. Schumanns Genoveva', *ZIMG*, xi (1909), 277

F. Strich: 'Byrons Manfred in Schumanns Vertonung', *Festgabe Samuel Singer* (Tübingen, 1930), 167

G. B. Shaw: *Music in London 1890–94*, iii (London, 1932), 107 [on *Genoveva*]

G. Abraham: 'The Dramatic Music', *Schumann: a Symposium*, ed. G. Abraham (London, 1952), 260

J. Horton: 'The Choral Works', *Schumann: a Symposium*, ed. G. Abraham (London, 1952), 283

H. Wolff: 'Schumanns "Genoveva" und der Manierismus des 19. Jahrhunderts', *Beiträge zur Geschichte der Oper* (Regensburg, 1969), 89

F. Cooper: 'Operatic and Dramatic Music', *Robert Schumann: the Man and his Music*, ed. A. Walker (London, 1972, rev. 2/1976), 324

L. Halsey: 'The Choral Music', *Robert Schumann: the Man and his Music*, ed. A. Walker (London, 1972, rev. 2/1976), 324

E. Sams: 'Schumann and Faust', *MT*, cxiii (1972), 543

L. Siegel: 'A Second Look at Schumann's *Genoveva*', *MR*, xxxvi (1975), 17

L. Zanoncelli: ' "*Manfred*" da Byron a Schumann: una metamorfosi', *Studi musicali*, x (1981), 121–54

ORCHESTRAL WORKS

P. Tschaikowski: 'Schumann als Symphoniker', *Musikalische Erinnerungen und Feuilletons* (Berlin, 1899)

F. Weingartner: *Ratschläge für die Aufführung klassicher Symphonien*, vii (Leipzig, 1918), 30–119

O. Karsten: *Die Instrumentation Robert Schumanns* (diss., U. of Vienna, 1922)

D. Tovey: 'Schumann', *Essays in Musical Analysis: Symphonies* (London, 1935), 45

——: 'Schumann: Concerto for Four Horns and Orchestra'; 'Introduction and Allegro appassionato op.92'; 'Piano Concerto in A minor op.54'; 'Violoncello concerto in A minor op.129', *Essays in Musical Analysis: Concertos* (London, 1936), 182; 184; 188

——: 'Schumann: Overture to Byron's "Manfred" ', *Essays in Musical Analysis: Illustrative Music* (London, 1936), 112

——: 'Schumann: Overture, Scherzo and Finale op.52', *Essays in Musical Analysis: Miscellaneous Notes* (London, 1939), 40

G. Abraham: 'The Three Scores of Schumann's D minor Symphony', *MT*, lxxxi (1940), 105; repr. in *Slavonic and Romantic Music* (London, 1968), 281

M. Carner: 'Mahler's Re-scoring of the Schumann Symphonies', *MR*, ii (1941), 97; repr. in *Of Men and Music* (London, 1944), 115

G. Abraham: 'On a Dull Overture by Schumann' [op.136], *MMR*, lxxvi

(1946), 238; repr. in *Slavonic and Romantic Music* (London, 1968), 288

B. Shore: 'Schumann'; 'Schumann's Symphony in D minor', *Sixteen Symphonies* (London, 1949), 99; 103

G. Abraham: 'Schumann's "Jugendsinfonie" in G minor', *MQ*, xxxvii (1951), 45; repr. in *Slavonic and Romantic Music* (London, 1968), 267

R. Réti: 'Schumann: Symphony in B-flat major', *The Thematic Process in Music* (New York, 1951, rev.2/1961), 295

M. Carner: 'The Orchestral Music', *Schumann: a Symposium*, ed. G. Abraham (London, 1952), 176–244

M. Lindsey: 'The Works for Solo Instrument and Orchestra', *Schumann: a Symposium*, ed. G. Abraham (London, 1952), 245 .

A. Zlotnik: 'Die beiden Fassungen von Schumanns D-Moll Symphonie', *ÖMz*, xxi (1966), 271

M. Maniates: 'The D minor Symphony of Robert Schumann', *Festschrift für Walter Wiora* (Kassel, 1967), 441

A. Gebhardt: *Robert Schumann als Symphoniker* (Regensburg, 1968)

A. Nieman: 'The Concertos', *Robert Schumann: the Man and his Music*, ed. A. Walker (London, 1972, rev. 2/1976), 241–76

B. Schlotel: 'The Orchestral Music', *Robert Schumann: the Man and his Music*, ed. A. Walker (London, 1972, rev. 2/1976), 277

S. Walsh: 'Schumann's Orchestrations: Function and Effect', *Musical Newsletter*, iii (1972), 3

H. Gál: *Schumann Orchestral Music* (London, 1979)

W. Boetticher: 'Zur Kompositionstechnik und Originalfassung von Robert Schumanns III. Sinfonie', *AcM*, liii (1981), 144

CHAMBER MUSIC

J. Fuller Maitland: *Schumann's Concerted Chamber Music* (London, 1929)

F. Davies: 'Some Notes on the Interpretation of Schumann's Chamber Music', *Cobbett's Cyclopedic Survey of Chamber Music* (London, 1929–30, rev., enlarged 2/1963), 390

G. Wilcke: *Tonalität und modulation in streichquartett Mendelssohns und Schumanns* (diss., U. of Rostock, 1932, Leipzig, 1933)

D. Tovey: 'Schumann: Quintet in E flat major op.44', *Essays in Musical Analysis: Chamber Music* (London, 1944), 149

A. Dickinson: 'The Chamber Music', *Schumann: a Symposium*, ed. G. Abraham (London, 1952), 138–75

O. Neighbour: 'Schumanns dritte Violinesonate', *NZM*, Jg.117 (1956), 423

A. Molnár: 'Die beiden Klavier-Trios in d-moll von Schumann (op.63) und Mendelssohn (op.49)', *Sammelbände der Robert-Schumann-*

Bibliography

Gesellschaft, i (Leipzig, 1961), 79

J. Gardner: 'The Chamber Music', *Robert Schumann: the Man and his Music*, ed. A. Walker (London, 1972, rev. 2/1976), 200–240

F. Reininghaus: 'Zwischen Historismus und Poesis: über die Notwendigkeit umfassender Musikanalyse und ihre Erprobung an Klavierkammermusik von Felix Mendelssohn Bartholdy und Robert Schumann', *Zeitschrift für Musiktheorie*, iv (1973), 22; v (1974), 34

AESTHETICS

R. Bouyer: 'Schumann et la musique à programme', *Le ménestrel* (1903), no.37, p.290

H. Kretzschmar: 'Robert Schumann als Ästhetiker', *JbMP 1906*, 49

A. Schmitz: *Untersuchungen über des jüngeren Robert Schumanns Anschauungen vom musikalischen Schaffen* (diss., U. of Bonn, 1919); extracts in *ZMw*, ii (1919–20), 535; iii (1920–21), 111

H. Kötz: *Der Einfluss Jean Pauls auf Robert Schumann* (Weimar, 1933)

'Robert Schumann': l'esthétique et l'oeuvre', *ReM* (1935), no.161 [special issue]

R. Jacobs: 'Schumann and Jean Paul', *ML*, xxx (1949), 250

M. Elssner: *Zum Problem des Verhältnisses von Musik und Wirklichkeit in den musikästhetischen Arbeiten der Schumann-Zeit* (diss., U. of Halle, 1964)

E. Lippmann: 'Theory and Practice in Schumann's Aesthetics', *JAMS*, xvii (1964), 310–45

T. Brown: *The Aesthetics of Robert Schumann* (New York, 1968)

OTHER SPECIAL STUDIES

R. Prochazka: 'Ernestine von Fricken: Robert Schumanns erste Braut', *Arpeggien: Musikalisches aus alten und neuen Tagen* (Dresden, 1897); repr. in *ÖMz*, xi (1956), 216

M. Kalbeck: 'Schumann und Brahms', *Deutsche Rundschau*, cxiv (1903), 232

J. Tiersot: 'Robert Schumann et la Révolution de 1848', *BSIM*, ix (1913), 5

F. Schnapp: *Heinrich Heine und Robert Schumann* (Hamburg, 1924)

P. Frenzel: *Robert Schumann und Goethe* (Leipzig, 1926)

W. Gurlitt: 'Robert Schumann in seinen Skizzen gegenüber Beethoven', *Beethoven-Zentenarfeier: Wien 1927*, 91

W. Boetticher: 'Robert Schumann in seinen Beziehungen zu Johannes Brahms', *Die Musik*, xxix (1936–7), 548

H. Redlich: 'Schumann Discoveries', *MMR*, lxxx (1950), 143, 182, 261; lxxxi (1951), 14

G. von Dadelsen: 'Robert Schumann und die Musik Bachs', *AMw*, xiv (1957), 46

U. Martin: 'Ein unbekanntes Schumann-Autograph aus dem Nachlass E. Krügers' [the copy Schumann made of *Die Kunst der Fuge*, in 1837], *Mf*, xii (1959), 405

R. Fritsch: *Schumanns Vater als Verleger* (Frankfurt, 1960)

O. Alain: 'Schumann und die französische Musik', *Sammelbände der Robert-Schumann-Gesellschaft*, i (Leipzig, 1961), 47

D. Schitormirski: 'Schumann in Russland', *Sammelbände der Robert-Schumann-Gesellschaft*, i (Leipzig, 1961), 19

D. Kämper: 'Zur Frage der Metronombezeichnungen Robert Schumanns', *AMw*, xxi (1964), 141

E. Sams: 'Did Schumann use Ciphers?', *MT*, cvi (1965), 584; see also cvii (1966), 392, 1051

M. Beaufils: 'Mythos und Maske bei Robert Schumann', *Sammelbände der Robert-Schumann-Gesellschaft*, ii (Leipzig, 1966), 66

E. Sams: 'Why Florestan and Eusebius?', *MT*, cviii (1967), 131

W. Boetticher: 'Robert Schumann und seine Verleger', *Musik und Verlag: Karl Vötterle zum 65. Geburtstag* (Kassel, 1968), 168

E. Sams: 'Politics, Literature, People in Schumann's Op.136', *MT*, cix (1968), 25

——: 'The Tonal Analogue in Schumann's Music', *PRMA*, xcvi (1969–70), 103; repr. in *Robert Schumann: the Man and his Music*, ed. A. Walker (London, 1972, rev. 2/1976), 390

——: 'A Schumann Primer?', *MT*, cxi (1970), 1096

B. Schlotel: 'Schumann and the Metronome', *Robert Schumann: the Man and his Music*, ed. A. Walker (London, 1972, rev. 2/1976), 109

M. H. Schmid: *Musik als Abbild: Studien zum Werk von Weber, Schumann und Wagner* (Tutzing, 1981)

B. Appel: 'Schumanns Davidsbund – Geistes- und sozialgeschichtliche Voraussetzungen einer romantischen Idee', *AMw*, xxxviii (1981), 1

FRANZ LISZT

Humphrey Searle

CHAPTER ONE

Life

I   Early years

Franz (Ferenc) Liszt was born on 22 October 1811 at
Raiding, near Sopron. He was the son of Adam Liszt
(1776–1827), an official in the service of Prince Nikolaus
Esterházy, and Maria Anna Lager (1788–1886), of
south Austrian origin. Through both parental lines he
descended from humble stock, largely Austrian migrant
farmers and peasants. Early writers occasionally linked
the family to the Hungarian nobility. Even though Liszt
at one time believed in a noble pedigree and tried without success to verify it, recent scholarship disproves the
possibility. A number of celebrated musicians, among
them Haydn, Cherubini and Hummel, occasionally
visited the prince's palace at Eisenstadt; Adam Liszt,
himself a talented amateur musician, played the cello in
the court orchestra, before his unhappy transfer in 1809
to Raiding for a new position as sheep inspector.

At the age of six Franz began to listen attentively to
his father's piano playing and also to show an interest
in both gypsy music and sacred music; and he developed
a strong religious sense that remained with him for the
rest of his life. Adam, who had been a Franciscan
novice at the age of 19 but had been dismissed from the
order because of his 'inconstant and variable character'
(as it was officially recorded), soon recognized his son's
musical talent and began teaching him the piano at the

age of seven. Franz began to compose in an elementary way when he was eight and may also have first played in public at Baden in the same year; he certainly appeared in concerts at Oedenburg and Pressburg in October and November 1820, after which a group of local Hungarian magnates put up a sum of money to provide for his musical education.

In spring 1822 Liszt's family moved to Vienna; Franz studied the piano with Czerny and composition with Salieri, then music director at the Viennese court. His first public concert in Vienna, on 1 December 1822, was a great success. He gave a second concert on 13 April 1823 which gave rise to the legend of Beethoven's kiss of consecration. Beethoven did not attend and the only documented meeting between them consisted of a private interview earlier in April. In 1875 Liszt recalled the interview, and the alleged kiss, with deep emotion; he seems to have regarded it as a kind of artistic christening. On the other hand, he denied ever having met Schubert, but biographers continue to endorse their personal acquaintance. He was asked to contribute a variation on a waltz by the publisher Diabelli to a symposium in which 50 Austrian composers took part, including Schubert, Czerny, Hummel and Moscheles (Liszt's variation was the 24th); Beethoven wrote his monumental set of 33 variations on the same theme.

In the autumn of 1823 Liszt moved with his family to Paris, giving concerts on the way in Munich, Stuttgart and other German towns. In Paris he was refused admission to the Conservatoire by Cherubini, on the grounds of being a foreigner; instead he studied theory with Reicha and composition with Paer. He made a success in Parisian society and played at many fashionable con-

ARGYLL ROOMS.

Master *LISZT's CONCERT*,
JUNE 21, 1824.

PART I.

Grand Sinfonia...*Haydn.*
Aria, Signor FABRI, " Non piu andrai."......*(Figaro.)*......*Mozart.*
Grand Concerto, (with Orchestral Accompaniments,)
    Piano Forte, Master LISZT..*Hummel.*
Duetto, Le Signore MARINONI ....*Rossini.*
Fantasia, Corno, Signor PUZZI...*Puzzi.*
Duetto, Madame CASTELLI and Signora MARINONI,
    " E ben per mia memoria."....*(La Gazza Ladra)*....*Rossini.*
Recit. and Air, Mr. BRAHAM, " Deeper and deeper still;"
    and " Waft her, Angels."............*(Jephthah.)*............*Handel.*
Variations, Mandolin and Piano Forte, Signor VIMERCATI
    and Master LISZT..*Mayseder.*
Terzetto, Madame CASTELLI, Signor MARINONI, & Signor GARCIA,
    " Cruda sorte."............*(Ricciardo e Zoraide.)*.........*Rossini.*

PART II.

Overture to *Prometheus*...*Beethoven.*
Ballad, Miss STEPHENS, " Gin living worth."
Fantasia, Harp, Monsieur LABARRE..................................*Labarre.*
Cavatina, Signor BOCCACINI, " Fra un istante."...............*Rossini.*
Variations, Piano Forte, (with Orchestral Accompaniments,)
    Master LISZT...*Czerny.*
Duetto, Signor GARCIA and Signor BOCCACINI,
    " Che bella vita."...*Generali.*

Extempore Fantasia on a written Thema, which Master LISZT
respectfully requests may be given to him by any Person in the Company.

Leader of the Band, Mr. MORI.

Conductor.........Sir GEORGE SMART.

(Master LISZT will perform on *Sebastian Erard's* New Patent
Grand Piano Forte

Mallett, Printer, 59, Wardour Street, Soho.

20. *Programme of Liszt's first public concert in London, at the Argyll Rooms, on 21 June 1824*

certs: his first, on 22 December 1823, was sensational. In the spring he visited England for the first time, playing privately and then in public on 21 June at the Argyll Rooms in London and on 29 June at Drury Lane, where he 'consented to display his inimitable powers on the New Patent Grand Piano Forte, invented by Sébastien Erard', as advertised on the playbill. The next spring he toured the French provinces and then again visited England, playing in London, in Manchester and before King George IV at Windsor; on 16 June his 'New Grand Overture', the prelude to a one-act opera *Don Sanche*, was performed in Manchester. Other compositions predating 1827 include bravura variations, rondos, and a number of lost works: two piano concertos, a trilogy of piano sonatas and chamber music. In 1826 he wrote the *Etude en douze exercices*, the sparkling forerunner of the Transcendental Studies of 1851. His most ambitious work, however, was the precocious *Don Sanche* (1825) for the Paris Académie Royale; it remained the only opera he completed.

II Contact with Parisian society

In 1826 Liszt again toured the French provinces and later Switzerland in the winter. This was followed by a third visit to England in 1827, but the constant touring was beginning to affect his health; and he expressed the desire to become a priest. He went to recover in Boulogne with his father, who died there suddenly of typhoid fever. Liszt returned to Paris and set up house with his mother. He gave up touring and earned a living as a piano teacher. In the next year he fell in love with one of his pupils, Caroline de Saint-Cricq, the daughter of Charles X's minister of commerce, but her father

insisted that the attachment be broken off. Liszt again became ill – there was even an obituary of him in a Paris newspaper – and he underwent a long period of religious doubts and pessimism. He again expressed a wish to join the church, but was dissuaded by his mother and his confessor the Abbé Bardin. From autumn and winter 1829 date his friendships with Wilhelm von Lenz and the sensitive, eccentric Chrétien Urhan.

The year 1830 marked a turning-point in Liszt's personal, intellectual and artistic development. He acquired a passion for literature, embracing a wide range of classics from Homer to Montaigne, as well as intoxicating newer publications by Hugo, Lamartine and Châteaubriand. Current theatrical and operatic productions also captivated his attention. It is often written that a sense of educational inadequacy motivated this cultural assault. However, a diary-like sketchbook of the period contains both profuse quotations and more personal soliloquies of elation and despair. His voracious reading, then, seems more closely related to his growth into intellectual maturity; attaining adulthood appears to have been for him an exhilarating, confusing rite of passage.

Liszt had composed little since 1827. Two events of 1829–30 stirred him into creative activity: a performance of Auber's *La fiancée*, which resulted in his first operatic paraphrase of extreme virtuosity, and the July Revolution of 1830, inspiring him to sketch his Revolutionary Symphony. These works mark his transition into a mature composer; they anticipate both his socalled *Glanzperiode* brilliance and the symphonic ambitions of the Weimar years.

In April 1832 Liszt went to Paganini's benefit concert

at the Opéra and also heard him in a private performance at the salon of Rothschild, the banker. Paganini's technique and charisma excited him into a period of feverish practice, literary absorption and composition. Determined to transfer Paganini's dazzling virtuoso effects to the piano, he drafted the demonic *Clochette* fantasy. In the next three years he produced a rich series of new works: piano transcriptions of Berlioz's *Symphonie fantastique* and the *Francs-juges* overture, the *Lélio* fantasy, his own poetic set of three *Apparitions*, the early *Harmonies poétiques et religieuses*, the *De profundis* for piano and orchestra, the *Konzertstück* for two pianos on Mendelssohn's *Lieder ohne Worte* and numerous operatic fantasies.

In October Liszt found himself attracted to the idealistic St Simonian cult, drawn into the fold by its leading aesthetician, Emile Barrault. In December he heard the première of Berlioz's *Symphonie fantastique*. The two composers had recently formed a friendship that would endure for more than 20 years. Through the rest of this period Liszt also developed close relationships with Chopin, Hugo, George Sand and Lamennais, and came into contact with most of the celebrated personalities of Paris.

Liszt had experienced his first mature liaison with the beautiful Adèle de Laprunarède, from 1831 to early 1833. But in January that year he was introduced to another married countess who was to play a far more important role in his life. In the Marquise Le Vayer's salon he met Marie d'Agoult; their stormy relationship encompassed the following 11 years and the birth of three children. In 1835 d'Agoult left her family to join Liszt in Geneva; their first daughter, Blandine-Rachel,

was born there on 18 December. Liszt taught at the newly founded Geneva Conservatory and wrote a manual of piano technique (which was later lost). With d'Agoult's help he began to establish a literary career by publishing essays on music in leading Parisian journals. The first of them argued for the raising of the artist's status from that of a kind of superior servant to a respected member of the community.

For the next four years Liszt and the countess lived together, mainly in Switzerland and Italy, though they occasionally visited Paris. During one of these visits, in the winter of 1836–7, Liszt took part in a Berlioz concert, gave chamber music concerts with Urhan and Alexandre Batta and, on 31 March, played at Princess Belgiojoso's house in the celebrated pianistic duel with Thalberg, whose fame as a pianist was beginning to rival Liszt's; Liszt was the victor. He continued work on the *Album d'un voyageur*, lyrical evocations of Swiss scenes which were later transformed into the first book of the *Années de pèlerinage*, and transcribed three Beethoven symphonies for the piano. In 1837 he went with the countess to Italy, where he wrote the original versions of the pieces that formed the second book of the *Années de pèlerinage*, the first version of the Paganini Studies and the 12 *grandes études* (a greatly enlarged and revised version of the *Etude en douze exercices*). Their second daughter, Cosima, was born on 24 December at Como. In 1838 Liszt gave concerts in various Italian cities and Vienna, where he popularized little-known works by Domenico Scarlatti and the largely forgotten lieder of Schubert, through his own incomparable transcriptions.

On 9 May 1839 Daniel, the only son of Liszt and the

countess, was born; but soon the recurring strain in their relationship became unbearable. When Liszt heard that the proposal to build a Beethoven monument in Bonn was in danger of collapse through lack of funds, he offered to make good from his own pocket the amount still needed. This meant a return to the life of a travelling virtuoso; and the countess returned with two of the children to Paris, while Liszt gave six concerts in Vienna. He then visited Hungary for the first time since childhood, playing at Poszony and Budapest, and was received with great acclamation; he also proposed the foundation of a national conservatory in Budapest. Once again he came in contact with gypsy music, which he began to transcribe for the piano; he was presented with a poem of homage by the Hungarian poet Mihály Vörösmarty and even with a sword of honour.

III Weimar

During the following eight years Liszt continued to tour the whole of Europe, from Ireland to Turkey, from Portugal to Russia. He spent his summer holidays with the countess and their children on the island of Nonnenwerth on the Rhine until 1844, when they finally separated and Liszt took the children to Paris to arrange for their education. This was Liszt's most brilliant period as a concert pianist: he was adulated everywhere, and honours were showered on him. He continued to compose throughout his so-called *Glanzperiode*, writing songs, choral works and piano music; although he dreamed of making his mark as an opera composer, none of his projects progressed beyond the sketching stage. He visited England again in 1840 and 1841, where he was warmly received by the court and

21. Franz Liszt: daguerreotype, c1841

the public, but not by the critics. The next year he was appointed Grand Ducal Director of Music Extraordinary at Weimar: he had conducted for the first time in Budapest in 1840, and from 1844 onwards he conducted regularly in Weimar and other German towns. In 1845 he wrote his first Beethoven cantata, for the Beethoven Festival in Bonn: his first work for chorus and orchestra.

In February 1847 Liszt played in Kiev, where he met the Princess Carolyne Sayn-Wittgenstein, who was to dominate most of the rest of his life. For a long time he had hoped to be able to abandon his constant touring in order to concentrate on composition. She encouraged the idea with unwavering moral support. After a tour of the Balkans, Turkey and Russia that summer, Liszt gave his final concert at Elisavetgrad in September and spent the winter with the princess at her estate at Woronince.

Liszt now decided to take up his conducting appointment in Weimar as a full-time job, and he moved there in February 1848. The princess visited the town in the same year in order to ask the Grand Duchess of Weimar to influence her brother, the Tsar of Russia, to grant her a divorce from her husband; when this was refused, the princess set up house with Liszt in the Altenburg the next year. Liszt now had ample time to compose, and during the next 12 years he wrote or revised most of the major works for which he is known: it was a period of remarkable productivity. In addition he conducted a number of stage works by contemporary composers, including *Tannhäuser*, *Lohengrin* (first performance, 1850), Schumann's *Genoveva* and incidental music to *Manfred*, Berlioz's *Benvenuto Cellini*, and operas by

Verdi and Donizetti. He was able to try out his own compositions with the Weimar orchestra and as a result made considerable revisions in his scores. He also attracted a number of pupils, the most important of whom were Hans von Bülow, Peter Cornelius and Carl Tausig, and Weimar became the Mecca of the avant-garde movement in Germany, known as the 'New German school' or the 'futurists'.

Naturally, this activity was not always to the taste of the more academic musicians. Joseph Joachim went to Weimar as the leader of Liszt's orchestra but left after two years, and when Liszt conducted his own works in Aachen and Leipzig in 1857 there was strong opposition in the press. Meanwhile both his daughters had been married, Cosima to Bülow and Blandine to Emile Ollivier (later prime minister of France). The Weimar court was now beginning to distrust Liszt because of his continued support of Wagner, who was living in Switzerland as a political refugee (Liszt had helped him to escape from Germany in 1849). The Tsar of Russia put pressure on the grand duchess to ban the princess from all court functions, and some of the more staid citizens of Weimar objected to her living openly with Liszt. Other intrigues (not yet fully understood) began to interfere with his artistic activities. Matters came to a head on 15 December 1858 at the first performance of Cornelius's *Der Barbier von Bagdad*, which Liszt conducted himself: the audience made a demonstration which Liszt took to be directed against himself, and he resigned his post.

Liszt remained in Weimar for a time, but his position there became more and more difficult. On 13 December 1859 his son died in Berlin at the age of 20: deeply

distressed, Liszt wrote *Les morts*, an 'oration for orchestra', in his memory. The princess had fallen out of favour with the Weimar court, and in May 1860 she went to Rome in the hope that the pope would grant her divorce. Meanwhile, a protest against the New German school signed by Brahms, Joachim, Grimm and Scholz had appeared in the Berlin *Echo*. On 14 September, in a troubled state, Liszt made his first will, in the form of a letter to the princess (he left a ring to his early love Caroline de Saint-Cricq, but the will does not mention Marie d'Agoult). In August 1861 Liszt left Weimar and travelled to Berlin and Paris, where he played before the emperor and empress and met many old friends; he arrived in Rome on 21 October, the day before his 50th birthday, on which he had hoped to be married to the princess. But at the last moment, and for reasons not yet clear, the pope revoked his sanction of her divorce, and the couple remained in Rome in separate establishments.

IV Rome and the last years

For the next eight years Liszt lived mainly in Rome, occupied chiefly with religious music; in Weimar he had written a *Missa solemnis* for the consecration of the basilica in Gran (Esztergom) and also a setting of Psalm xiii, and he had begun work on two large oratorios, *Die Legende von der heiligen Elisabeth* and *Christus*. He soon completed these and also wrote a number of smaller religious works, including a *Missa choralis* based partly on Gregorian themes. In 1862 his elder daughter Blandine died at the age of 26. This inspired Liszt to write the variations on the passacaglia theme from Bach's Cantata no.12 ('Weinen, Klagen'): the work ends

with the chorale 'Was Gott tut, das ist wohlgetan'. In 1863 Liszt entered the Oratorio della Madonna del Rosario, on Monte Mario, where he was visited by the pope: the princess's husband died in 1864, but there was no further talk of marriage, and in 1865 Liszt took the four minor orders of the Catholic Church (though he never became a priest). He spent some time at the Villa d'Este at Tivoli, where Cardinal Hohenlohe provided him with a suite of rooms. Here, in the middle of writing religious pieces, he composed 'quelques traits sur le piano pour la jonglerie indienne de l'Africaine', a fantasia on themes from Meyerbeer's last opera. He also wrote a mass for the coronation of Emperor Franz Josef of Austria as King of Hungary, thus renewing his links with his native land, and two orchestral works of a semi-autobiographical character, *La notte* and *Le triomphe funèbre du Tasse*. Meanwhile his younger daughter Cosima had begun an affair with Wagner and had borne him two illegitimate children: the press embraced the affair as a public scandal, causing Liszt embarrassment. The ensuing rift with his daughter and Wagner continued well into 1872.

In 1869 Liszt was invited back to Weimar to give master classes in piano playing, and two years later he was asked to do the same in Budapest. From then until the end of his life he made regular journeys between Rome, Weimar and Budapest – his 'vie trifurquée', as he described it. He was visited by numerous composers, including Anton Rubinstein, Albéniz, Borodin, Saint-Saëns and Fauré. He also taught a number of pianists, including Frederic Lamond, Sophie Menter, Moriz Rosenthal, Emil von Sauer and José Vianna da Motta, as well as Eugen d'Albert and Felix Weingartner. After

his reconciliation with Wagner in 1872 he attended the Bayreuth Festival; he also occasionally appeared as a pianist in charity concerts. In 1882, while staying with the Wagners in Venice and working on his last oratorio *Die Legende vom heiligen Stanislaus*, a strange presentiment caused him to break off work on it and to write two pieces entitled *La lugubre gondola*, inspired by the funeral processions that he had seen on the Venetian canals. Two months later Wagner died in Venice, and his body was borne from the Palazzo Vendramin by gondola.

In 1885 Debussy visited Liszt in Rome and was advised by him to hear the music of Palestrina and Lassus at the church of S Maria dell'Anima. The next year, his 75th, Liszt set out on his last tour. He attended concerts of his works in Budapest, Liège and Paris and then visited London for the first time since 1841; he was enthusiastically received, and several concerts of his works were given, including a performance of *Elisabeth* under Alexander Mackenzie. He also attended a performance of *Elisabeth* in Paris and visited Antwerp, Weimar and Sondershausen. He had become very weak, and his doctor diagnosed dropsy. On 3 July he visited Bayreuth for the wedding of his granddaughter Daniela von Bülow; then he went to Colpach in Luxembourg to stay with his friend the Hungarian painter Munkácsy. He made his last appearance as pianist at a concert of the Musical Society of Luxembourg, playing his first *Liebestraum*, one of his arrangements of Chopin's Polish songs and no.6 of his *Soirées de Vienne*. He returned to Bayreuth on 21 July and saw *Parsifal* on 23 July and *Tristan und Isolde* two days later. His illness developed into pneumonia, and he died painlessly at

250

22. 'Fantaisie brilliante sur Liszt': caricature from 'La vie parisienne' (3 April 1886)

11.30 p.m. on 31 July. The Princess Sayn-Wittgenstein died the following year on 8 March, having completed the 24th volume of her book *Causes intérieures de la faiblesse extérieure de l'Eglise en 1870.*

CHAPTER TWO

Character and personality

Liszt had a profound effect on all who knew him. Marie d'Agoult described him on their first meeting as having

an excessively tall and thin figure, a pale face with sea-green eyes which shone with rapid flashes like waves in flames . . . an indecisive walk in which he seemed to glide rather than set foot on the ground, a distracted and unquiet appearance like that of a ghost about to return to the darkness.

This is the young Liszt as the painters Devéria and Ingres saw him. 40 years later his young American pupil Amy Fay wrote of him:

His mouth turns up at the corners, which gives him a most crafty and Mephisophelean expression when he smiles, and his whole appearance and manner have a sort of Jesuitical elegance and ease. . . . He is all spirit, but half the time at least, a mocking spirit. . . . He is rather tall and narrow. . . . he made me think of an old-time magician more than anything, and I felt that with a touch of his wand he could transform us all.

These two writers have seized on many of Liszt's essential qualities – his romantic abstraction and other-worldliness combined with elegant, worldly manners and a diabolism and feeling of magic. The historian Gregorovius, seeing him in Rome in 1865 shortly after he had taken the four minor orders of the Catholic Church, described him as 'Mephistopheles disguised as an abbé'. Liszt was certainly a mass of contradictions, but he nevertheless remained a unique and comprehen-

253

sive personality with a strength of vision that lasted all his life. There were some who found a source of irritation in these apparent changes of character: Chopin came to dislike his hob-nobbing with the aristocracy, Mendelssohn complained of his textual alterations to the classics during his period as a travelling virtuoso, and Joachim often said that it was a wonderful experience to play through sonatas or other chamber music works with him once, but that afterwards he would double simple passages in octaves or 3rds, convert ordinary trills into trills in 6ths, and so on. What these habits revealed above all was intellectual restlessness; for he could also be a purist, as is shown by his straightforward, unadorned transcriptions of Bach organ fugues for the piano.

One of the most important factors in Liszt's character was his generosity to others. From his early meeting with Berlioz until the end of his life, he helped countless composers and performers both artistically and financially, and no-one more than Wagner. After giving up his career as a virtuoso he was not well off: his Weimar post brought him only a small stipend, he never took fees for teaching, and his only public appearances from then on were at charity concerts. He continued to give generous praise even to former friends who had quarrelled with or drawn apart from him, including Chopin, Berlioz and, for a time, Wagner. He was probably too self-effacing, even suggesting in later years that proposed performances of his works should be cancelled. It is true that his symphonic poems had met with some opposition, and when *Elisabeth* was more warmly received he was somewhat wry about it, declaring:

For years, in his symphonic poems, masses, piano works and songs,

Liszt has written only confused and objectionable stuff; with the *Elisabeth* he seems to behave slightly more reasonably. However, . . . I am in no way inclined to issue a general *peccavi* for my compositions.

And very few of his later experimental works were published in his lifetime: he seems to have wished not to annoy his public further.

His triple character as gypsy, Franciscan and original creator was symbolized to some extent by his travels between Budapest, Rome and Weimar, respectively, in the last years of his life. He needed the anchor of the Catholic Church, but at the same time was passionately attracted to gypsy music (though his neglect of genuine Hungarian folk music is the main reason for his failure to found a truly Hungarian school of composers). But his creativity remained constant in spite of all difficulties: he invariably worked long hours and with enormous energy, often requiring the aid of brandy to sustain him (Cosima's ban on alcohol during his last illness may well have hastened his death). Throughout his life Liszt remained extremely attractive to women, and two of them at least, Marie d'Agoult and Princess Sayn-Wittgenstein, influenced his life considerably. Yet over the years it has become increasingly clear that in spite of his inconsistencies he remained, throughout, the same character, one of a unique power, strength and originality.

From 1834 to 1859 Liszt published widely: articles for the Paris *Revue et gazette musicale*, books on Chopin and gypsy music, and numerous polemic pieces. Even during his lifetime rumours circulated that his writings were not his own work. The controversy has not yet been resolved by modern scholarship, a vexing point since his publications form an important corpus

of source material. It is clear that both Marie d'Agoult and later the Princess Sayn-Wittgenstein collaborated heavily with Liszt on these literary endeavours. But to what degree the contents accurately reflect his own ideas remains unclear. Only the early piece on the future of church music and *De la fondation Goethe à Weimar* survive in Liszt's hand; manuscripts, drafts and proofs related to the rest of his literary output have vanished.

CHAPTER THREE

Liszt the pianist

It is generally accepted that Liszt was the greatest pianist of his time, indeed quite possibly of all time: certainly from the point of view of keyboard technique his achievements have hardly ever been rivalled, let alone surpassed (the 1839 version of the Transcendental Studies is one of the most difficult sets of pieces ever written for the piano). Berlioz wrote of Liszt's music about 1840: 'Regrettably, one cannot hope to hear music of this kind often: Liszt created it for himself, and no-one else in the world could flatter himself that he could approach being able to perform it'. Liszt's unusually long fingers gave him an enormous reach: he was able to play 10ths as easily as most pianists could play octaves, and he even wrote rapid consecutive 10ths in several works, including his transcription of Schubert's *Divertissement à l'hongroise*.

The extraordinary breadth of his repertory is shown by the list of works he played in concerts between 1838 and 1848, drawn up by August Conradi and revised by Liszt himself. In addition to transcriptions of overtures by Beethoven, Berlioz, Mozart, Rossini and Weber and symphonies by Beethoven (nos.5–7) and Berlioz, he performed concertos by Bach, Beethoven (nos.1, 3 and 5), Chopin (nos.1–2), Mendelssohn (nos.1–2), Henselt, Hummel and Moscheles, and Weber's *Konzertstück*. His chamber music repertory included Hummel's D

minor Septet, Spohr's C minor Quintet and Beethoven's
E♭ Quintet, piano trios, violin sonatas and cello sonatas.
Besides his own piano compositions and transcriptions,
Liszt played ten of Beethoven's sonatas (including all the
late ones) and the Diabelli Variations; Schumann's
Carnaval, C major Fantasia (dedicated to Liszt) and F♯
minor Sonata; Schubert's *Wandererfantasie*, Weber's
four sonatas, Mendelssohn's *Variations sérieuses*,
numerous works by Chopin and by Czerny; sonatas and
fantasias by Hummel, sonatas and studies by Moscheles,
and the composite work *Hexaméron*, written by six
composers (including Liszt) for a benefit concert in
Paris in 1837. He was one of the few pianists of his day
to play Bach's fugues, both those from *Das wohltemperirte Clavier* and (in his own transcriptions) the
organ fugues; his 18th-century repertory also included
the Goldberg Variations, Handel's suites and fugues and
Domenico Scarlatti's sonatas. Such an exhaustive
repertory would certainly daunt most modern pianists.

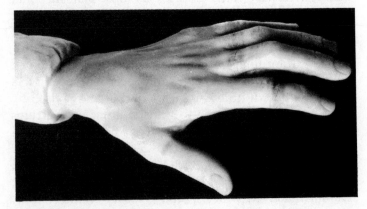

23. Cast of Liszt's left hand

One can get some idea of the effect of Liszt's playing on his contemporaries from the following two descriptions, one from near the beginning and one from the end of his career. Charles Hallé, then a student in Paris and an accomplished pianist himself, first heard him in 1836 and wrote:

Such marvels of executive skill and power I could never have imagined. He was a giant, and Rubinstein spoke the truth when, at a time when his own triumphs were greatest, he said that in comparison with Liszt all other pianists were children. Chopin carried you with him into a dreamland, in which you would have liked to dwell for ever; Liszt was all sunshine and dazzling splendour, subjugating his hearers with a power that none could withstand. For him there were no difficulties of execution, the most incredible seeming child's play under his fingers. One of the transcendent merits of his playing was the crystal-like clearness which never failed for a moment even in the most complicated and, to anybody else, impossible passages; it was as if he had photographed them in their minutest detail upon the ear of his listener. The power he drew from his instrument was such as I have never heard since, but never harsh, never suggesting 'thumping'.

50 years later J. A. Fuller Maitland heard Liszt play in London in the last year of his life and later wrote (in *Grove 3*): 'His playing was a thing never to be forgotten or approached by later artists. The peculiar quiet brilliance of his rapid passages, the noble proportion kept between the parts and the meaning and effect which he put into the music were the most striking points'.

Liszt must be credited with the invention of the modern piano recital. Before his day pianists usually took part in a kind of variety bill with other artists; it was Liszt who introduced a whole evening of serious piano music as a regular feature of public concert life.

CHAPTER FOUR

Style and structure

Two elements of style give Liszt's compositions a distinctly personal stamp: experiment with formal structure, and thematic transformation. Many of his smaller works show considerable ingenuity of design. In larger-scale pieces he often made daring experiments in extending the Classical one-movement sonata form, in unifying multi-movement works and in using large-scale tonal areas for new dramatic effect.

A good many of Liszt's works, including most of the Transcendental Studies, are monothematic, and it is usually in the longer works that he introduced more than one theme. This monothematicism led Liszt to develop the concept of 'transformation of themes', a process by which one or more short ideas are subjected to various techniques of alteration (change of mode; change of rhythm, metre or tempo; ornamentation; change of accompaniment etc) to form the thematic basis of an entire work. One of the main ideas of the *Faust-Symphonie*, for instance, appears during the work in various guises, all of which are shown in ex. 1. Liszt's technique does not require that the form of a work be dependent entirely on pictorial or dramatic elements; in this respect it differs from Wagner's use of leitmotifs, in which similar transformations are guided by the necessities of the drama. Liszt described the *Faust-Symphonie* simply as 'three character-pieces', and the element of actual story-

telling is very small. Often he found a title for a work after he had written the music, for example the individual Transcendental Studies and the symphonic poem *Les préludes*, which offers one of the best illustrations of the technique of transformation of themes. Even such atmospheric pieces as *Pastorale* and *Le mal du pays* from the

Ex.1 *Eine Faust-Symphonie*

261

Swiss book of *Années de pèlerinage* were first published without titles.

In general, the purpose of Liszt's new piano technique was to exploit all the resources of the instrument – to make it sound like an orchestra, when necessary – and at the same time be practical for the player. This 'transcendental technique', as it was called, can best be shown by two music examples, the first (ex.2), from *Mazeppa*, showing a three-handed effect, the second (ex.3), from the transcription of the overture to *Tannhäuser*, illustrating an orchestral effect achieved by purely pianistic means: with the descending string passages written in octaves followed by single notes, the pianist can pass his hand over the thumb; the bass is filled out with repeated chords to sustain the sonority. In his last piano works Liszt adopted an austere, almost skeletal style of keyboard writing, though the transcendental technique was still used when needed.

Liszt's harmonic style up to about 1870 was similar to the early Romantic harmonies found in the works of Chopin and the late works of Schubert. There was also some influence of the cantilena of Italian opera, notably in *Ricordanza* from the Transcendental Studies. As early as *Il penseroso* from the Italian book of the *Années de pèlerinage* there is some remarkable chromatic har-

Style and structure

Ex.2 Transcendental study no.4 *Mazeppa*

Allegro

Ex.3 Piano transcription of Wagner's overture to *Tannhäuser*

Andante maestoso

mony; this tendency increased over the years, with a copious use of augmented and diminished chords, especially the tritone, which had a considerable influence on Wagner's mature style. The whole-tone scale appears throughout the recitation *Der traurige Mönch* (1860), though such early works as the single piece *Harmonies poétiques et religieuses*, the *Réminiscences de La juive*, the Fantasia on themes from *Niobe*, and the *Grand galop chromatique* have passages partly based on whole-tone scales. Other works, including *Funérailles* and *Héroïde funèbre*, have themes based on the 'gypsy' scale with its characteristic augmented 2nds. The opening theme of the *Faust-Symphonie* includes all the notes of the chromatic scale, and in several passages in the work the tonality is suspended altogether.

CHAPTER FIVE

Solo piano music

Works for the piano naturally make up the greater part of Liszt's output and may be divided into two categories: original works, and transcriptions and fantasias. The earliest original extant pieces written before 1830 are strikingly effective for the keyboard; the Czernian influence is strong, but they also contain adventurous harmonic details. The most interesting are the *Etude en douze exercices* and the little G minor Scherzo with its wide leaps and free use of the diminished 7th. Two pieces in the Hungarian style, both entitled *Zum Andenken*, were probably written some time in the 1830s to celebrate Liszt's meeting with the Hungarian violinist Ebner in Paris in 1828: they are simple arrangements of melodies by the Hungarian musicians Fáy and Bihari.

Between 1829 and 1933 Liszt had already sketched ideas for *Mazeppa* and the E♭ Concerto as well as a number of other works completed years later. His first mature original piano piece was probably the single work *Harmonies poétiques et religieuses* (1833, revised 1835), which takes its title from a collection of poems by Lamartine, to whom it is dedicated. It is a kind of free improvisation, mostly without key signature or time signature, and based on a single theme that recurs throughout in various different forms – an early example of the transformation of themes. The mood of the piece is poetic,

though there are some more agitated passages, and it ends with an Andante religioso. The three *Apparitions*, also written in 1834, have a similar poetic atmosphere.

The years 1835–9 were mainly spent on the Transcendental Studies, the Paganini Studies and the first two books of the *Années de pèlerinage*. All these works were revised and given their present form during the Weimar period, but their essential inspiration dates from Liszt's 20s. The Transcendental Studies – 24 were announced but only 12 were completed – take as their starting-point the early *Etude en douze exercices*, but it is hardly possible to recognize it in this new greatly expanded and technically more difficult form. The 1839 version of the studies had no titles, but in 1847 Liszt published no.4 separately as *Mazeppa*, with a slightly altered ending to fit the well-known story of the exiled Polish nobleman. In the 1852 version titles were added to all but two of the others, and some of the more formidable difficulties were smoothed out and some cuts were made. They range widely in mood, from the lyrical *Paysage*, *Ricordanza* and *Harmonies du soir* and the delicate *Feux follets* to the brilliant A minor and F minor studies and the furious *Mazeppa* and *Wilde Jagd*; one of the most impressive studies, *Chasse-neige*, gives the feeling of a landscape being relentlessly covered with snow. Liszt's original plan was to write two studies in each of the major and minor keys; they are in fact paired by key signature – C major, A minor; F major, D minor; and so on. The six Paganini Studies, which are more or less straightforward transcriptions of five of the 24 Caprices op.1 and part of *La campanella* from the B minor Violin Concerto, are lighter in both texture and content. The earlier version (1838) again is much more

difficult than the later (1851); Busoni, in his edition of the studies, felt that Liszt had pruned away too much in the later version and so he restored some passages from the original.

In complete contrast to the brilliance of the studies are the first two books of *Années de pèlerinage*, mainly lyrical evocations of nature scenes or of works of art. The first (Swiss) book, chiefly concerned with nature, was originally published in 1842 as *Album d'un voyageur* and consisted of many small pieces, apart from the extended *Vallée d'Obermann* and *Lyon*. Most of these pieces were later revised and included in the first volume of *Années de pèlerinage* published in 1855, and three other pieces were republished in 1877 as *Trois morceaux suisses*. The whole collection gives a feeling of purity and freshness: Liszt absorbed the atmosphere of his surroundings without attempting to write 'picturesque' music. The first piece in the volume, *La chapelle de Guillaume Tell*, is a fine evocation of the Swiss national hero; *Au lac de Wallenstadt* and *Au bord d'une source* are charming pastoral pieces; *Orage* gives a graphic description of mountain storm; *Eglogue* again reflects the freshness of the countryside and *Le mal du pays* conjures up the atmosphere implied in its title by very simple, almost folklike means. The longest piece in the collection, *Vallée d'Obermann*, represents a transcendental experience described in the novel *Obermann* by Senancour, a difficult feat to bring off in music, but here carrying conviction throughout.

The second (Italian) book of *Années de pèlerinage* is concerned more with art than with nature: pastoral scenes are replaced by the legacy of the past, with a consequent change of atmosphere. The first piece,

24. Liszt at the piano: painting (1840) by Joseph Danhauser; (from left to right) Alexandre Dumas, Victor Hugo, George Sand, Paganini, Rossini, Liszt and the Countess Marie d'Agoult

Sposalizio, inspired by Raphael's *Lo sposalizio della vergine*, is a lyrical piece with a free use of notes foreign to the harmony that anticipates Debussy. *Il penseroso*, inspired by Michelangelo's statue in the Medici Chapel in Florence, has a mood of brooding melancholy throughout and ends with a remarkable passage of chromatic harmony that looks forward to the style of *Tristan und Isolde* of 20 years later. The *Canzonetta del Salvator Rosa* is a cheerful, straightforward setting of a song attributed to the celebrated painter; the three Petrarch Sonnets, originally written as songs but very effective in their piano arrangement, are mainly quiet and lyrical. But the final piece in the collection, *Après une lecture du Dante*, an impression of Dante's *Inferno*, contains many violent passages: much of it is deliberately confused and chaotic, though it, too, contains moments of great lyrical beauty. As a supplement to the Italian volume of *Années de pèlerinage* Liszt wrote a small collection called *Venezia e Napoli*, which exists in two versions. The first, written about 1840, consists of four pieces; one of these is based on the same Venetian gondolier's song which Liszt later used in *Tasso*. The collection was revised and reissued in 1859: its three pieces, *Gondoliera*, *Canzone* and *Tarantella* are all based on other composers' themes, the second using the Gondolier's song from Rossini's *Otello*.

The remaining original piano works of the 1830s are fairly slight in character. They include the brilliant *Grande valse di bravura* and the charming *Valse mélancolique*, as well as the *Rondeau fantastique sur un thème espagnol, El contrabandista* which inspired George Sand's story *Le contrebandier*. A curiosity is the composite piece *Hexaméron*, written in 1837 for a charity

concert given at the house of Princess Belgiojoso and consisting of variations by six composers on the March from Act 2 of Bellini's *I puritani* – a patriotic call for freedom that was probably chosen by the staunchly republican princess. The composers, Liszt, Thalberg, J. P. Pixis, Herz, Czerny and Chopin, were all musicians of non-French birth who were living in Paris at the time (except Czerny, who was there on a visit). Liszt wrote the introduction (which includes a theme of his own as a counterpoise to Bellini's), the piano arrangement of the march theme, the linking passages between the variations and the finale. Although the work was not completed until several months after Belgiojoso's soirée, and thus did not receive its première on that occasion, Liszt later performed it frequently. Another favourite piece from this time was the *Grand galop chromatique* (1838).

During the period of Liszt's travels as a virtuoso pianist in the 1840s he completed fewer original works, but a good many pieces were rewritten and issued in revised form in the 1850s. He revised a study written in 1840 under the title *Morceau de salon* for a piano primer edited by Fétis and Moscheles and republished it in 1852 as *Ab irato*. It is a short and violent piece, very effective, with some thematic links with *Les préludes*. There is also an excellent Galop in A minor which has all the gaiety of Offenbach. Liszt's reawakening interest in Hungarian music is shown by the Heroic March in the Hungarian Style (1840), a fine, compact piece with which Liszt reciprocated Vörösmarty's gesture in writing a poem of homage to him; the march was later expanded to form the basis of the symphonic poem *Hungaria*. Between 1840 and 1847 Liszt published 20

pieces based on Hungarian gypsy music, which, together with four unpublished works, became the basis of the first 15 Hungarian rhapsodies of 1851–3. Liszt did not carry out very deep research into the origins of Hungarian folk music, being inclined to overestimate the part played by the gypsies in its development; moreover, many of the themes he used were not traditional gypsy melodies but themes by various amateur composers quite well known by name – Liszt merely applied the gypsy style of ornamentation to them. Nevertheless, he was successful in transcribing for the piano the sound of the gypsy orchestra of solo violin, clarinet, cimbalom and strings. In the hands of a virtuoso with an affinity for the gypsy style, the rhapsodies can be stunningly effective, since they embody sensual dance rhythms, harmonic daring, exotic melodies and compelling technical demands.

Towards the end of this period Liszt began the series of pieces which eventually became the collection of *Harmonies poétiques et religieuses* (1853): this is a mixed group, three of the ten pieces being transcriptions of choral works written about 1846 and one being an arrangement of a work by Palestrina; the single piece *Harmonies poétiques et religieuses* reappears here under the title *Pensée des morts* in a revised form that also includes a passage from the unfinished *De profundis* of the same year (the theme to which Psalm cxxx was set in that work). The set does contain three other fine works, the expressive *Cantique d'amour* (the collection is dedicated to the Princess Sayn-Wittgenstein), the dramatic *Funérailles* (written after the defeat of the Hungarian Revolution, in memory of three patriots

271

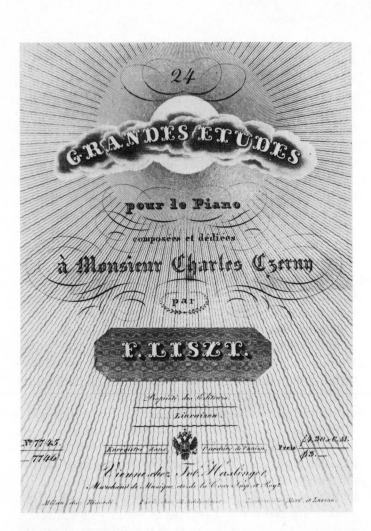

25. Title-page of Liszt's 'Vingt-quartre grandes études' of 1837, dedicated to Czerny and published by Haslinger and others in 1839

who had died), and the *Bénédiction de Dieu dans la solitude*, in which Liszt expressed a mood of mystical contemplation similar to that of Beethoven in his last period.

About 1848 Liszt wrote the brilliant, evocative *Trois études de concert*. About this time too he composed a number of pieces in Chopinesque forms: two ballades, two polonaises, a berceuse (two versions), a *Mazurka brillante* and revised versions of some of his earlier waltzes. The second ballade, in particular, is intensely dramatic. The polonaises are on the bombastic side, but the two versions of the berceuse are interesting: the first, in the same key as Chopin's (D♭) and similarly based on a tonic pedal, is written simply but effectively, while the second version presents the same material in a fantastically ornamented form.

Three of Liszt's greatest piano works date from this period: the *Grosses Konzertsolo* (1849), the Scherzo and March (1851) and the Sonata (1852–3). The *Grosses Konzertsolo* contains a theme which is strikingly similar to one of the main themes of both the Sonata and the *Faust-Symphonie*. It is in a 'three-movements-in-one' form, a rearrangement of the elements of the Classical sonata form, which normally consisted of four sections – exposition and development; slow movement; further development of the first section and recapitulation; coda, including a part repeat of the slow movement. The *Grosses Konzertsolo*, a fine dramatic piece, was arranged for piano and orchestra about 1849 and he revised and rearranged it for two pianos in 1856 under the title *Concerto pathétique*. The Scherzo and March contrasts the grim, chattering

TABLE 1: Formal, thematic and tonal outline of Liszt's Piano Sonata

Bars	Formal unit	Themes	Tempo, time signature	Key
1–204	Exposition			
1–7	introduction	[I	Lento assai, **C**	[(g)
8–81	1st subject	II	Allegro energico, **¢**	b
81–104	modulation	I		
105–19	2nd subject	III	Grandioso, 3/2	D
120–52	transition	II	[Allegro], **C**	D
153–204	3rd subject	II′		
205–330	Development section, part 1	II (I, III, II′)		
331–459	Slow movement			
331–48	1st subject	IV	Andante sostenuto, 3/4	F♯
349–62	2nd subject	II′	Quasi adagio, **C** (bar 347)	A
363–94	middle section	III, I	3/4	F♯–g–
394–452	recapitulation	IV, II′	3/4–**C**	F♯
453–9	coda	[I		[(f♯)
		II	Allegro energico, **¢**	b♭
460–530	Development section, part 2			
531–672	Recapitulation			
673–760	Coda			B
673–710		I, II, III	Presto–Prestissimo, **¢**–3/2	
711–28		IV	Andante sostenuto, 3/4	
729–60		II, I	Allegro moderato–Lento assai, **C**	

humour of the scherzo section with a broader, central march section, after which the scherzo returns in a condensed form. Both themes are combined in the coda, which Louis Kentner described as 'religion doing battle with the Devil'.

The B minor Sonata, one of the masterpieces of 19th-century piano literature, has maintained its place in the repertory. It is a work of great dramatic power and lyrical expression, with frequent changes of mood; but it does not attempt to tell a story, and its construction is logical on purely musical grounds. It carries the principle of transformation of themes to its limits, and does so extremely successfully. Its formal structure is so complex that no single analytical interpretation has achieved widespread acceptance. Theories of its design include an extended one-movement sonata form, a three-movement cycle, a four-movement cycle and general programmatic treatments. The use of a limited number of themes and their transformations shows an extraordinary sense of economy (exx.4*a–e*). Study of the manuscript reveals a rich compositional genesis with a completely sketched early version of the Andante sostenuto differing significantly from the final version.

Ex.4 Piano Sonata
   (a) Theme I

275

(b) Theme II

(c) Theme III

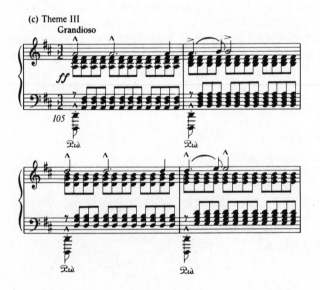

(d) Theme II′

(e) Theme IV

The piano works of the 1860s are mainly religious in character. One of the most important is the set of variations on the passacaglia theme from Bach's Cantata no.12 ('Weinen, Klagen'). Liszt had written a short prelude on the same theme three years before, in 1859,

277

and the variations show an advanced use of chromaticism, probably derived from some passages in Bach's work and based mainly on chords of the diminished 7th, here used in such a way that a sense of tonality is severely strained. There are also two legends for piano, *St François d'Assise: la prédication aux oiseaux* and *St François de Paule marchant sur les flots*, illustrative pieces that nevertheless have a definite musical shape; the first is most delicately written, while the second evokes the atmosphere of the story. A curiosity is *A la Chapelle Sixtine: Miserere d'Allegri et Ave verum corpus de Mozart*. In the first part of the piece Liszt treated the *Miserere* (which Mozart is reputed to have written down from memory as a child) in a distinctly 19th-century manner, including a good deal of chromatic harmony, so that the work almost becomes an original piece based on Allegri's themes; but the transcription of the Mozart that follows is quite straightforward. The material from the first part returns and works up to a climax, but the work ends quietly with the *Ave verum*. Two secular works of this period are the concert studies *Waldesrauschen* and *Gnomenreigen*, the first a lyrical piece and the second a kind of Mendelssohnian scherzo. Liszt also wrote the first two of five little piano pieces, for his friend Olga von Meyendorff, which again show his skill as a miniaturist; a *Rhapsodie espagnole*, a brilliant fantasy on Spanish themes and one of his best works in this form, and a funeral march for Emperor Maximilian of Mexico (executed in 1867), an impressive piece which was later included in the third book of *Années de pèlerinage*.

The rest of the pieces in this collection were written

in the 1870s: Liszt's style had now begun to change and to become starker and more austere, with long passages in single notes and a considerable use of whole-tone chords. Definite cadences are usually avoided, and the tonality is often in doubt; there is little brilliance, and often a curiously resigned atmosphere prevails – Liszt was now writing for himself and no longer for his public. Of these later pieces, the most interesting are those connected with the Villa d'Este, two being inspired by its cypresses (said to be the tallest in Italy) and one by its fountains. The two cypress pieces, called threnodies, have a powerful, rather gloomy atmosphere; *Les jeux d'eau*, with its impressionistic harmony, had a considerable effect on the style of Debussy and Ravel. The collection also contains an interesting *Sunt lacrymae rerum*, which makes some use of the Hungarian scale, and *Sursum corda*, in which the whole-tone scale figures prominently.

The 'Christmas Tree Suite' (*Weihnachtsbaum*), written about 1875, is a mixed collection of pieces, some of which are arrangements of carols. There are also two 'bell' pieces, *Carillon* and *Abendglocken*, and the collection includes a nostalgic waltz *Jadis* which is a recollection of the first meeting of Liszt and Princess Sayn-Wittgenstein nearly 30 years before; a march in the Hungarian style represents Liszt himself and a mazurka portrays the princess. The other principal collection of late pieces, the seven *Historische ungarische Bildnisse*, contains portraits of various Hungarian politicians and artists; two of them, depicting the composer Mihály Mosonyi and the poet Sándor Petőfi, were written in the 1870s and published separately, and the portrait of the statesman Ladislaus Teleky was

279

composed in 1885 and published posthumously as *Trauermarsch*. These three pieces are all fine, grave elegies, and the *Trauermarsch* is based on an ostinato figure which produces some remarkable harmonic clashes. The other four pieces, also composed in 1885, are in Liszt's Hungarian style, fiery but not quite so original. Liszt's orchestral version seems to be lost.

Other Hungarian works of this period. include the Hungarian rhapsodies nos.16–19, less brilliant than their predecessors and much more experimental. There are also three *csárdás*, of which the more interesting are the *Csárdás obstiné* and the *Csárdás macabre*. The latter is unusual for its period in that it consists mainly of bare parallel 5ths; the theme of its two contrasting trio sections is a Hungarian folksong, the text of which reads: 'The hut is burning while I make love to a gypsy girl'. Liszt made arrangements of five Hungarian folksongs during this period: they are simple settings that anticipate Bartók's work.

There are a number of other late piano pieces that express a definite mood, such as the charming nocturne *En rêve* and the rather more disturbed *Schlaflos, Frage und Antwort*. The two Elegies (1874 and 1877) depend little on instrumental colour, and are thus more 'abstract' than the earlier piano works, which could be effective only on the piano. The second version of *La lugubre gondola* was originally written for violin or cello and piano, and the elegy on the death of Wagner, *Am Grabe Richard Wagners*, exists in versions for string quartet and harp and for piano solo; the latter is based on a theme from Liszt's *Excelsior!* (the prelude to his oratorio *Die Glocken des Strassburger Münsters*), which Wagner later used in *Parsifal*, and also on the bell

motif from *Parsifal* itself. Both 'gondola' pieces have a strange, withdrawn atmosphere, as does Liszt's other piece written in Wagner's memory, *R. W.–Venezia*. The Third Mephisto Waltz also dates from this period: it is angry and violent throughout, with little relief, but is a fine piece with some unusual use of harmony. A lighter mood is represented by a Toccata (1879–81, pubd 1970), and by the *Carousel de Mme Pelet-Narbonne*, occasioned by the sight of a fat lady riding on a merry-go-round – probably Liszt's only humorous piece. Another recently discovered work, the *Bagatelle sans tonalité*, originally intended as a Fourth Mephisto Waltz, shows an advanced use of chromaticism, and though it is not atonal in any Schoenbergian sense, there is certainly no feeling of key.

The piano works on other composers' themes fall into two classes, the 'partitions de piano', which are usually fairly straightforward transcriptions, and the para-phrases and fantasias, which are original works based on material from elsewhere; Liszt wrote works in both categories during most of his life. As early as 1829 he wrote a fantasy on the Tyrolienne from Auber's *La fiancée* and followed it three years later with the complex and very difficult Fantasia on Paganini's *La campanella*: both works show a masterly command of virtuoso technique, the treatment of Paganini's simple theme being most ingenious. In contrast to these, the transcriptions of Berlioz's *Symphonie fantastique* and *Francs-juges* Overture are entirely straightforward: Liszt helped Berlioz to gain recognition by playing them at his concerts and even paid for the publication of the symphony transcription. The arrangements are aston-ishingly well done: Liszt did not simply transcribe the

notes but recast the texture so as to obtain an orchestral effect on the piano.

Liszt continued his long series of operatic fantasias with those on themes from Halévy's *La juive*, Pacini's *Niobe* and Donizetti's *Lucia di Lammermoor*, all written in 1835. He has often been attacked for writing such works, and admittedly some of them are superficial and over-written. But some are extremely fine works, and many of them give a kind of conspectus of the action of the opera in a comparatively small space. Apart from further fantasies on operatic themes, he also wrote two fantasies on themes from Rossini's *Soirées musicales*, little songs which Rossini used to write for evening parties with his friends; the two fantasias are complex virtuoso pieces, but Liszt later transcribed 12 of the songs for piano in a simple and straightforward manner. His *L'idée fixe*, a short piece on the main theme of the *Symphonie fantastique*, has a charming lyrical atmosphere.

Liszt's next transcriptions were of Beethoven's symphonies, nos.5–7. One might hardly think that Beethoven needed championing, but in the Paris of 1837 he was by no means a popular composer. Liszt had to work for him, as he did for Schubert, transcribing parts of his *Divertissement à l'hongroise* and several of his songs in the following year. It is true that in some of the song transcriptions Liszt added extra material not always at one with the original: but he had a genuine feeling for Schubert, who was then more or less unknown in Paris. The Beethoven transcriptions are masterly, as one can see by comparing them with normal routine arrangements, and they are also entirely faithful to the original. A few lighter transcriptions of

this period include the overture to Rossini's *Guillaume Tell* and some delightful songs by Donizetti and Mercadante.

The transcriptions and fantasias of the 1840s continued on much the same lines, including further works by Beethoven (*Adelaide*, the Gellert songs and the Septet) and Schubert (lieder and some of the marches for piano duet). Liszt also transcribed a group of songs by Mendelssohn, Weber's *Freischütz*, *Oberon* and *Jubel* overtures and some other works, and Chernomor's Circassian March from Glinka's *Ruslan and Lyudmila* (transcribed in 1843, the year after the first performance of the opera): this was his first contact with the Russian nationalist school, occasioned by his visit to Russia in 1843. Another important enterprise was the transcription of some organ preludes and fugues of Bach, who was still comparatively unknown in spite of Mendelssohn's efforts: Liszt's arrangements are simple and straightforward and quite in keeping with the spirit of the music.

The operatic fantasias of the 1840s are among the most important in this genre. That on *Norma* summarizes the musical content of the opera in a concentrated form, and those on *La sonnambula*, Donizetti's *Dom Sébastien* and Auber's *La muette de Portici* are far more than mere salon pieces. The fantasia on *Don Giovanni* is more open to question, but it does summarize three aspects of the story, justice, seduction and carefree enjoyment (in that order), and is a satisfying piece. It presents Liszt's own view of Mozart's opera (a parallel might be Berio's transcription of Mahler in his *Sinfonia*). The *Figaro* fantasia, which Liszt left very nearly complete (it was finished and first published by

26. *Autograph MS from Liszt's 'Réminiscences de Simone Boccanegra', composed December 1882*

Busoni), is a slighter work based on Figaro's two arias from Act 1 and does not attempt a similar interpretation of the opera. This period also saw the composition of several works based on national themes, apart from the Hungarian Rhapsodies: a fantasia on *God Save the Queen*; an over-complex and over-written fantasia on Spanish themes which is less successful than the later *Rhapsodie espagnole*; two Pyrenean folksongs, written

on his meeting with Caroline de Saint-Cricq at Pau; transcriptions of two Russian songs; and the *Glanes de Woronince*, based on gypsy themes that Liszt heard at Princess Sayn-Wittgenstein's estate in the Ukraine: these folksong arrangements are concise and have a charming atmosphere.

In spite of his other preoccupations Liszt continued his series of transcriptions and fantasias during the Weimar period: the former included works of some of the younger German composers, such as Draeseke, Raff and Nicolai. In addition, Liszt transcribed six of Chopin's Polish songs – a charming collection – and wrote nine *Soirées de Vienne*, based on Schubert's waltzes, which successfully evoke the Romantic atmosphere of Vienna in Schubert's time. Apart from three 'illustrations' from Meyerbeer's *Le prophète*, the most important arrangements of operas in this period were those of the works of Wagner and Verdi: in the Wagner works, Liszt was again making propaganda for an underestimated composer. Most of his Wagner transcriptions are straightforward attempts to reproduce the sound of Wagner's music on the piano – the *Phantasiestück* on themes from *Rienzi* is one exception – and many are extremely successful in this, notably the brilliant transcription of the *Tannhäuser* overture and the subtle treatment of Isolde's *Liebestod*. Liszt continued to transcribe Wagner's music right up to the end of his life, his last Wagner transcription being that of the March to the Grail from *Parsifal* (1882): he did much the same for Verdi, writing straightforward transcriptions of passages from many of his operas, except for two fantasias on *Ernani* and one on *Simon Boccanegra*, Liszt's last work in this genre. But perhaps the most

interesting transcription of this period is that of the Waltz from *Faust*, made soon after the first production of the opera (1859), which invests Gounod's tune with a diabolic quality, creating more the atmosphere of an orgy than a simple fair.

The transcriptions and fantasias of the 1860s are few, but in the last years of his life Liszt produced some interesting arrangements: one of these, of Saint-Saëns's *Danse macabre*, takes considerable liberties with the original (much to its advantage). Liszt's sympathy with the new Russian nationalism was shown by his transcriptions of tarantellas by Cui and Dargomïzhsky, and he also arranged the Polonaise from Tchaikovsky's *Eugene Onegin* for piano. These transcriptions are mostly in the austere style of his late works, as in the *Boccanegra* fantasia with its whole-tone harmonies.

CHAPTER SIX

Orchestral and other instrumental music

By the time Liszt began composing music for orchestra alone, he had already completed quite a number of works for piano and orchestra. The *Grande fantaisie symphonique* (1834), based on themes from Berlioz's *Lélio*, is the first of these, though the scoring seems not to have been Liszt's own; the work is in two sections, the first a meditation on the *Chant du pêcheur*, the second an Allegro on the *Chanson des brigands*. The unfinished score of the instrumental psalm *De profundis* of the same year contains only a rough sketch of the orchestral part in Liszt's hand. The origins of the *Malédiction*, for piano and strings, are unclear; its title applies only to the striking opening theme: later themes are marked 'Pride', 'Tears–Anguish–Dreams' and 'Raillery'.

Both piano concertos also originated in the 1830s and were frequently revised up to the Weimar period. The first borrows its four-movements-in-one form from Schubert's *Wandererfantasie*, the four sections being also interlinked thematically. The second concerto, less brilliant and more poetical, has a freer, highly original form. The last work for piano and orchestra, *Totentanz* (planned 1839, sketched 1849, revised 1859), is a set of variations on the *Dies irae* inspired by Orcagna's frescoes *Il trionfo della morte*, which Liszt saw in the Campo Santo, Pisa, in 1838.

As late as the mid-1840s, when at work on the first of his Beethoven cantatas, Liszt was worried about his lack of skill as an orchestrator, and when he began to write for orchestra alone he enlisted the help of collaborators for his scores. The first of these, August Conradi, worked with Liszt in Weimar in 1848–9. Liszt usually made a sketch on a small number of staves with some indications of the orchestration, and from this Conradi prepared a full score. This was then revised by Liszt (often a new score was prepared) and the whole process could be repeated several times; an example of Conradi's work is given in Raabe's Liszt biography. However, though technically proficient, Conradi was not a man of great imagination, and his place was soon taken by the more original Joachim Raff, who went to Weimar for this purpose and remained until 1854. Raff was able to give Liszt many helpful suggestions, as may be seen from the reproductions in Raabe's biography. He went too far, however, when he claimed to have scored Liszt's orchestral works: his role was more that of a superior copyist, and after Liszt dispensed with his services, at the insistence of Princess Sayn-Wittgenstein, he never used collaborators again. The final versions of all his orchestral works were revised by Liszt himself and represent his own intentions.

The first 12 'symphonic poems' were written in Weimar between 1848 and 1858. Liszt invented the term 'sinfonische Dichtung' to describe works that did not obey Classical forms strictly and were based to some extent on a literary or pictorial idea. But many of them were originally written for other purposes, for instance as overtures (*Tasso*, *Les préludes*, *Orpheus*, *Prometheus*, *Hamlet*) or were expanded versions of

earlier works (*Mazeppa*, *Héroïde funèbre*, *Hungaria*), while the first and last of these 12 works are really symphonies in a three-movements-in-one form. Liszt took some time to write many of these works, and though his rehearsals of them with the Weimar orchestra led to alterations and improvements, it should be remembered that the forces at his disposal were very meagre – only 21 string players, double woodwind, four horns, two trumpets, one trombone, one tuba and one timpanist. This was in 1851: later in the year Liszt asked the court for two more trombones, harp, organ and some percussion instruments, thus bringing the orchestra up to the size that he used for most of his orchestral works.

Liszt's conception of programme music was rather different from that of Berlioz and also some later composers who wrote symphonic poems, like Smetana, Dvořák, Saint-Saëns and Strauss. He tried to express general ideas in music rather than to use pictorial realism: even a battle-piece like *Hunnenschlacht* is treated symbolically rather than realistically for half of its length. The prefaces affixed to the symphonic poems are misleading: most of them were written by the Princess Sayn-Wittgenstein or by Bülow; though intended to give the audience some idea of the thoughts behind the music, they often do not really correspond to Liszt's actual compositions. To him musical construction was always more important than scene-painting.

The first symphonic poem, *Ce qu'on entend sur la montagne*, is described as 'after Victor Hugo', and Hugo's poem of the same name is prefaced to the score. Liszt also wrote a short synopsis in which he spoke of the voice of nature, 'splendid and full of order', contrasting with that of humanity, 'hollow, full of pain'; in fact

he did not follow the exact form of the poem, taking from it two main ideas, Hugo's 'broad, immense, confused sound' (this is the rumbling figure with which the work begins) out of which the voice of Nature appears, and the contrast between the voices of Nature and Humanity (each with its own theme) often at war with each other. Written in July 1829, a year before the Paris Revolution, Hugo's poem expresses the thoughts of a man longing to escape from political turmoil; a quieter middle section in the symphonic poem, 'Song of the Anchorites', portrays this desire, but the stormy material returns after it and leads finally to the triumph of the voice of Nature. The poem ends on a pessimistic note: 'Why does the Lord eternally mix in a fatal marriage the song of Nature with the cries of the human race?', but Liszt, a practising Catholic, gave his work a more definitely religious ending by bringing back the anchorite chorale.

Liszt's second symphonic poem, *Tasso: lamento e trionfo*, was first sketched during the 1840s as an overture to Goethe's play *Torquato Tasso* and revised several times (with an added middle section) until 1854. In Tasso, Liszt saw a symbolic figure, 'the genius who is misjudged by his contemporaries and surrounded with a radiant halo by posterity'; and he felt, with good reason, that he might himself suffer the same fate. The preface to the score relates the music to various events in Tasso's life, but was in fact written afterwards. The opening lament is based mainly on a theme to which Liszt heard a Venetian gondolier sing the opening lines of *Gerusalemme liberata*; there is a more violent Allegro passage symbolizing Tasso's sufferings. The middle section, light, graceful and in the style of a minuet, sup-

posedly represents Tasso at the court of Ferrara, but was probably added as a bridge section between the opening lament and the final triumphant section.

Les préludes, the best-known of the symphonic poems, was written as an introduction to the unpublished choral work *Les quatre élémens*, on a text by the Provençal poet Joseph Autran, consisting of four choruses, *Les aquilons*, *Les flots*, *Les astres* and *La terre*. It was orchestrated by Conradi; the introduction was scored by Raff and revised by Liszt between 1852 and 1854. Liszt then decided to use the introduction as a separate work and was obliged to find a 'programme' for it: the nearest he could get was Lamartine's long poem *Les préludes*, a meditation in which pastoral and warlike elements are closely linked together. This is the only feature the poem has in common with Liszt's music, which was written in France and Spain and was in fact inspired by the Mediterranean atmosphere. The first main theme, derived from *Les astres*, is treated typically along the lines of Liszt's 'transformation of themes' method, appearing in many guises throughout the work and holding it together. The second main theme, which appears later on horns and violas and plays an almost equally important part, is taken from *La terre*; the trumpet figure, which appears in the stormy central section, is from *Les flots*. *Les préludes* probably owes its popularity to the tightness of its construction and the avoidance of recitative passages, which perhaps hold up the action in many of Liszt's other symphonic poems.

Orpheus (1853-4) was written as an introduction to the first Weimar performance of Gluck's *Orphée et Euridice*. The preface to the work refers to an Etruscan vase in the Musée du Louvre which shows Orpheus

taming the wild beasts by his music – a symbol of art as the civilizing influence on man's more brutal instincts. *Prometheus* (1850) was also originally an overture, to Herder's *Der entfesselte Prometheus*; Liszt also set the choruses in the play, and the entire work, orchestrated by Raff from Liszt's indications, was first performed at the unveiling of a memorial to Herder in Weimar. Five years later Liszt rescored the overture and the choruses, and in this form the overture was first performed as a symphonic poem, here the symbolism being that of suffering for the sake of the enlightenment of mankind.

Mazeppa, an expanded version of the fourth of the Transcendental Studies, was first scored in 1851 with Raff's help and revised by Liszt in 1854. Hugo's poem *Mazeppa* is prefaced to the score, and here again the symbolism of artistic creation appears; the poem speaks of 'mortal man tied to the saddle of genius', triumph being able to come only after defeat and collapse: 'He runs, he flies, he falls and stands up king!' The first part of the work follows the piano study without great alteration, but the section depicting Mazeppa's fall is considerably expanded, and in the final section, showing Mazeppa as hetman of the Cossacks, Liszt introduced two new themes, one from his own *Arbeiterchor* of 1848, the other a 'Cossack' theme with an oriental flavour. *Festklänge* (1853) was written in anticipation of Liszt's forthcoming marriage with the Princess Sayn-Wittgenstein and contains a section in polonaise rhythm, referring to the princess's Polish origins. In 1861 Liszt published some 'variants' to the work which considerably lengthen it; he also marked an optional cut of 45 pages in the original version. Prone to accepting advice about his works, not always to his own advan-

tage, he seems here to have been particularly doubtful as to whether he wanted a full-scale symphonic work or a short 'festival overture', which was his original intention. Neither *Festklänge* nor *Mazeppa* is among the more inspired of the symphonic poems.

Héroïde funèbre (1849–57), however, is an interesting work. Its origins go back to the planned Revolutionary Symphony of 1830, and after the revolutionary uprisings in Europe from 1848 onwards Liszt intended to take up the idea again in the form of a five-movement symphony based partly on national themes and partly on settings of biblical texts (he had intended to incorporate national themes in the early Symphony). But in the end he only wrote one movement (which eventually became *Héroïde funèbre*), a fine, dignified funeral march. The theme of the outer sections has a distinctly Hungarian flavour, and the middle section contains a theme similar to part of the *Marseillaise*: Liszt had intended to use the tune in the 1830 symphony, and he also planned to base a whole movement on it in the 1849 version. Together with *Funérailles* and the cantata *Hungaria 1848*, the symphonic poem *Hungaria* (1854) was intended as a reply to the poem of homage that Vörösmarty had dedicated to him in 1840. It is a much expanded version of the Heroic March in the Hungarian Style of the same year, but it contains other themes as well and is somewhat episodic. There is an impressive funeral march towards the end, based on the second theme of the Heroic March, and this is followed by a crescendo to an exciting climax: Liszt was no doubt thinking of the defeat of Kossuth's revolution in Hungary and symbolizing the idea that Hungary would one day be liberated by its own people. *Hun-*

garia was an enormous success at its first performance in Budapest in 1856.

Hamlet, the last of the first 12 symphonic poems to be composed (1858) and also performed (1876), was planned as an overture to Shakespeare's play, which Liszt had seen in Weimar in 1856. He was greatly impressed by Bogumil Dawison's interpretation of the part of Hamlet, not as an indecisive dreamer but as a gifted prince waiting for the right moment to complete his revenge. One of Liszt's finest, most concise symphonic poems, *Hamlet* is a remarkable psychological portrait of the prince, not an attempt to depict the action of the play; it is a short work, with a slow introduction, a violent central Allegro and a slow final section that refers to the introduction and ends with a funeral march. The Allegro contains two short passages of a 'shadowy' character referring to Ophelia; these were afterthoughts on the part of Liszt, who did not see Ophelia as central to the action of the play.

Hunnenschlacht (1857) was suggested by an enormous mural by Wilhelm von Kaulbach representing the battle in the Catalaunian fields in 451 between the Huns under Attila and the forces of the Christian Emperor Theodoric for the possession of Rome. At the time Liszt was considering the idea of writing a series of symphonic poems after Kaulbach's frescoes which was to include the Tower of Babylon, Nimrod, Jerusalem and the Glory of Greece. The battle depicted in this painting is said to have been so fierce that the spirits of the fallen were seen continuing the fight in the sky, and Liszt had originally intended to show this in his music; later he decided to end the work with a full-scale treatment of the chorale *Crux fidelis* to represent the victor-

ious Christians. The battle itself thus occupies only half the work; but it is one of Liszt's most imaginative creations, beginning with ghostly string passages interrupted by ferocious horn-calls, and continuing with a confrontation between the rival themes that becomes wilder and wilder until the Christian chorale eventually triumphs. The second half of the work, rather more episodic, is mainly a meditation on this chorale theme, beginning quietly on the organ, which plays a prominent part in the work, and finally building up to a big climax.

Together with the *Faust-Symphonie*, *Die Ideale* (1857) was first performed in Weimar at a concert in honour of the laying of the foundation stone for a memorial to Goethe's patron, the Grand Duke Karl August, and the unveiling of memorials to Goethe, Schiller and Wieland. Various extracts from Schiller's poem of the same name are printed in the score as a guide to the mood of each section of the work; but Liszt altered the order of the extracts from the poem for musical reasons, and towards the end of the work wrote: 'I have allowed myself to add to Schiller's poem by repeating the motifs of the first movement joyously and assertively as an apotheosis'. The ending is hardly in keeping with Schiller, and the entire work is long and episodic.

The other main orchestral works of the Weimar period, the *Faust-Symphonie* and the Dante Symphony, were planned much earlier. As early as 1830 Liszt was introduced to Goethe's *Faust* (in the French translation of Gérard de Nerval) by Berlioz, who dedicated his *La damnation de Faust* (1846) to Liszt (Liszt replied by dedicating the *Faust-Symphonie* to Berlioz). He seems to have started sketching out the work during his travels

295

27. Autograph score of part of the third movement ('Mephistopheles') from Liszt's 'Faust-Symphonie', composed 1854–7

in the 1840s, but was at first rather hesitant about his task, writing: 'Anything having to do with Goethe is dangerous for me to handle'. However, the performance of Berlioz's *La damnation de Faust* in Weimar in 1852 stimulated him to take up the work again, and the entire first version was completed in the astonishingly short time of two months, from August to October 1854. But Liszt was still not happy about it: the first version, scored for a very small orchestra, without trumpets, trombones or percussion, does not contain the martial music of the first and last movements. Liszt tried out the work with the Weimar orchestra and made a number of alterations, adding the extra brass and percussion and, in 1857, a final chorus. Even after the score was published in 1861, Liszt continued to make alterations, and as late as 1880 added 12 bars to the slow movement. The symphony, considered by many to be Liszt's masterpiece, makes no attempt to tell the story of Goethe's play. Instead, the three movements are character studies of Faust, Gretchen and Mephistopheles. The Faust movement is a supreme example of Liszt's method of transformation of themes, lasting nearly half an hour though based on only five short phrases. It is one of Liszt's most masterly creations, and with its extraordinarily vivid, acutely perceived range of moods gives all the appearance of a self-portrait.

The Gretchen movement is scored with marvellous delicacy and is a fine example of Liszt's 'chamber music for full orchestra' which differentiates his orchestral style from the thicker textures favoured by many of his contemporaries. Mephistopheles, the spirit of negation (Goethe's 'Geist der stets verneint'), is allowed virtually no themes of his own in the third movement, but merely

297

parodies the Faust themes from the first movement. This ingenious idea is brilliantly carried out, and the whole movement has an atmosphere of sardonic mockery. There is only a short phrase (taken from the early *Malédiction*) for him, and it makes a highly appropriate appearance here. At the end of the movement there is a tremendous struggle, in which the Devil is defeated and the music sinks down; the final section is a setting of the *Chorus mysticus* which ends the second part of *Faust* for tenor, male chorus and orchestra.

Liszt used to read Dante with Marie d'Agoult in the 1830s, and after sketching out *Après une lecture du Dante* began to think of an orchestral work based on Dante. By 1847 he was able to play Princess Sayn-Wittgenstein some of the themes, and at this time he had the idea of illustrating the music with lantern slides and using a wind machine in the final section of the *Inferno* movement, but both ideas were dropped when he settled down to composing the work in 1855–6. He originally planned to write the symphony in three movements, corresponding to the three books of Dante's *Divina commedia*, but Wagner (to whom the work is dedicated) managed to persuade him that no human could truly express the joys of Paradise: the work therefore remains in two movements, *Inferno* and *Purgatorio*, to which Liszt added a choral *Magnificat*. The balance of the symphony is thus destroyed, and at the end the listener is left, like Dante, gazing up at the heights of Heaven and hearing its music from afar.

The last important orchestral works of the Weimar years were the two episodes from Nikolaus Lenau's *Faust*, both written about 1860. Lenau (1802–50) was a partly Hungarian poet who wrote in German; his *Faust*

(1836), a long poem, contains many episodes that differ from Goethe's version of the story. The second episode that Liszt set, *Der Tanz in der Dorfschenke*, is the well-known First Mephisto Waltz, whose combination of excitement and sensuality is typical of Liszt's music at this time; some of the harmonic progressions in the quieter middle section anticipate Skryabin.

The principal orchestral works of the 1860s are *Salve Polonia*, based on the folksong *Boze cos Polske*, which Liszt later intended to use for an interlude in the unfinished *Stanislaus* oratorio, and the autobiographical *Trois odes funèbres*. The first of the odes, *Les morts*, written in 1860 in memory of his son Daniel, is described as an oration for orchestra, and throughout the score is written a prose passage by Lamartine which begins: 'They, too, have lived in this earth; they have passed down the river of Time; their voices were heard on its banks, and then were heard no more. Where are they now? Who shall tell? But blessed are they who die in the Lord'. In 1866 Liszt added a male chorus, which sings part of this text in Latin, 'Beati qui in Domino moriuntur'. The second ode, *La notte* (1863–4), an extended version of *Il penseroso* from the second book of the *Années de pèlerinage*, is prefixed with Michelangelo's quatrain containing the line: 'So long as injustice and shame remain on earth I count it a blessing not to see or feel'. Liszt wrote a new middle section for this ode, ending with the *bokázó* or so-called 'Hungarian cadence' found chiefly in the rhapsodies, to which he affixed a quotation from Virgil about an Argive companion of Aeneas remembering his native land at the moment of his death in battle in Italy (Liszt composed this ode in Rome at a time when he, too, felt that he

might die far from his homeland). Liszt wanted both *Les morts* and *La notte* played at his funeral, but his wish was not fulfilled; they were first performed only in 1912. The third ode, *Le triomphe funèbre du Tasse* (1866), was written as an epilogue to *Tasso* and uses some of its themes. Here Liszt returned to the idea of the true worth of the artist being recognized only after his death, but unlike the earlier work this ode remains dignified and restrained throughout. It is dedicated to Leopold Damrosch, a violinist in the Weimar orchestra for many years and an early champion of Liszt's music. At Liszt's request, Abbate Pier-Antonio Serassi's account of Tasso's splendid funeral was printed in the programme of the first performance and in the published score.

Two important orchestral works date from Liszt's last years. The Second Mephisto Waltz (1880–81) shows the same mixture of violence and sensuality as its predecessor, but is more sparingly written and has an unexpected ending: after building to a climax in the main key of E♭, it suddenly falls on to the tritone B–F and ends with these two notes only. It is dedicated to Saint-Saëns, a somewhat doubtful compliment in view of Liszt's 'diabolical' transcription of *Danse macabre* five years earlier. The last symphonic poem, *Von der Wiege bis zum Grabe* (1881–2), inspired by a painting by Count Michael Zichy, is a restrained and austere work that looks towards the future both musically and in its subject matter.

Liszt's few chamber works were written mostly at the beginning or the end of his career. An early duo (or sonata) for violin and piano, based on Chopin's Mazurka in C♯ minor op.6 no.2, is a substantial four-movement

work in a Classical style probably typical of his early lost chamber works. Another work for violin and piano, the *Grand duo concertant* on Lafont's *Le marin* (1835), was presumably written for Liszt to play with the well-known violinist Charles Lafont. In the 1860s Liszt arranged some of his works for chamber combinations, including *La notte* and two numbers from the Hungarian Coronation Mass for violin and piano and *Pester Karneval* (the ninth Hungarian rhapsody) for piano trio. The rest of his chamber works date from the last 15 years of his life: these include the two elegies, the first of which was composed in 1874 in memory of Countess Marie Mukhanov and originally scored for the unusual combination of cello, piano, harp and harmonium; *Epithalam* (1872) for violin and piano, written for the wedding of the violinist Ede Reményi (for whom he also began a violin concerto); and *Die Wiege* (1881), a work for four violins that was later transformed into the first section of *Von der Wiege bis zum Grabe*. Liszt also sketched out a piece for string quartet, *Die vier Jahreszeiten*, about 1880, but completed only 'Winter' and part of 'Frühling'.

Liszt is not known to have written a single original piece for piano duet. His Prelude to Borodin's 'Chopsticks' Polka (1880) is actually for piano solo. The *Notturno* in E is probably by Gottschalk, not Liszt. And the substantial *Festpolonaise*, written for the marriage of Princess Marie of Saxony in 1876, was originally drafted for piano solo; the duet arrangement was completed the next day, as Liszt clearly indicated on both manuscripts. There are also only two original works for two pianos, the unpublished *Grosses Konzertstück* of 1834, based on Mendelssohn's *Lieder ohne Worte*, and the *Concerto*

pathétique, a revised and somewhat expanded version of the *Grosses Konzertsolo* for piano.

The most important of Liszt's organ works are the Fantasy and Fugue on *Ad nos, ad salutarem undam* (1850) and the Prelude and Fugue on B–A–C–H (1855). The former is a large work, lasting over half an hour in performance, and is in three sections played without a break. The opening fantasia, a brilliant, exciting section with many changes in mood, is based on only the first half of the main theme of the work, from Meyerbeer's *Le prophète*: there it is sung by the three Anabaptists who call on the people to seek rebaptism in the healing water (the theme is Meyerbeer's own and not a traditional chorale). The following Adagio meditates calmly on the entire chorale melody. A violent *fortissimo* passage introduces the fugue, which leads quickly to a heroic apotheosis. The work appears to have been revised in the 1870s, but the second version is lost. The B–A–C–H Prelude and Fugue also begins with a brilliant opening section, in which the four-note motif is treated in a number of ingenious ways. The fugue begins with a passage of sliding chromaticism in which tonality is suspended for a while: later it becomes more fantasia-like, and at the end the four-note theme appears as an ostinato under staccato chords.

These two pieces are Liszt's only important original organ works, though he composed a number of smaller works such as the quiet and gentle *Ora pro nobis* (1864) and the *Salve regina* (1877), based on the Gregorian chant and harmonized modally. There are also a short mass and a requiem for organ solo, intended for liturgical use with the spoken Mass; the latter has connections with the male-voice Requiem of 1867–8. Liszt

28. Franz Liszt: photograph

arranged *Les morts* for organ under the title *Trauerode*, and *Sposalizio* under the title *Zur Trauung* (1883): here the thundering octaves and *fortissimo* chords of the piano version are replaced by quiet held chords over a rippling accompaniment and a dynamic range that never goes above *piano*.

CHAPTER SEVEN

Sacred choral music

Liszt's first sacred work was a *Tantum ergo* (now lost), written at the age of 11. From the 1840s he produced many sacred choral works, ranging from the oratorios *Die Legende von der heiligen Elisabeth* and *Christus* to small motets. His aim was to free church music from the cloying sentimentality prevalent at the time – he regarded his male-voice mass of 1848, simple and liturgical in style, as a step towards this goal, and hoped above all that it would express 'religious absorption, Catholic devotion and exaltation. . . . Where words cannot suffice to convey the feeling, music gives them wings and transfigures them'. The other side of his approach to church music may be seen in his large religious works for chorus and orchestra, such as Psalm xiii, the *Missa solemnis* and the Hungarian Coronation Mass and the two oratorios.

As early as 1834, in his essay 'Über die zukünftige Kirchenmusik', Liszt had dreamed of a religious music that would unite 'the theatre and the Church on a colossal scale', and many of these works attempted to use all the resources of secular music in the service of the Church. Thus the setting of Psalm xiii for tenor solo, chorus and orchestra is intensely dramatic, almost operatic in style, while Psalm xviii is set in a style that Liszt described as 'very simple and massive – like a monolith'. Both these works date from the Weimar

years, as does the *Missa solemnis*; this is written in a full-blooded Romantic style, which led to accusations that he was trying to 'smuggle the Venusberg into church music'. Yet it is an impressive and well-constructed work, bearing some resemblances to Beethoven's *Missa solemnis*, though containing many original ideas. The Hungarian Coronation Mass is rather simpler in style and makes occasional use of the Hungarian idiom; the Credo is written in a plainsong style, and there are some beautiful lyrical passages, especially in the Benedictus with violin solo.

Liszt received the initial inspiration for *Elisabeth* and *Christus* during the Weimar period, and he completed both oratorios in Rome in the 1860s. Princess Elisabeth of Hungary was born in 1207 and taken to the Wartburg at the age of four as the future bride of Landgrave Ludwig of Thuringia: they were married in 1220. Elisabeth's acts of charity to the poor at first aroused her husband's anger, but he was converted by the Miracle of the Roses (whereby a basket of bread for the poor was turned into red roses) and went on the sixth crusade in 1227, on which he was killed. In the oratorio Elisabeth is driven from the Wartburg by her mother-in-law and after living in poverty she finds asylum in the house of her uncle, the Bishop of Bamberg. She died in 1231 and was canonized four years later. *Elisabeth* is thus partly oratorio, partly drama; but it is not an opera, and Liszt refused to attend a stage performance mounted at Weimar in 1881. The music is based on several leitmotifs, a plainchant usually sung on the feast of St Elisabeth, a Hungarian popular melody to symbolize Hungary, a pilgrims' hymn that probably dates from the time of the crusades, and a

Hungarian hymn to St Elisabeth. Here Liszt was using
Wagner's methods of construction, though the music is
not chromatic in a Wagnerian manner. There are some
effective ceremonial scenes, such as the arrival of the
child Elisabeth at the Wartburg, the departure of the
crusaders (dominated by the three-note motif F–G–B♭
that Liszt took from plainchant and used as the symbol
of the Cross in many of his sacred works) and the
solemn ceremony of Elisabeth's burial. The Miracle of
the Roses is delicately scored, and there is some
violently dramatic music in the scene of Elisabeth's
expulsion from the castle; but the oratorio as a whole is
patchy and is not one of Liszt's best achievements.

For *Christus* (completed in 1866) Liszt had no
need to write ceremonial scenes, and, in spite of its
length, it is a much more intimate and personal work.
The text is based on the Bible, the Catholic liturgy and
some Latin hymns. It is in three parts, a Christmas
Oratorio, a collection of scenes from the life of Christ,
and a Passion and Resurrection. The first part is based
on Latin hymns and is mainly pastoral in character: its
first three numbers are scored with the delicacy of
chamber music, though the fourth, the March of the
Three Holy Kings, uses the full orchestra to admirable
effect. The Beatitudes, for baritone solo, chorus and
organ, which begins the second part of the work, had
been written as early as 1855, and shows both intensity
and restraint. Nietzsche, who heard this piece at
Meiningen in 1867, thought that in it Liszt had 'found
the character of the Indian Nirvana excellently'. The
Miracle, also from the second part, which describes
Christ walking on the waves, is more than a conventional storm piece, and all its effects form part of a

logically ordered musical whole. It is a magnificent
orchestral piece with some short vocal interludes; at the
climax the disciples cry out: 'Lord, we perish!' and after
a dramatic silence Jesus answers: 'How fearful ye are, O
ye of little faith'. The beginning of the third part, *Tristis
est anima mea* for baritone and orchestra, uses the voice
sparingly, thus increasing the effectiveness of its entries.
This number begins with a gloomy and powerful
introduction and ends with a big orchestral climax, with
a serene epilogue after the words 'Not what I will, but
what thou wilt': it is certainly one of the finest works
inspired by the Agony in the Garden.

Of the other religious works of the 1860s, the most
important are the *Cantico del sol di S Francesco d'Assisi* (1862) and the *Missa choralis* (1865), a simple and
restrained work that makes a considerable use of
Gregorian themes and avoids chromaticism and the
Romantic style. The male-voice Requiem of 1867–8,
however, uses whole-tone harmonies and themes and an
extra chromaticism that anticipates *Parsifal*: it is concise and tightly constructed. Like the late piano music,
much of Liszt's later choral music shows a tendency
towards experimental harmonic effects. *Qui seminant in
lacrimis* (1884) is advanced in its chromaticism; *Ossa
arida* (1879) gradually builds up all the notes of the
diatonic scale in 3rds in the organ part and sounds them
all *fortissimo* at the entrance of the male chorus – a
formidable effect. The setting of Psalm cxxix (1881) is
restrained and austere throughout, with an original use
of chords frequently based on major 7ths: it contrasts
strongly with the earlier, more Romantic psalm settings.
Liszt intended to insert this into *Stanislaus*, a dramatic
oratorio planned in six scenes along the lines of

29. Autograph MS from Liszt's oratorio 'Christus', completed in 1866

Elisabeth, which he worked on intermittently until his death. The most interesting of the late choral works is *Via crucis* (1878–9), whose text was arranged by Princess Sayn-Wittgenstein from biblical quotations, Latin hymns and German chorales. While remaining restrained and devout throughout, there is a consistent use of experimental harmony that gives the work a curious atmosphere; yet it is deeply felt and very moving. But when Liszt sent it with two other sacred works to the chief publisher of Catholic church music in Germany it was refused. Liszt failed in his attempt to revolutionize liturgical music, and this was chiefly due to the hidebound attitude of the ecclesiastical authorities of his time. But his large-scale works for chorus and orchestra did have a certain amount of success in his lifetime, perhaps because they were well suited to performance in the concert hall.

CHAPTER EIGHT

Opera and secular vocal music

Liszt completed no mature opera. *Don Sanche*, put
on mostly for propaganda purposes, has an 'effective'
libretto containing almost every theatrical device in the
repertory – peasant dances, descending Cupids, a sleep
aria, a storm scene, a fight, a funeral march and a final
ballet. It is well written in a way, and shows considerable melodic gifts, but the music belongs to Liszt's
period of immaturity. His later attempts at opera came
to nothing, apart from 111 pages of sketches for an
Italian opera *Sardanapale* (1846–51), intended for the
Kärntnertor-Theater in Vienna, together with another
opera that was to have been either *Marguerite* or
Richard en Palestine after Scott. The text of
Sardanapale, adapted from Byron, is in the old sensational operatic manner and the music leans towards
Bellini, Meyerbeer and the Wagner of *Rienzi*. Liszt still
had operatic aspirations at the end of the 1850s, but
after that he apparently decided to abandon the genre.

The secular choral works date mainly from the
1840s. The first Beethoven Cantata (1845) was glow-
ingly praised by Berlioz after its first performance, but it
does not really show Liszt at his most characteristic.
During this period he wrote a number of short works
for male chorus and piano, and some of these have
connections elsewhere, such as *Les quatre élémens* and
the *Arbeiterchor*. *Hungaria 1848*, a cantata inspired by

the Hungarian Revolution of 1848, contains in its main theme references to the Rákóczy March, and its music is in Liszt's 'Hungarian' style. The other important works in this genre are the choruses from Herder's *Der entfesselte Prometheus*, first performed as part of the play and later in concert form with a linking narrative by Pohl, and *An die Künstler*, a setting of lines from a poem by Schiller on the duty of artists to preserve their integrity; the version for male chorus and orchestra was highly praised by Wagner. The second Beethoven Cantata (1869–70), like the first, contains an orchestration of the Adagio from Beethoven's Archduke Trio op.97.

Liszt wrote some 70 songs on French, German, Italian and Hungarian poems and one Tennyson setting in English. Many of these were written in the 1840s and revised in the Weimar period; the earlier versions tend to overemphasize the piano part at the expense of the vocal line, and Liszt was inclined to overdramatize the simpler poems; these defects mostly disappear in the later versions. The early songs include several delightful settings of Hugo, and also of poems by Goethe and Heine (Liszt interpreted Heine with particular feeling and subtlety). The original version of the three Petrarch Sonnets is firmly in the Italian operatic style, the vocal part going as high as $c\sharp'''$; the revised version of 1861, for baritone, is in a much more subdued manner. The songs of the 1870s show the same tendencies as Liszt's other works of this period – austerity, restraint and a certain world-weariness: titles like *J'ai perdu ma force et ma vie* and *Verlassen* symbolize Liszt's prevailing mood. *Und wir dachten der Toten* contains some remarkable harmonic progressions, and a setting of Tennyson's *Go not, happy day*, though simple and

restrained, is charming. Liszt's melodramas (or recitations), with piano, are also interesting, especially *Der traurige Mönch* (1860), which is based entirely on the whole-tone scale.

CHAPTER NINE

Harmonic development

During his last 15 years Liszt's harmonic style changed
considerably, producing some extraordinarily imaginative results. While he still used augmented and diminished chords and frequently used the whole-tone scale,
he created new contrapuntal effects by playing themes
and accompaniments against each other without any
regard for the rules of harmony; a simple example of
this is given in ex.5a, from the first version of *La lugubre
gondola*. In *Ladislaus Teleky* (ex.5b–c) the use of an ostinato bass leads to similar clashes of tonality, while in
Unstern! (ex.5d) the left-hand part moves against a fixed
augmented triad in the right hand. The ending of *Nuages*

Ex.5

(a) *La lugubre gondola*, 1st version

Harmonic development

(b) Historische ungarische Bilder, no.4 *Ladislaus Teleky*

(c) ibid

(d) *Unstern!*

(e) *Nuages gris*

gris (ex.5*e*) consists of three juxtaposed contrapuntal elements, an ostinato figure in the bass B♭–A, a chromatically descending figure in the left-hand part and a chromatically ascending figure in the right-hand part.

Another important feature of Liszt's late harmonic style is his construction of chords, particularly those derived from a single interval. The opening of the Third Mephisto Waltz (ex.6*a*) is based on a chord in 4ths; in

Ex.6

(a)   Third Mephisto Waltz

ex.6*b* the second chord is simply the fourth inversion of the first chord. In the Hungarian Rhapsody no.17, a

316

chord in 4ths is derived from the 'Hungarian' intervals F–G♯ and B♭–C♯ (ex.6*c*). The opening of the First Mephisto Waltz (ex.6*d*) is based on chords in 5ths. In the motet *Ossa arida* there is a striking example of a chord in 3rds from *B* to *e'''* (ex.6*e*), and in Psalm cxxix a motif built from major 7ths is made into a chord of the

(c) Hungarian Rhapsody no.17

(d) First Mephisto Waltz

(e) *Ossa arida*

317

(f) Psalm cxxix

Langsam ♩ = 40

(g) Hungarian Rhapsody no.16

Allegro

(h) *Qui seminant*

Qui se - mi - nant in la - cri - mis, in

la - cri - mis.

11th (ex.6*f*). Finally, Liszt's experimentation in these late works was accompanied by a much bolder approach to chromaticism, whether of the type derived from the melodic inflections in Hungarian gypsy music (ex.6*g*) or of that of the Romantic style, with its sinuous counterpoint and enharmonic ambiguities (ex.6*h*).

CHAPTER TEN

Conclusion

In many ways Liszt may be considered to epitomize the 19th century, and the various aspects of his character are naturally reflected in the many different kinds of music that he wrote. He carried a walking-stick on which were carved the heads of St Francis of Assisi, Gretchen and Mephistopheles – an apt symbol of his longing for spiritual and ethical meaning in life, his love and reverence for women and his cynical diabolism. His enjoyment of purely physical sensation led to the glitter of the studies and rhapsodies; yet the late pieces have a speculative, inward quality (though even at the end of his life he was capable of brilliance, as in the *Boccanegra* fantasia). He was always searching for the new in music, not only in his own compositions but also by helping others, as conductor, arranger, pianist or writer: the New German school was entirely Liszt's conception, not Wagner's. He absorbed the standard Classical forms of music, and went on to develop new forms, partly – though not wholly – inspired by literature, painting or nature. It is a procedure with its dangers, and not all Liszt's attempts in this direction were equally successful; but he did feel it his mission to heighten man's experience and at the same time embody it in all its manifestations – the quest for the spiritual, the knowledge of the diabolical, the ceaseless exploration in spite of loneliness and insecurity. Liszt stands at the centre of the 19th

century, not only in music and the arts, but as a person- ality; like Mahler, he felt that music should embrace the world, and he cast his net as wide as possible. Princess Sayn-Wittgenstein wrote prophetically: 'He has hurled his lance much further into the future than Wagner'.

WORKS

Editions: F. Liszt: *Musikalische Werke*, ed. F. Busoni, P. Raabe and others (Leipzig, 1907–36/*R*) [B]
Liszt Society Publications (London, 1950–) [LS]
Opernïe transkriptsii dlya fortep'yano, ed. V. Belov and K. Sorokin (Moscow, 1958–) [RS]
F. Liszt: *Neue Ausgabe sämtlicher Werke/New Edition of the Complete Works*, 1st ser., ed. Z. Gárdonyi, I. Sulyok and I. Szelényi (Kassel and Budapest, 1970–) [NA]

Unless otherwise stated printed works were published in Leipzig; for fuller lists, including works planned, works by other composers edited by Liszt, and additional information on sources and early performances of works, see H. Searle: 'Liszt, Ferencz', *Grove 5* [S], and P. Raabe: *Franz Liszt* (Stuttgart, 1931, rev. 2/1968) [R] ('a' etc denotes supplementary material)

* – *autograph*   † – *autograph not traced*   fs – *full score*   vs – *vocal score*

Numbers in the right-hand column denote references in the text.

S	R	Title, genre	Composition; first performance	MSS; publication; dedication	Edition	
		STAGE				
1	476	Don Sanche, ou Le château d'amour (opera, 1, Mme Théaulon and de Rancé, after Claris de Florian)	1824–5, collab. Paer; Paris, Opéra, 17 Oct 1825	*F-Po, A-Wn*; lib pubd Paris, ?1825		244, 311 240, 311
		SACRED CHORAL				305–10
2	477	Die Legende von der heiligen Elisabeth (oratorio, O. Roquette), S, A, 3 Bar, B, chorus, orch, org	1857–62; Budapest, 15 Aug 1865; excerpts arr. pf 4 hands, 578	†; vs 1867, fs 1869; ded. Ludwig II of Bavaria	NA xvi (3 pf arrs.)	248, 250, 254–5, 305, 306–7, 310
3	478	Christus (oratorio, 3, Bible and Catholic liturgy), S, A, T, Bar, B, chorus, org, orch	planned 1853, 1855, 1859, rev. 1862–6; Weimar, 29 May 1873; excerpts arr. pf 4 hands, 579, org, 664; see also 25, 29, 36 and 261	*GB-Lbm, D-WRgs, H-Bn* (no.12 only); fs and vs 1872	NA xvi (2 pf arrs.)	248, 305, 306, 307–8, 309
4	479	Cantico del sol di S Francesco d'Assisi, Bar, male vv, orch, org	1862, rev. 1880–81; Rome, 1862; arr. pf, 499; see also 175/1, 183/1, 665, 677	Bl (sketches); fs and vs 1884; ded. A. Freiherr Senfft von Pilsach	B v/5	308
5	480	Die heilige Cäcilia (Mme E. de Girardin), legend, Mez, chorus ad lib, orch/pf	1874; Weimar, 17 June 1875; 2 earlier settings, 1845 and 1868–9, lost	*F-Pc, H-Bn* (frag.), Bl; fs and vs 1876; ded. L. Haynald, Archbishop of Kálocsa		

6	482	Die Glocken des Strassburger Münsters (Longfellow), Mez, Bar, chorus, orch	1874; Budapest, 10 March 1875; prelude arr. pf, 500, pf 4 hands, 580, org, 666; see also 135	fs and vs 1875; ded. Longfellow	—	280
7	481	Cantantibus organis, antiphon for the feast of St Cecilia, solo vv, chorus, orch	1879; Rome, 1880	†; fs Rome, 1880, vs Rome, 1881	B v/5	
8/1	485	Mass, 4 male vv, org 1st version	1848, rev. 1859; Weimar, 15 Aug 1852	1853; ded. Pater Albach	—	305
8/2		2nd version	1869; Jena, June 1872	*F-Pc; Paris, 1869	B v/3	
9	484	Missa solemnis zur Einweihung der Basilika in Gran, S, A, T, B, chorus, orch	1855, rev. 1857–8; Gran [Esztergom], 31 Aug 1856	US-Wc, *H-Bn (frag.); fs Vienna, 1859	—	248, 305, 306
10	486	Missa choralis, chorus, org	1865; Lwów, 1869	1869	B v/3	248, 308
11	487	Hungarian Coronation Mass, S, A, T, B, chorus, orch	1867; Budapest, 8 June 1867; excerpts arr. pf, 501, pf 4 hands, 581, vn, org, 678, vn, pf, 381, vn, orch, 362, org, 667; see also 15a	*and other MSS in D-WRgs, H-Bn, A-Wn, frag. in private collection of Mrs Franklin Geist (New York); fs 1869	—	249, 301, 305, 306
12	488	Requiem, 2 T, 2 B, male vv, org, brass ad lib	1867–8; Lwów, 1869; see also 45, 266	Paris, 1869	B v/3	302, 308
13	489	Psalm xiii, T, chorus, orch	1855, rev. 1858, 1862; Berlin, Singakademie, 6 Dec 1855	D-WRgs; fs (with vs) 1864, vs 1878; ded. P. Cornelius	—	248, 305
14	490	Psalm xviii, male vv, orch/org/ (ww, brass)	Aug 1860; Weimar, 25 June 1861	*and other MSS in WRgs, US-SPmoldenhauer; fs and vs 1871; ded. Princess Sayn-Wittgenstein	—	305
15/1 15/2	491	Psalm xxiii, 2 versions for T/S, harp/pf, org/harmonium for T/S, male vv ad lib, harp/pf, org/harmonium	1859, rev. 1862	*D-WRgs; 1864 *and other MSS in WRgs		
15a	—	Psalm cxvi, male vv, pf	1869; added to 11 as grad, 1869	*private collection of Mrs Franklin Geist (New York)	—	
16/1	492	Psalm cxxix, 2 versions for Bar, male vv, org	Nov 1881; added to 688	*and other MSS in H-Bn; pubd in NZM, 1880s, suppl.	—	303, 318–19
16/2		for B/A, pf/org	1880	*H-Bn, US-Wc; 1883		
17	493	Psalm cxxxvii, S, female vv, vn, harp, org	1859, rev. 1862	*D-WRgs; 1864	—	

S	R	Title, genre	Composition; first performance	MSS; publication; dedication	Edition	
18	506	Five Choruses: 1 Qui m'a donné, 3 equal vv; 2 L'Eternel est son nom (Racine), mixed vv; 3 Chantons, chantons l'auteur, mixed vv; 4 untexted, mixed vv; 5 Combien j'ai douce souvenance (Chateaubriand), mixed vv	1840s; a 6th chorus probably lost	*WRgs (nos.1, 3, 4), *F-Pc (nos.2, 5); unpubd	—	
19	508	Hymne de l'enfant à son réveil (Lamartine), female vv, harmonium/pf, harp ad lib	sketched as 173/6, Nov 1847, rev. 1862, 1865, 1874; Weimar, 17 June 1875	*D-WRgs, H-Bn and private collection of J. Vallier (London); Budapest, 1875; ded. Liszt Choral Society (Budapest)	B v/5	271
20/1	496	Ave Maria I, 2 versions	1846	Vienna, 1851 with 21/1	B v/6	271
20/2		B♭, for chorus, org	c1852; see also 173/2 and 264	†; 1852 with 21/2; ded. Pater Albach	B v/6	
		A, for 4vv, org				
21/1	518a	Pater noster II, 2 versions	1846	†; Vienna, 1851 with 20/1	—	271
21/2	518b	for male vv	c1848; arr. pf, 173/5	†; 1852 with 20/2; ded. Pater Albach	B v/6	
		for 4 equal vv, org				
22	520	Pater noster IV, mixed vv, org	1850	*D-WRgs; unpubd	B v/5	
23	504	Domine salvum fac regem, T, male vv, org/orch	1853, orchd Raff	fs B	—	
24	—	Te Deum II, male vv, org	?1853	*F-Pc; B	B v/7	
25	529	Die Seligkeiten, Bar, mixed vv, org	1855–9, Weimar, 2 Oct 1859; added to 3	1861; ded. Princess Sayn-Wittgenstein	B v/6	
26	505	Festgesang zur Eröffnung der zehnten allgemeinen deutschen Lehrerversammlung (Hoffmann von Fallersleben), male vv, org	1858; Weimar, 27 May 1858	†; 1859; ded. German School-teachers' Association	B v/6	
27	533	Te Deum I, mixed vv, org, brass and drums ad lib	May 1867	*US-PHci; B	B v/7	
28	494	An den heiligen Franziskus von Paula, prayer, solo male vv, male chorus, harmonium/org, 3 trbn and timp ad lib	before 1861, rev. c1874; see also 175/2	*Wc; Budapest, 1875	B v/5	

29	519	Pater noster I, 4vv, org	before 1861; Dessau, 25 May 1865; added to 3	1864	B v/6
30	526	Responses and antiphons, 4vv, ? org acc.	1860	B	B v/7
		Christus ist geboren I (T. Landmesser), Christmas carol, 2 versions			
31/1	536a	for mixed vv, org	?1863	Berlin, 1865	B v/6
31/2	536c/1	for male vv, org	?1863	†; Berlin, 1865	B v/6
		Christus ist geboren II, 3 versions			
32/1	536b	for mixed vv, org	?1863; arr. pf, 502	Berlin, 1865	B v/6
32/2	536c/2	for male vv unacc., org postlude	?1863	†;Berlin, 1865	
32/3	536d	for SSA unacc.	?1863	†; Berlin, 1865	
33	531	Slavimo slavno slaveni! (U. Pucić), male vv, org	1863, rev. 1866; Rome, 3 July 1863; arr. pf, 503, org, 668	*D-WRgs, CS-Pnm; B	B v/6
		Ave maris stella, 2 versions			
34/1	499	for mixed vv, org	?1865–6	†; 1870	B v/6
34/2		for male vv, org/harmonium	1868; arr. pf, 506, 1v, pf/harmonium, 680, org, 669	*F-Pc; Paris, 1868	
35	501	Crux! (Guichon de Grandpont), sailors' hymn, male vv unacc. or female/children's vv, pf	1865	*Pc, *US-Wc (frag.); Brest, 1865	B v/6
36	502	Dall'alma Roma, 2vv, org	1866; based on 'Die Gründung der Kirche' from 3	*D-WRgs, GB-Lbm (no.2), unpubd	—
37	513	Mihi autem adhaerere (from Ps lxxiii), male vv, org	1868	†; 1871, with 34/1–2, 38, 40, 41/1, 42/1, 43, 44 and 45 as 9 Kirchenchorgesänge	B v/6
38	497	Ave Maria II, mixed vv, org	1869; arr. pf/harmonium, 504, 1v, org/harmonium, 681; see also 264	Regensburg, 1870 with 41/1; ded. Mme J. Laussot	B v/6
39	510	Inno a Maria Vergine, mixed vv, harp, org/(pf 4 hands, harmonium)	1869	B	B v/5
40	516a	O salutaris hostia I, female vv, org	1869	*D-WRgs; Regensburg, 1871 with 42/1; ded. F. X. Haberl	B v/6
		Pater noster III, 2 versions			
41/1	521a	F, for mixed vv, org/pf	1869	*and other MSS in WRgs 1871, in 9 Kirchenchorgesänge	B v/6
41/2	521b	B♭, male vv, org/harmonium/pf		B	B v/6
	532	Tantum ergo	1869		B v/6

S	R	Title, genre	Composition; first performance	MSS; publication; dedication	Edition
42/1		for male vv, org		Regensburg, 1871 with 40; ded. F. X. Witt	
42/2		for female vv, org		B†; 1871, in 9 Kirchenchorgesänge	B v/6
43	516b	O salutaris hostia II, mixed vv, org	?c1870	*private collection of J & J Lubrano S. Lee, Mass., 1871, in 9 Kirchenchorgesänge	B v/6
44	500	Ave verum corpus, mixed vv, org ad lib	1871	*H-Bl; 1871, in 9 Kirchenchorgesänge	—
45	511	Libera me, male vv, org	1870; added to 12		
	495	Anima Christi sanctifica me, male vv, org			
46/1		1st version	June 1874	*F-Pc; B; ded. Pater J. Mohr †; 1882, with 34/1–2, 37, 38, 40, 41/1, 42/1, 43, 44 and 45 as 12 Kirchenchorgesänge	B v/6
46/2		2nd version	?c1874		B v/6
47	483	Sankt Christoph, legend, Bar, female vv, pf, harmonium, harp ad lib	1881	D-WRgs; unpubd	—
48	503	Der Herr bewahret die Seelen seiner Heiligen, Festgesang zur Enthüllung des Carl-August-Denkmals in Weimar	1875; Weimar, 3 Sept 1875	†; 1887, with 92	B v/6
49	535	O heilige Nacht, Christmas song, T, female chorus, org/harmonium	after 1876; Rome, 25 Dec 1881; arr. of 186/2	Berlin, 1882	B v/6
50	72	Zwölf alte deutsche geistliche Weisen	?1878–9	*WRgs; B (nos.1–7 only), others unpubd; *nos.8, 9, 12, in Stargardt auction catalogue (Marburg, 1975); MSS (nos.1, 2, 9), private collection of Friedrich Schnapp, Eschenberg	B v/7 [nos. 1–7]; NA x
		1 Es segne uns Gott, mixed vv, org			
		2 Gott sei uns gnädig, mixed vv, org	see 51		
		3 Nun ruhen alle Wälder, chorus/pf			
		4 O Haupt voll Blut, chorus/pf	added to 53		
		5 O Lamm Gottes, chorus/pf	arr. pf 4 hands, 582		
		6 Was Gott tut, chorus/pf			
		7 Wer nur den lieben Gott, chorus/pf			

No.	S	Title, scoring	Composition / notes	Sources, remarks	Edn	bb
		8 Vexilla regis, chorus/pf / 9 Crux benedicta, chorus/pf / 10 O Traurigkeit, chorus/pf / 11 Nun danket alle Gott, chorus/pf / 12 Jesu Christe, chorus/pf	added to 53 / added to 53 / added to 53 / see 61	*WRgs; unpubd; ded. R. Pohl	—	
51	507	Gott sei uns gnädig und barmherzig (Meine Seel' erhebt den Herrn!), mixed vv, org	1878; based on 50/2			
52	530	Septem sacramenta, responsories, Mez, Bar, mixed vv, org: 1 Baptisma, 2 Confirmatio, 3 Eucharistia, 4 Penitentia, 5 Extrema unctio, 6 Ordo, 7 Matrimonium	1878; Weimar, 10 July 1879 (nos.3 and 7, Vienna, 8 April 1879)	*of no.7 at Editio Musica, Budapest; B	B v/7	
53	534	Via crucis, les 14 stations de la croix, solo vv, chorus, org/pf	sketched 1866, rev. 1878–9 Budapest, Good Friday 1929, arr. pf, 504a, org, 674a pf 4 hands, 583	*H-Bn, Bl, D-WRgs; B	B v/7	310
54	514	O Roma nobilis, mixed vv, org ad lib, or 1v, org	1879; arr. pf, 546a	*I-Rsc, D-WRgs; B	B v/7	
55	517	Ossa arida, unison male vv, org 4 hands/pf 4 hands	1879	B	B v/6	308, 317
56	527	Rosario: 1 Mysteria gaudiosa, 2 Mysteria dolorosa, 3 Mysteria gloriosa, all mixed vv, org/harmonium, 4 Pater noster, Bar/unison male vv, org/harmonium	Nov 1879; nos.1–3 arr. org. 670	B	B v/7	
57	509	In domum Domini ibimus, mixed vv, org, brass, drums	after 1880; prelude arr. pf, 505, org, 671	*H-Bn; B	B v/5	
58	515	O sacrum convivium, A solo, female vv ad lib, org/harmonium	after 1880	D-WRgs; B	B v/6	
59	523	Pro Papa: I Dominus conservet eum, mixed vv, org, 2 Tu es Petrus, unison male vv, org	?1880	†; Rome, 1881; ded. Pope Leo XIII	B v/6	

S	R	Title, genre	Composition; first performance	MSS; publication; dedication	Edition	
60	498	Zur Trauung (Ave Maria III), org/harmonium, unison female vv ad lib	1883; arr. of 161/1	*and other MS in *W-Rgs*; 1890	B v/6	304
61	408	Nun danket alle Gott, org (male/mixed vv, brass, drums ad lib)	1883	†; 1884; ded. C. Hase	B v/7	
62	512	Mariengarten (Quasi cedrus), SSAT, org	before 1885	†; B	B v/6	
63	525	Qui seminant in lacrimis, mixed vv, org	1884	*Göllerich collection; B	B v/6	308, 318–19
64	522	Pax vobiscum!, male vv, org	1885	Zurich, 1885-6; ded. Strasbourg Male Choir	B v/6	
65	524	Qui Mariam absolvisti, Bar, unison mixed vv, org/harmonium	1885	1886, in Der Chorgesang	B v/6	
66	528	Salve regina, mixed vv	1885	*W-Rgs*; B	B v/6	
		SECULAR CHORAL				244, 311–2
67	537	Festkantate zur Enthüllung des Beethoven-Denkmals in Bonn (O. L. B. Wolff), 2 S, 2 T, 2 B, chorus, orch	1845; Bonn, 13 Aug 1845; arr. pf, 507, pf 4 hands, 584	*D-W-Rgs*; unpubd		246, 288, 311
68	538	Zur Säkularfeier Beethovens (2nd Beethoven cantata) (A. Stern and F. Gregorovius), S, A, T, B, double chorus, orch	1869–70; Weimar, 29 May 1870	*?Bds*, *frags. in *GB-Lbm*, *H-Bn*; fs and vs 1870; ded. Grand Duchess Sophie of Saxe-Weimar	—	246, 288, 312
69	539	Chöre zu Herders Entfesseltem Prometheus, S, A, 2 T, 2 B, double chorus, orch	1850, orchd Raff, rev. 1855; Weimar, 24 Aug 1850; Pastorale arr. pf, 508, pf 4 hands, 585	fs Weimar, 1855	—	292, 312
70	540	An die Künstler (Schiller), 2 T, 2 B, male chorus, orch	1853, orchd Raff, 1st rev. late 1853, 2nd rev. 1856; Karlsruhe, June 1853; see also 114	fs and vs Berlin, 1854 (1st rev.), fs and vs Leipzig, 1856 (2nd rev.)	—	312
71	541	Gaudeamus igitur, humoresque, solo vv ad lib, male/mixed chorus, orch	1869; Jena, 1870; arr. pf, 509, pf 4 hands, 586	†; 1871; ded. Dr C. Gille	—	

72	542	Vierstimmige Männergesänge	1841	Berlin, 1843 *Bn*; ded. J. Lefebvre	
		1 Rheinweinlied (Herwegh), with pf	Jena, 30 Nov 1841		
		2 Studentenlied aus Goethes Faust, unacc.	Leipzig, 6 Dec 1841	†; ded. W. Speier	
		3 Reiterlied (Herwegh), 1st version with pf	Paris, 1842 (1st perf. of no.3 or no.4)	ded. S. Teleky	
		4 Reiterlied, 2nd version, unacc.	see 72/3 for 1st perf.	†	
73	543	Es war einmal ein König (Goethe), B, male vv, pf	1845	*D-WRgs*; unpubd	
74	545	Das deutsche Vaterland (E. M. Arndt), 4 male vv	1839; Leipzig, Dec 1841	*US-NYp, Phci* (later version), *D-WRgs*; Berlin, 1843; ded. Friedrich Wilhelm IV of Prussia	
75/1	544	Über allen Gipfeln ist Ruh (Goethe), 2 versions for male vv	1842; arr. 1v, pf, 306/1	ded. Prince F. W. Constantin von Hohenzollern-Hechingen Cologne, 1844, with 76, 90/3 and 90/12	
75/2		for male vv, 2 hn	1849	Hamburg, 1849, with 84, 85, 227 and 303 as Festalbum zur Säkularfeier von Goethes Geburtstag	
76	546	Das düstre Meer umrauscht mich, male vv, pf	1842	†; Cologne, 1844, with 75/1, 90/3 and 90/12	
77	551	Die lustige Legion (A. Buchheim), male vv, pf ad lib	1846	†; arr. T, pf, chorus, pubd Vienna, 1848	
78	550	Trinkspruch, male vv, pf	1843	†; pubd in *ZfM*, x (1929); ded. male-voice quartet in the Stubenvoll (Munich)	
79	549	Titan (F. von Schober), Bar, male vv, pf	1842, rev. 1845 and 1847, orchd Conradi c1848	*and other MSS in *WRgs*; unpubd	
80	547	Les quatre élémens (J. Autran), male vv, pf/orch; 1 La terre, 2 Les aquilons, 3 Les flots, 4 Les astres	1839 (no.2), 1845 (nos.1, 3, 4), orchd Conradi 1848; no.2 perf. Marseilles, Aug 1844; see also 97	*and other MSS in *WRgs*; unpubd	291, 311
81	548	Le forgeron (Lamennais), male vv, pf/orch	1845, orchd Conradi 1848	vs ed. I. Szelényi (Budapest, 1962)	
81a	564a	A patakhoz [To the brook] (J. Garay), male vv	1846	†; pubd in Apollo, iii (1874), 13	

S	R	Title, genre	Composition; first performance	MSS; publication; dedication	Edition	
82	552	Arbeiterchor (? trans. of Lamennais), B solo, 4 male vv, male chorus, pf	before 1849; arr. pf, 510, pf 4 hands, 587	†; ed. Budapest, 1954	—	292, 311
83	553	Hungaria 1848 (Schober), cantata, S, T, B, male vv, pf/orch	1848, orchd Conradi; orch version perf. Weimar, 21 May 1912	†; vs ed. I. Szelényi (Budapest, 1961)	—	293, 311–12
84	554	Licht, mehr Licht (?Schober), male vv, brass	1849	*private collection of Mrs Franklin Geist (New York); pubd in Leipziger illustrierte Zeitung (25 Aug 1849) and as 75/2	—	
85	555	Chor der Engel (Goethe: Faust), mixed vv, harp/pf	1849	pubd as 75/2	—	
86	556	Festchor zur Enthüllung des Herder-Denkmals in Weimar (A. Schöll), male vv, pf/orch	1850, orchd Raff; Weimar, 25 Aug 1850	vs pubd in Leipziger illustrierte Zeitung (2 Nov 1850)	—	
87	557	Weimars Volkslied (P. Cornelius), 1 male vv, wind, 2 male vv, pf ad lib, 3 male vv, 4 male vv, org, 5 S,A/TB, 6 3vv	1857, based on 357; Weimar, 3 Sept 1857; arr. 1v, pf, 313, pf, 542, pf 4 hands, 588, org, 672	*H-Bn (no.5), D-WRgs (no.4); 1871	—	
88	558	Morgenlied (Hoffmann von Fallersleben), female vv	1859	†; pubd in Mädchenlieder (Weimar, 1861)	—	
89	559	Mit klingendem Spiel, children's vv	c1859	pubd in Vaterländisches Liederbuch, ed. A. W. Gottschalg and others (Weimar, 1860)	—	
90	560	Für Männergesang	1842–59	*(nos.1–7, 10) and other MSS (nos.8, 11) in W-Rgs; *(no.9), H-Bn; entire collection pubd 1861	—	
		1 Vereinslied (Hoffmann von Fallersleben)	1856			
		2 Ständchen (Rückert), with T solo	before 1858			
		3 Wir sind nicht Mumien (Hoffmann von Fallersleben), orig. with pf acc.	1842	Cologne, 1844, with 75/1, 76 and 90/12		
		Geharnischte Lieder (T. Meyer), orig. with pf acc.: 4 Vor der Schlacht ('Es rufet Gott'), 5 Nicht gezagt, 6 Es rufet Gott	1845, rev. 1860; arr. pf, 511	?Leipzig, 1845; 4 ded. C. Brenner, 5 ded. A. Müller, 6 ded. 'Herr Architekt Heimlicher'		

S	R	Title, genre	Composition: performance	Arrangements	MSS: publication: dedication	Edition	
		7 Soldatenlied aus Goethes Faust, tpts and timp ad lib	6 July 1844				
		8 Die alten Sagen kunden, with 4 solo vv	c1845				
		9 Saatengrün (Uhland)	?c1845				
		10 Der Gang um Mitternacht (Herwegh), with T solo	?c1845		pubd in *Leipziger illustrierte Zeitung* (12 Nov 1859)		
		11 Festlied zu Schillers Jubelfeier (Dingelstedt), with Bar solo	1859		†; Cologne, 1844, with 75/1, 76 and 90/3		
		12 Gottes ist der Orient (Goethe)	1842				
91	561	A lelkesedés dala – Das Lied der Begeisterung	1871, rev. 1874 (2nd version)		*US-Wc*; Budapest, 1871: ded. Hungarian Choral Society	—	
92	562	Carl August weilt mit uns, Festgesang zur Enthüllung des Carl-August-Denkmals in Weimar, male vv, brass, drums and org ad lib	1875; Weimar, 3 Sept 1875		*and other MSS in *D-WRgs*; 1887, with 48	—	
93	563	Magyar király-dal – Ungarisches Königslied (K. Ábrányi), male vv, or mixed vv, or male vv, pf, or mixed vv, pf/orch, or orch, male/mixed vv ad lib, or children's vv	1883; Bratislava, 1884; see also 340		*frag. and other MSS in *H-Ba(mi)*; Budapest, 1884		
94	564	Grüss, male vv	?1885		†(photo of autograph in *D-WRgs*); pubd in *Der Chorgesang*, i (Leipzig, 1885); ded. Riga Liedertafel		254, 288–300

ORCHESTRAL

S	R	Title, genre	Composition: performance	Arrangements	MSS: publication: dedication	Edition	
95	412	Ce qu'on entend sur la montagne ('Bergsymphonie'), sym. poem, after Hugo	1848–9, orchd Raff, 1st rev. 1850, orchd Raff, 2nd rev. spr. 1854; Weimar, Feb 1850	pf 4 hands, 589, 2 pf, 635	*Breitkopf & Härtel archive, Wiesbaden, *D-WRgs*; 1857; ded. Princess Sayn-Wittgenstein	B i/1	289–90

S	R	Title, genre	Composition: performance	Arrangements	MSS: publication: dedication	Edition	
96	413	Tasso: lamento e trionfo, sym. poem, after Byron	1841–5, orchd Conradi, 1st rev. 1850–51, orchd Raff, 2nd rev. 1854; Weimar, 28 Aug 1849	pf 4 hands, 590, 2 pf, 636	*WRgs (sketches); 1856; ded. Princess Sayn-Wittgenstein	B i/1	269, 288, 290–91, 300
97	414	Les préludes, sym. poem, after Lamartine	1848, as introduction to 80, rev. before 1854; Weimar, 23 Feb 1854	pf 4 hands, 591, 2 pf, 637	1856; ded. Princess Sayn-Wittgenstein	B i/2	261, 270, 288, 291
98	415	Orpheus, sym. poem	1853–4; Weimar, 16 Feb 1854	pf 4 hands, 592, 2 pf, 638, chamber ens, 723a	*H-Bn (frag.); 1856; ded. Princess Sayn-Wittgenstein	B i/2	288, 291–2
99	416	Prometheus, sym. poem	1850, orchd Raff, rev. 1855; Weimar, 24 Aug 1850	pf 4 hands, 593, 2 pf, 639	1856; ded. Princess Sayn-Wittgenstein	B i/3	288, 292
100	417	Mazeppa, sym. poem, after Hugo	1851, orchd with Raff, rev. before 1854; Weimar, 16 April 1854; based on a pf study (see 136–9)	pf 4 hands, 594, 2 pf, 640	*D-WRgs, F-Pc; 1856; ded. Princess Sayn-Wittgenstein	B i/3	289, 292, 293
101	418	Festklänge, sym. poem	1853; Weimar, 9 Nov 1854	pf 4 hands, 595, 2 pf, 641	1856; ded. Princess Sayn-Wittgenstein	B i/4	292–3
102	419	Héroïde funèbre, sym. poem	1849–50, orchd Raff, rev. c1854; Breslau, 10 Nov 1857; based on 1st movt of 690	2 pf, 642, pf 4 hands, 596a	†; 1857; ded. Princess Sayn-Wittgenstein	B i/4	264, 289, 293
103	420	Hungaria, sym. poem	sum. 1854; Budapest, National Theatre, 8 Sept 1856	pf 4 hands, 596, 2 pf, 643	1857; ded. Princess Sayn-Wittgenstein	B i/5	270, 289, 293–4
104	421	Hamlet, sym. poem	1858; Sondershausen, 2 July 1876	pf 4 hands, 597, 2 pf, 644	*D-WRgs, H-Bn (frag.); 1861; ded. Princess Sayn-Wittgenstein	B i/5	288, 294
105	422	Hunnenschlacht, sym. poem, after W. von Kaulbach	completed 10 Feb 1857; Weimar, 29 Dec 1857	2 pf, 645, pf 4 hands, 596b	1861; ded. Princess Sayn-Wittgenstein	B i/6	289, 294–5
106	423	Die Ideale, sym. poem, after Schiller	1857; Weimar, 5 Sept 1857; see also 114	2 pf, 646, pf 4 hands, 596c	*D-WRgs; 1858; ded. Princess Sayn-Wittgenstein	B i/6	295
107	424	Von der Wiege bis zum Grabe–Du berceau jusqu'à la tombe, sym. poem	1881–2; 1st part based on 198	pf, 512, pf 4 hands, 598	*F-Pc (2nd part only); Berlin, 1883; ded. Count Michael Zichy	B i/10	289, 300, 301

108	425	Eine Faust-Symphonie in drei Charakterbildern, after Goethe, T, male vv, orch	planned 1839, 1854, 1857; Weimar, 5 Sept 1857	pf, 513 (2nd movt), 2 pf, 647	*H-Bn, private collection of Mrs Jephta Drachman (Stevenson, Maryland) (frag. 2nd movt); 1861; ded. H. Berlioz	B i/8 9	260–61, 264, 273, 295–8
109	426	Eine Symphonie zu Dantes Divina commedia	planned 1839, 1855–6; Dresden, 7 Nov 1857	2 pf, 648	*US-Wc (frag.); 1859; ded. R. Wagner	B i/7	295, 298
110	427	Two episodes from Lenau's Faust: 1 Der nächtliche Zug, 2 Der Tanz in der Dorfschenke (First Mephisto Waltz)	1860–61; Weimar, 8 March 1861 (no.2 only)	pf, 514 (no.2), pf 4 hands, 599	*private collection of Baron von Vietinghoff-Scheel (Berlin) (no.2 only), D-WRgs; ded. C. Tausig	B i/10	298, 317
111	428	Second Mephisto Waltz	1880–81; Budapest, 9 March 1881	pf, 515, pf 4 hands, 600	H-Bn, Berlin, 1881; ded. C. Saint-Saëns	B i/10	300
112	429	Trois odes funèbres 1 Les morts (Lamennais), male vv ad lib	1860–66; Weimar, 21 May 1912	org, 268/2, pf, 516, pf 4 hands, 601	*D-WRgs, F-Pc; B: ded. C. von Bülow	B i/12	299–300 248, 299, 304
		2 La notte, after Michel-angelo	1863–4, based on 161/2; Weimar, 6 Dec 1912	vn, pf, 377a, pf 4 hands, 602, pf, 699	H-Ba(mi); B	B i/12	249, 299, 301
		3 Le triomphe funèbre du Tasse	1866, as epilogue to 96; New York, March 1877	pf, 517, pf 4 hands, 603	1877; ded. L. Damrosch	B i/2	249, 300
113	430	Salve Polonia	1863, as interlude for 688; Weimar, 1884	pf, 518, pf 4 hands, 604	*US-Wc; 1884	—	299
114	432	Künstlerfestzug zur Schillerfeier 1859	1857, based on themes from 70 and 106; Weimar, 8 Nov 1860	pf, 520, pf 4 hands, 605	1860	B i/11	
115	433	Festmarsch zur Goethejubiläumsfeier	1849, orchd Conradi, 1st rev. orchd Raff, 2nd rev. 1857; Weimar, 28 Aug 1849; based on 227	pf, 521, pf 4 hands, 606	†; 1859	B i/11	
116	436	Festmarsch nach Motiven von E. H. zu S.-C.-G., on themes from Duke Ernst of Saxe-Coburg-Gotha's Diana von Solange	before 1860	pf, 522, pf 4 hands, 607	†; 1860, as Coburg Festmarsch no.2, with 115	—	

S	R	Title, genre	Composition: performance	Arrangements	MSS; publication; dedication	Edition	
117	439	Rákóczy March	1865, rev. 1867, based on themes from 242/13 and 1st version of 244/15; Budapest, 17 Aug 1875	pf, 2nd version of 244/15, pf 4 hands, 608	*SM; 1871		312
118	438	Ungarischer Marsch zur Krönungsfeier in Ofen-Pest am 8. Juni 1867	1870	pf, 523, pf 4 hands, 609	*Wc; 1871	B i/12	
119	437	Ungarischer Sturmmarsch	1875, based on 232	pf, 524, pf 4 hands, 610	*H-Bl; Berlin, 1876; ded. Count S. Teleky	B i/12	
		PIANO AND ORCHESTRA					287
120	453	Grande fantaisie symphonique, on themes from Berlioz's Lélio	1834; Paris, April 1835	—	D-WRgs; unpubd	—	242, 287
121	452	Malédiction, pf, str orch	1833	—	*and other MSS in WRgs; B	B i/13	287, 298
122	454	Fantasie über Motive aus Beethovens Ruinen von Athen	c1837, rev. 1849; Budapest, 1 June 1853	pf, 389, 2 pf, 649	*WRgs, US-Wc; 1865; ded. N. Rubinstein	—	287
123	458	Fantasie über ungarische Volksmelodien	?1852; Budapest, 1 June 1853	—	*US-Wc(frag.); 1864; ded. H. von Bülow	—	
124	455	Concerto no.1, E♭	sketched Jan 1832, rev. 1849, 1853, 1856; Weimar, 17 Feb 1855	2 pf, 650	*and other MSS in D-WRgs, US-Wc, NYpm; Vienna, 1857; ded. H. Litolff	B i/13	265, 287
125	456	Concerto no.2, A	1839, revs 1849–61; Weimar, 7 Jan 1857	2 pf, 651	D-WRgs; Mainz, 1863; ded. H. von Bronsart	B i/13	287
126	457	Totentanz	planned 1839, 1849, rev. 1853 and 1859; 1st rev. related to 691; The Hague, 15 April 1865	pf, 525, 2 pf, 652	*and other MSS in WRgs, US-NYpm; 1865; ded. H. von Bülow	B i/13	243, 287
126a	—	Piano Concerto in the Hungarian Style	1885; Russia, 1892	—	†; New York, 1909, as Hungarian Gypsy Songs, arr. S. Menter, orchd P. Tchaikovsky		

CHAMBER MUSIC

S	R	Title	Composition	Publication	Autograph: dedication; remarks	Edition	
							300–02
127	461	Duo (Sonata), vn, pf	c1832–5, based on Chopin's Mazurka op.6 no.2	—	D-WRgs; ed. New York, 1964	—	300–01
128	462	Grand duo concertant, on Lafont's Le marin, vn, pf	1835, rev. 1849	—	*H-Bl, D-WRgs; Mainz and Paris, 1852	—	301
129	466	Epithalam zu E. Reményis Vermählungsfeier, vn, pf	1872	pf, 526, pf 4 hands, 611	†; Budapest, 1873	—	301
130	471	[Erste] Elegie, vc, pf, harp, harmonium/(vc, pf)/(vn, pf)	1874; Weimar, 17 June 1875	pf, 196, pf 4 hands, 612	†: 1875–6; in memory of M. Mukhanov	—	301
131	472	Zweite Elegie, vn/vc, pf	1877	pf, 197	*US-NYp; 1878; ded. L. Ramann	—	301
132	467	Romance oubliée, va/vn/vc, pf	1880	pf, 527	*H-Bn; Hanover, 1881; ded. H. Ritter	—	
133	475	Die Wiege, 4 vn	?1881	—	†: unpubd	—	301
134	468	La lugubre gondola, vn/vc, pf	1885	pf, 200/2	D-WRgs; ed. Budapest, 1974	—	250, 280, 281
135	474	Am Grabe Richard Wagners, str qt, harp ad lib	1883, based on 6	pf, 202, org, 267	*US-NYpm; LS	LS ii	280

PIANO SOLO

S	R	Title	Composition	Publication	Autograph: dedication; remarks	Edition	
							244, 265–86
		studies and exercises					
136	1	Etude en douze exercices	1826	Paris, 1826, as op.6	†; ded. L. Garella; pubd orig. as Etude en quarante-huit exercices	B ii/1	240, 243, 265, 266
137	2a	Vingt-quatre grandes études	1837	Berlin and elsewhere, 1839	*H-Bn (nos.1, 7); ded. C. Czerny; based on 136, only 12 studies composed	B ii/1	243, 272
138	2c	Mazeppa	sketched 1829, 1840	Berlin and Vienna, 1847	*D-WRgs; ded. V. Hugo; based on 137/4	B ii/1	265, 266

S	R	Title	Composition	Publication	Autograph; dedication; remarks	Edition	
139	2b	Etudes d'exécution transcendante [Transcendental Studies]: 1 Preludio, 2 in a, 3 Paysage, 4 Mazeppa, 5 Feux follets, 6 Vision, 7 Eroica, 8 Wilde Jagd, 9 Ricordanza, 10 in f, 11 Harmonies du soir, 12 Chasse-neige	1851	1852	ded. Czerny; based on 137–8, see also 100	B ii/2; NA i	240, 260, 261, 262, 263, 266, 292
140	3a	Etudes d'exécution transcendante d'après Paganini: 1 in g, 2 in E♭, 3 La campanella, 4 in E, 5 La chasse, 6 in a	1838–9	Vienna, 1840	*H-Bn (frag.); ded. C. Schumann; based on Paganini's 24 Caprices op.1 and B minor Vn Conc., last mvt	B ii/3	243, 257, 266–7
141	3b	Grandes études de Paganini	1851	1851	*D-WRgs; ded. C. Schumann; based on 140	B ii/3; NA ii	266–7
142	4a	Morceau de salon, étude de perfectionnement	1840	Berlin, 1841	†; for Fétis's Méthode des méthodes	B ii/3	270
143	4b	Ab irato	1852	Berlin and Vienna, 1852	†; based on 142	B ii/3; NA ii	270
144	5	Trois études de concert, A♭, f, D♭	c1848	1849	*WRgs (no.1 only); ded. E. Liszt	B ii/3; NA ii	273
145	6	Zwei Konzertetüden: 1 Waldesrauschen, 2 Gnomenreigen	?1862–3	Stuttgart, 1863	*WRgs; ded. D. Pruckner; for Lebert and Stark's Klavierschule	B ii/3; NA ii	278
146	7	Technische Studien, 12 bks	1868–c1880	1886, ed. A. Winterberger	*WRgs; other technical studies lost	—	
147	26	Variation über einen Walzer von Diabelli	1822	Vienna, 1823	†; no.24 in Diabelli's Vaterländischer Künstlerverein	B ii/7	242
148	27	Huit variations	c1824	Paris, 1825, as op.1	†; ded. S. Erard	B ii/7; NA ix	

149	28	Sept variations brillantes sur un thème de Rossini	c1824	Paris and London, 1824, as op.2	†; ded. Mme Panckoucke; based on a theme from Ermione	—
150	29	Impromptu brillant sur des thèmes de Rossini et Spontini	1824	Vienna and elsewhere, 1825, as op.3	†; ded. Countess Eugénie de Noirberne; based on themes from La donna del lago, Armida, Olympie and Fernand Cortez	RS i/1
151	30	Allegro di bravura	1824	Paris and Vienna, 1825	WRgs; ded. Count Thaddeus Amadé; pubd in Vienna with 152 as op.4 arr. orch, 701a	B ii/7; NA xi
152	31	Rondo di bravura	1824	Paris and Vienna, 1825	†; ded. Count Amadé; pubd in Vienna with 151 as op.4	B ii/7; NA xi
153	19	Scherzo, g	27 May 1827	in AMz, xxii–xxiii (1896)	*auctioned by Stargardt (Marburg), 1975	B ii/9	265

miscellaneous

154	13	Harmonies poétiques et religieuses	1833, rev. 1835	Gazette musicale de Paris (7 June 1835), suppl.	*WRgs; ded. A. de Lamartine; orig. conceived for pf, orch, see also 173/4	B ii/5; NA ix, appx	242, 264, 265–6, 271
155	11	Apparitions: 1 Senza lentezza quasi allegretto, 2 Vivamente, 3 Molto agitato ed appassionato	1834	Berlin and elsewhere, 1834	†; 1 ded. Countess Clara de Rauzan, 2 ded. Viscountess Frédéric de Larochefoucauld, 3 ded. Mme la marquise de Camaran; 3 based on a waltz by Schubert, used also in 427/4	B ii/5; LS ii [nos.1, 2]; NA ix	242, 266
156	8	Album d'un voyageur, 3 bks	1835-8	Vienna and Berlin, 1842 (complete)		B ii/4; LS ii [no.1]; LS v [nos. 3–4, 7–9]; NA vi [bks I, II]	243, 267

S	R	Title	Composition	Publication	Autograph; dedication; remarks	Edition
		I Impressions et poésies: 1 Lyon, 2a Le lac de Wallenstadt, 2b Au bord d'une source, 3 Les cloches de G ...., 4 Vallée d'Obermann, 5 La chapelle de Guillaume Tell, 6 Psaume	?Sept 1837–Jan 1838	Paris, c1839	†; 1 ded. F. de Lamennais, 2b ded. F. Denis, 3 ded. Blandine Liszt, 4 ded. E. Pivert de Senancour, 5 ded. V. Schölcher; orig. pubd as Première année de pèlerinage, Suisse	
		II Fleurs mélodiques des Alpes: 7a Allegro, C, 7b Lento, e/G, 7c Allegro pastorale, G, 8a Andante con sentimento, G, 8b Andante molto espressivo, g, 8c Allegro moderato, E♭, 9a Allegretto, A♭, 9b Allegretto, D♭, 9c Andantino con molto sentimento, G	?Jan–May 1838	Paris, 1840	†; ded. Mme H. Reiset; orig. pubd as Deuxième année de pèlerinage	
		III Paraphrases: 10 Improvisata sur le ranz de vaches de F. Huber, 11 Un soir dans les montagnes, 12 Rondeau sur le ranz de chèvres de F. Huber	Nov 1835–1836	Paris, 1836	†; 10 ded. Mme A. Pictet, 11 ded. Countess Marie Potocka, 12 ded. Count Theobald Walsh	
157	9	Fantaisie romantique sur deux mélodies suisses	1836	Paris, 1836, as op.5 no.1	†; ded. V. Boissier	B ii/5; LS v
158	[10b]	Tre sonetti del Petrarca, orig. version for pf	1844–5	Vienna and elsewhere, 1846	*WRgs, H-Bn (no.1); based on 270/1; see also 161/4–6	B ii/5
159	10d	Venezia e Napoli, orig. version: 1 Lento, 2 Allegro, 3 Andante placido, 4 Tarantelles napolitaines	c1840	B	set up for publication, Vienna, c1840 (proof copy in D-WRgs)	B ii/5; NA vii 269

160	10a	Années de pèlerinage, première année, Suisse: 1 La chapelle de Guillaume Tell, 2 Au lac de Wallenstadt, 3 Pastorale, 4 Au bord d'une source, 5 Orage, 6 Vallée d'Obermann, 7 Eglogue, 8 Le mal du pays, 9 Les cloches de Genève	1848–54	Mainz, 1855	*USSR-Lsc; no.1 based on 156/5, no.2 based on 156/2a, no.3 based on, 156/7c, no.4 based on 156/2b, no.6 based on 156/4, no.8 based on 156/7b, no.9 based on 156/3	B ii/6; NA vi	243, 261–2, 266, 267
161	10b	Années de pèlerinage, deuxième année, Italie: 1 Sposalizio, 2 Il penseroso, 3 Canzonetta del Salvator Rosa, 4 Sonetto 47 del Petrarca, 5 Sonetto 104 del Petrarca, 6 Sonetto 123 del Petrarca, 7 Après une lecture du Dante, fantasia quasi sonata	1837–49	Mainz, 1858	* and other MSS in *D-WRgs*; no.1 arr. org, vv ad lib, 60, no.2 arr. orch, 112/2, vn, pf, 377a, pf 4 hands, 602. nos.4–6 based on 158; see also 270/2	B ii/6; NA vii	243, 262, 266, 267, 269, 299, 304
							298
162	10c	Venezia e Napoli, rev. version: 1 Gondoliera, 2 Canzone, 3 Tarantella	1859	Mainz, 1861, as suppl. to 161	*WRgs*; based on 159/3–4 and themes by Peruchini, Rossini and G. L. Cottrau or G. Bononcini	B ii/6; NA vii	269
163	10e	Années de pèlerinage, troisième année: 1 Angelus!, 2 Aux cyprès de la Villa d'Este (3/4), 3 Aux cyprès de la Villa d'Este (4/4), 4 Les jeux d'eau à la Villa d'Este, 5 Sunt lacrymae rerum, 6 Marche funèbre, 7 Sursum corda	1867–77	Mainz, 1883	no.1, *GB-Lbm, *H-Ba(mi)*, *I-Ria*, *US-Wc*, 1877, rev. 1880 and 1882, ded. D. von Bülow	B ii/6; NA viii	278–9
						no.2, †, 1877		
						no.3, *Wc*, 1877		279
						no.4, †, 1877		279
						no.5, *Wc*,1872, ded. H. von Bülow		
						no.6, *D-WRgs*, 1867, in memory of Emperor Maximilian I of Mexico		278
						no.7, *US-N Ypm*, 1877		279

S	R	Title	Composition	Publication	Autograph: dedication; remarks	Edition
164	64/1	Albumblatt, E	c1841	1841	*H-Bn (frag.); based on 210	—
165	62	Feuilles d'album, A♭	1841	Mainz, 1844	*D-WRgs; ded. G. Dubousquet	B ii/10
166	63	Albumblatt in Walzerform, A	1841	in Göllerich (1908), suppl.	*WRgs	B ii/10
167	64/2	Feuille d'album, a	1842	1843, with 164	*WRgs; based on 274/1	NA xvii
167a 167b	113a —	Ruhig Miniatur Lieder	?1883–6 —	unpubd —	*US-Wc MS sold at Christie's, London, 22 Oct 1980; 24 bars marked 'als Gottespfennig für Schubert (aber nicht zu publizieren)'	—
168	75	Elégie sur des motifs du Prince Louis Ferdinand de Prusse	1842, rev. c1851	Berlin and Paris, ?1843	†; ded. Princess Augusta of Prussia	NA xi
168a	64a	Andante amoroso		in Album offert aux abonnés de la Revue et gazette musicale (Paris, 1847)	†	—
169	66a	Romance	1848	Leipzig and Warsaw, 1849	D-WRgs; ded. J. Koscielska; based on 301a	LS vii NA xi
170	15	Ballade no.1, D♭	1845, rev. 1848	1909	*F-Pc, *D-WRgs; ded. Prince Eugen Wittgenstein	B ii/8; 273 NA ix
171	16	Ballade no.2, b	1853	1854	*D-MZsch; ded. Count Karl von Leiningen	B ii/8; 273 NA ix
171a	[12]	Madrigal	1844	unpubd	*private collection of E. Helm (Asolo, nr. Treviso); ded. M. Ziegsar; early version of 172/5	—
172	12	[6] Consolations	1844 (no.5), 1848 (nos.1, 2, 4, 6)	1850	*D-WRgs (nos.1, 2, 4, 6)	B ii/8; NA ix

173	14	Harmonies poétiques et religieuses: 1 Invocations, 2 Ave Maria, 3 Bénédiction de Dieu dans la solitude, 4 Pensée des morts, 5 Pater noster, 6 Hymne de l'enfant à son réveil, 7 Funérailles, 8 Miserere, d'après Palestrina, 9 Andante lagrimoso, 10 Cantique d'amour	1840 (no.2), 1845–52	1853	*A-Wgm (no.9); *D-WRgs (nos.1, 3, 6, 7–9); *US-Wc (no.8); H-Ba(mi) (no.10); ded. Princess Sayn-Wittgenstein; no.2 based on 20/2; no.4 based on 154 and 69I; no.5 based on 21/2; no.6 based on 19; no.10 arr. harp, 1856. *in private collection of R. Pohl	B ii/7; NA ix	264, 271, 273, 293
173a	—	Hymne de la nuit; Hymne du matin	1847	—	*D-WRgs; orig. intended for inclusion in 173	NA ix	
174		Berceuse					
	57a	1st version	1854	in Elisabeth-Fest-Album (Vienna, 1854)	*F-Pc	B ii/9	273
	57b	2nd version	1862	1865	*D-WRgs; ded. Princess M. Czartoryska	NA ix	
175	17	Légendes: 1 St François d'Assise: la prédication aux oiseaux, 2 St François de Paule marchant sur les flots	1863	Budapest and Paris, 1866; Budapest, 1976 (simplified version of no.2)	*H-Bn (simplified version of no.2); ded. C. von Bülow; arr. orch, 354	B ii/9; NA x (simplified version of no.2) in appx	278
176	18	Grosses Konzertsolo	?1849	1851	*and other MS in D-WRgs; *private collection of I. B. Grenier, Buenos Aires ded. A. Henselt, arr. 2 pf, 258, pf, orch, 365 †; ded. T. Kullak	B ii/8; NA v	273, 275, 302
177	20	Scherzo und Marsch, b	1851	Brunswick, 1854	*WRgs (frag.), *US-NYpm; ded. R. Schumann	B ii/8	273
178	21	Sonate, b	sketched 1851, 1852–3	1854		B ii/8; NA v	273, 274–7
179	23	'Weinen, Klagen, Sorgen, Zagen', Präludium	1859	Berlin, 1863	*Wc; ded. A. Rubinstein; based on theme from Bach's Cantata no.12	B ii/9	277
180	24	Variationen über das Motiv von Bach	1862	Berlin, 1864	*D-Bds, H-Bl, ded. A. Rubinstein; on the bass line from the 1st movt of Cantata no.12	B ii/9	248–9, 277–8

S	R	Title	Composition	Publication	Autograph; dedication; remarks	Edition
181	25	Sarabande und Chaconne, from Handel's Almira	1879	1880	*D-WRgs; ded. Walter Bache	RS i/1
182	67	Ave Maria ('Die Glocken von Rom')	1862	Stuttgart, 1863	*WRgs	B ii/9, NA xi
183	68	Alleluja et Ave Maria (d'Arcadelt)	1862	1865	*WRgs (Alleluja only); Alleluja based on 4 Ave Maria arr. org, 659	NA xi
184	69	Urbi et orbi, bénédiction papale	1864	unpubd	*WRgs	NA xii
185	70	Vexilla regis prodeunt	1864	unpubd	*WRgs; arr. orch, 355	NA xii
186	71	Weihnachtsbaum–Arbre de Noël: 1 Psallite, 2 O heilige Nacht!, 3 Die Hirten an der Krippe, 4 Adeste fideles, 5 Scherzoso, 6 Carillon, 7 Schlummerlied, 8 Altes provençalisches Weihnachtslied, 9 Abendglocken, 10 Ehemals!–Jadis, 11 Ungarisch, 12 Polnisch	sketched 1866, rev. 1876	Berlin, 1882	*WRgs (sketches); *F-Pc; *US-Wc (no.12 only) ded. D. von Bülow; arr. pf 4 hands, 613 no.2 arr. vv, org, 49	B ii/9, NA x
187	73	Sancta Dorothea	1877	B	*Göllerich collection	B ii/9, LS vii, NA xiii
187a	388	Resignazione	1877	pubd in Göllerich (1908)	*Wc; see also 263	NA xii
188	74	In festo transfigurationis Domini nostri Jesu Christi	1880	B	*D-WRgs	B ii/9, LS vii, NA xii
189	44a	Piano piece, Ab	May 1866	unpubd	*private collection of O. Haas (London)	—
189a	—	Piano piece, Ab	1845	? pubd in Hung. trans. of Mil'shteyn (1956)	*USSR-Mcl; *WRgs	NA ix, appx

279

No.	S	Title	Composed	Publication	Notes	Sources	p.
190	65	La marquise de Blocqueville, portrait en musique	1868	in Le Figaro (14 April 1886)	*F-AU; pubd as the 3rd of 3 pieces by H. Herz, F. Planté and Liszt	NA xii	
191	59	Impromptu ('Nocturne')	1872	1877	*US-Wc, D-WRgs; ded. Baroness Olga von Meyendorff	B ii/9, NA xii	
192	60	Fünf kleine Klavierstücke	1865–79	B (nos.1–4), no.5 ed. Kassel, 1970	*US-Wc (no.5), D-WRgs; ded. Baroness Olga von Meyendorff	B ii/9 and LS i [nos. 1–4], NA x	278
193	61	Klavierstück, F♯	after 1860	B	*WRgs	B ii/10, LS vii, NA xi	
194	110	Mosonyi gyászmenete–Mosonyis Grabgeleit	Nov 1870	Budapest, 1871	†; see also 205/7	LS iii	
195	111	Petofi szellemének–Dem Andenken Petőfis	1877	Budapest, 1877	US-Wc; arr. pf 4 hands, 614, see also 205/6	LS iii	
196	76	[Erste] Elegie	1874	1875	H-Ba(mi); in memory of M. Mukhanov; arr. chamber ens, 130	B ii/9, LS iii, NA x	280
197	77	Zweite Elegie	1877	1878	*D-WRgs, US-NYp; ded. L. Ramann; arr. chamber ens, 131	B ii/9, LS iii, NA x	280
197a	60a	Toccata	1879–81	ed. Kassel, 1970	*US-Wc	NA xii	281
198	58	Wiegenlied–Chant du berceau	May 1880	ed. London, 1958	*A-Wn; ded. A. Friedheim; see also 107	NA xii	
199	78	Nuages gris	1881	B	*D-WRgs	B ii/9, LS i, NA xii	314–6
	81	La lugubre gondola				B ii/9, LS i, NA xii	250,280
200/1		1st version (6/8)	Dec 1882	pubd in Kinsky (1916)	*WRgs	NA xii	314
200/2		2nd version (4/4)	1885	1886	†; based on 134	NA xii	
201	82	R. W.–Venezia	1883	B	*Stargardt's auction catalogue (Marburg, 1975)	B ii/9, LS i, NA xii	281

S	R	Title	Composition	Publication	Autograph; dedication; remarks	Edition	
202	85	Am Grabe Richard Wagners	22 May 1883	LS	*US-NYpm; arr. chamber ens, 135	LS ii, NA xii	280, 281
203	79	Schlaflos, Frage und Antwort, nocturne	March 1883	B	H-Bn; after a poem by T. Raab	B ii/9, LS iii, NA xii	280
204	86	Recueillement	1887	in Pel monumento a V. Bellini, Naples, ?1884	†	B ii/9, NA xii	279
205	112	Historische ungarische Bildnisse–Magyar történelmi arcképek: 1 Stephan Széchényi, 2 Josef Eötvös, 3 Michael Vörösmarty, 4 Ladislaus Teleky, 5 Franz Deák, 6 Alexander Petőfi, 7 Michael Mosonyi	Nov 1870 (no.7), 1877 (no.6), 1885 (nos.1–5)	ed. I. Szelényi, Budapest, 1956 (complete)	US-Wc (nos.1, 3, 5) Antiquariat (Marburg); D-WRgs (no.5); US-NYpm (no.6); nos.4, 6 and 7 same as 206/2, 195 and 194, respectively	LS i [no.4], iii [nos. 6–7] NA x	314–5
206/1 206/2	83 84	Trauervorspiel und Trauermarsch Trauervorspiel Trauermarsch	April 1885 Sept 1885	1887	†; ded. A. Göllerich see also 205/4	NA xii LS i, NA xii	280
207	87	En rêve, nocturne	wint. 1885	Vienna, 1888	†; ded. A. Stradal	B ii/9, LS i, NA xii	280
207a	—	Prelude to Borodin's 'Chopsticks' Polka	1880	in Paraphrases, St Petersburg, 2/1880	*Facs. edn. in Tcherepnin: Paraphrases (Bonn, 1959); written for a set of variations on the same theme by Borodin, Cui, Rimsky-Korsakov, Lyadov	—	301
208	80	Unstern: sinistre, disastro	after 1880	B	*Göllerich collection	B ii/9, LS i, NA xii	314–5

		works in dance form					
208a	—	Waltz, A		in The Musical Gem, London, 1832 in Album musical, Leipzig, 1836	†	—	273
209	32a	Grande valse di bravura ('Le bal de Berne')	1836		*D-WRgs (sketches); ded. P. Wolf and C. Ludlow (in different edns.); arr. pf 4 hands, 615, see also 214/1	B ii/10	269
210	33a	Valse mélancolique	1839	Vienna and elsewhere, 1840	*US-Wc (frag.); see also 164, 214/2	B ii/10, LS iv	269
211	34	Ländler, A♭	1843	in Neue Musikzeitung, xlii (1921)	*D-DO	—	
212	35	Petite valse favorite ('Souvenir de Pétersbourg')	1842	Vienna and Paris, 1843	ded. M. von Kalergis; see also 213	B ii/10	
213	36	Valse impromptu Trois caprices-valses	c1850 c1850	1852 Vienna and Berlin, 1852	†; based on 212 †	B ii/10 B ii/10 [nos. 1–2]	
214/1 214/2 214/3	32b 33b 155	no.1 no.2 no.3			based on 209 based on 210 based on 401		
214a	60b	Carousel de Mme Pelet-Narbonne	1875–81	ed. Kassel, 1970	*US-Wc	NA xii	281
215	37	Quatre valses oubliées	1881 (no.1), 1883 (nos.2–4)	Berlin, 1881 (no.1), Berlin, 1884 (nos.2–3), ed. Bryn Mawr, Penn., 1954 (no.4)	*Wc (nos.1, 4), *NYp (no.2), *F-Pc (no.3), US-NYpm (no.2); nos.2–3 ded. O. von Meyendorff	B ii/10 [nos. 1–3], LS iv [nos. 2–3]	281, 316
216	38	Third Mephisto Waltz	1883	Berlin, 1883	†; ded. M. Jaëll	B ii/10, LS i	
216a	60c	Bagatelle ohne Tonart – Bagatelle sans tonalité	1885	ed. I. Szelényi, Budapest, 1956	intended as Fourth Mephisto Waltz	—	281
217	39	Mephisto Polka	1883	Berlin, 1883	*D-WRgs; ded. L. Schmalhausen	B ii/10, LS v	
218	40	Galop, a	?1841	B	*WRgs	B ii/10, LS iv	270
219	41	Grand galop chromatique	1838	Breslau, 1838	†; ded. Count Rudolf Apponyi; arr. pf 4 hands, 616	B ii/10	264, 270

S	R	Title	Composition	Publication	Autograph; dedication; remarks	Edition	
220	42	Galop de bal	c1840	?Paris, ?c1840	†	—	
221	43	Mazurka brillante	1850	Leipzig and Paris, 1850	†; ded. A. Koczuchowski	B ii/10, LS v	273
222		deleted, same as 212					
223	44	Two Polonaises, c, E	1851	1852	*US-Wc (no.1)	B ii/10 LS i	273
224	46	Csárdás macabre	1881–2	LS; longer version ed. I. Szelényi, Budapest, 1954	*GB-Lbm; arr. pf 4 hands, 617		280
225	45	Two Csárdás: 1 Allegro, 2 Csárdás obstiné	1884 1886	Budapest and Vienna, 1886	*H-Bn (no.2), US-Wc (no.1); no.2 arr. pf 4 hands, 618	LS iii [no.2]	280
226	47	Festvorspiel–Prélude	1856	in Das Pianoforte, Stuttgart, 1857	*CS-Pnm; ded. K. Tausig; arr. orch, 356, orig. entitled Preludio pomposo	NA xi	
227	48a	Festmarsch zur Saekularfeier von Goethes Geburtstag	1849	pubd as 75/2	†; ded. Grand Duke Carl Friedrich of Saxe-Weimar; see also 115	B ii/10	
228	49	Huldigungsmarsch	1853	Berlin, 1858	†; ded. Grand Duke Carl Alexander; arr. orch, 357	NA xv	
229	50	Vom Fels zum Meer, deutscher Siegesmarsch	1853–6	Berlin, 1856	D-WRgs; ded. Kaiser Wilhelm I of Prussia; arr. orch. 358, pf 4 hands, 618a	NA xv	
230	52	Bülow-Marsch	1883	Berlin, 1884	*WRgs; ded. Meiningen Court Orchestra; arr. pf 4 hands, 619, 2 pf, 8 hands, 657b	B ii/10	
230a	296	Festpolonaise	14 Jan 1876	—	*US-NYpm; arr. pf 4 hands, 634b; for wedding of Princess Marie of Saxony	NA xvii	
231	53	Heroischer Marsch im ungarischen Styl	1840	Hamburg, 1840	†; ded. King Ferdinand of Portugal	—	270, 293
232	54a	Seconde marche hongroise–Ungarischer Sturmmarsch	1843	Berlin and Paris, 1843	†; ded. S. Teleky; see also 119	NA xvi, appx	
233	56	Ungarischer Geschwindmarsch–Magyar Gyors induló	1870	Bratislava, 1871	*H-Bn	—	

No.	S.	Title	Composed	Publication	Sources, remarks	pp.
233a	56a	Siegesmarsch–Marche triomphale	c1882	unpubd	*Bn	—
233b	—	March, E♭		pubd in Mil'shteyn (1956)	*USSR-Lsc	—

works based on national themes

AUSTRIAN

No.	S.	Title	Composed	Publication	Sources, remarks	pp.
233c		renumbered as 385a				

CZECH

| 234 | 100 | Hussitenlied, on a melody by J. Krov | 1840 | Prague and Paris, 1840 | *CS-Pnm; ded. Count Chotek of Chotkowa and Wognin | — |

ENGLISH

| 235 | 98 | God Save the Queen | 1841 | Leipzig and Paris, ?1841 | † | 284 |

FRENCH

236	93–4	[2 pieces:] Faribolo Pastour, Chanson du Bearn	1844	Mainz and elsewhere, 1845	†; ded. C. D'Artigaux	284–5
237	95	La Marseillaise	c1850	1872		
238	96	La cloche sonne	?1870–80	ed. London, 1958; LS		
239	97	Vive Henri IV		LS	D-WRgs, US-Wc	LS vii

GERMAN

| 240 | 99 | Gaudeamus igitur, concert paraphrase | 1843 | Breslau, 1843 | † | — |

HUNGARIAN

| 241 | 107 | Zwei Sätze ungarischen Charakters (Zum Andenken) | ?1831–7 | pubd in Gárdonyi (1935) | *D-Bds; arrs. of works by Fáy and Bihari | 264 |
| 242 | 105a | [21 Hungarian themes and rhapsodies] Magyar dallok–Ungarische National-melodien; i: 1 Lento, c, 2 Andantino, C, 3 Sehr langsam, D♭, 4 Animato, C♯, 5 Tempo giusto, D♭, 6 Lento, g; ii: 7 Andante cantabile, E♭; iii: 8 Lento, f, 9 Lento, a; iv: 10 Adagio sostenuto, D, 11 Andante sostenuto, B♭ | 1839–47 | Vienna, 1840–47; no.20 ed. O. Beu, Vienna, 1936, nos.18, 19 and 21 RS | *D-WRgs (sketches), *A-Wgm (no.20), H-Bn (no.13), *USSR-Lsc (no.19), MSS of nos. 18,21 in A-Wst, nos. 1–6 ded. Count Leo Festetics, no.7 ded. Count Casimir Esterházy, no.9 ded. Count Alberti, no.13 ded. Count Festetics and others; nos.14–15 ded. Baron Fery Orczy, no.16 ded. B. Egressy; see also 243–4 | LS vii [nos. 1–3, 6 8–10]; 270–71, 284 |

S	R	Title	Composition	Publication	Autograph; dedication; remarks	Edition
	105b	Magyar rhapsodiák – Rhapsodies hongroises v: 12 Mesto, e ('Héroïde élégiaque') vi: 13 Tempo di marcia, a viii: 14 Lento a capriccio, a vii: 15 Lento, d ix: 16 in E x: 17 Andante sostenuto, a				284
	105c	[Other works:] 18 Adagio, c♯/C♯, 19 Lento patetico, f♯, 20 Allegro vivace, g, 21 Tempo di marcia funebre, e				
243	105d	[3] Ungarische Nationalmelodien	c1840	Vienna, c1840	†: ded. Count Anton Apponyi; no.1 same as 242/5, no.2 same as 242/4, no.3 based on 242/11	—
244	106	[19] Hungarian Rhapsodies			*1, 2, *4–9, 10, *11–15; D-WRgs	B ii/12; 271, 280, 284
		1 Lento, quasi recitativo, c♯	1846	Leipzig, 1851	H-Bn; ded. E. Zerdahélyi	NA iii iv
		2 Lento a capriccio, c♯	1847	Leipzig and Milan, 1851	*(frag.) and other MS in Bn; ded. Count László Teleky; arr. orch. 359/4, pf 4 hands, 621/4	
		3 Andante, B♭		Vienna, 1853	†; ded. Count Leo Festetics; based on 242/11	
		4 Quasi adagio, alticeramente, E♭		Vienna, 1853	ded. Count Casimir Esterházy; based on 242/7	
		5 Héroïde-élégiaque, e		Vienna and Paris, 1853	ded. Countess Sidonie Revicsky; arr. orch. 359/5, pf 4 hands, 621/5; based on 242/12	
		6 Tempo giusto, D♭		Vienna, 1853	ded. Count Anton Apponyi; arr. orch. 359/3, pf 4 hands, 621/3; based on 242/4	

No.	Title	Year	Place, date	Notes	Pages
7	Lento, d		Vienna, 1853	ded. Baron Fery Orczy; based on 242/15	
8	Lento a capriccio, f♯		Mainz, 1853	ded. Baron Anton Augusz; based on 242/19	301
9	Pester Karneval, E♭ 1st version 2nd version		Vienna and Paris, 1848 Mainz, 1853		
10	Preludio, E		Mainz, 1853	ded. H. W. Ernst; arr. orch, 359/6, pf trio, 379, pf 4 hands, 621/6 †; ded. B. Egressy; based on 242/16 and a work by Egressy	
11	Lento a capriccio, a		Berlin, 1853	H-Bl; ded. Baron Orczy; based on 242/14	
12	Mesto, c♯		Berlin, 1853	ded. J. Joachim; arr. orch, 359/2, pf 4 hands, 621/2; based on 242/18 and 20	
13	Andante sostenuto, a		Berlin, 1853	ded. Count Festetics; based on 242/17	
14	Lento quasi marcia funebre, f		Berlin, 1853	Bl; ded. H. von Bülow; arr. pf, orch, 123, orch, 359/1, pf 4 hands, 621/1; based on 242/21	312
15	Rákóczy March, a 1st version		Leipzig, 1851	H-Bn, Bo; based on 242/10 and 13	
	2nd version		Leipzig, 1871	*Bn; arr. pf 4 hands, 608; based on 117	
16	Allegro, a	1882	Budapest, 1882	*US-Wc; ded. M. Munkácsy; arr. pf 4 hands, 622	280, 318-19
17	Lento, d	1886	Budapest, 1886	*NYpm	280, 316
18	Adagio, c♯	1885	Budapest, 1885	*H-Bn (frag.), *Ba(mi) (frag.); arr. pf 4 hands, 623	280
19	Lento, d	1885	Budapest, 1886	*Bn; arr. pf 4 hands, 623a, based on Ábrányi's Csárdás nobles	280

S	R	Title	Composition	Publication	Autograph: dedication; remarks	Edition	
245	108	Fünf ungarische Volkslieder	Jan 1873	Budapest, 1873	*Radio-Televisione Italiana (Milan); based on arrs. by Ábrányi for 1v, pf	LS iii	
246	113	Puszta-Wehmut—A Puszta keserve	after 1880	Budapest, after 1884	*CS-BRsa; arr. of work by L. Zámoyská, after a poem by N. Lenau; arr. vn, pf, 379a	—	
		ITALIAN					
247		renumbered as 252a					
248	92	Canzone napolitana	1842	Dresden, 1843	†; ded. Mlle C. de Groeditzberg	—	
		POLISH					
249	101	Glanes de Woronince: 1 Ballade d'Ukraine, dumka. 2 Mélodies polonaises, 3 Complaintes, dumka	1847–8	1849	*US-NYpm; ded. Princess Marie von Sayn-Wittgenstein	—	285
		RUSSIAN					
250	102	Deux mélodies russes ('Arabesques'): 1 Le rossignol, 2 Chanson bohémienne	1842	Hamburg and elsewhere, 1842	no.1 based on a song by Alabyev, no.2 based on a song by Bulakhov	—	285
251	104	Abschied	1885	Leipzig, 1885	†; ded. A. Siloti	—	
		SPANISH					
252	88	Rondeau fantastique sur un thème espagnol ('El contrabandista')	1836	Vienna, 1837	†; ded. George Sand; based on a song by M. Garcia	—	269
252a	91	La romanesca	c1832	Gazette musicale de Paris (6 April 1833), suppl.	†; ded. Mme H. Seghers	LS vii	
253	89	Grosse Konzertfantasie über spanische Weisen	Feb 1845	1887	*D-WRgs; ded. L. Ramann	—	278, 284
254	90	Rhapsodie espagnole ('Folies d'Espagne et jota aragonesa')	c1863	1867	*WRgs	—	278, 284

255	296	renumbered as 634b				—	301
256	297	renumbered as 207a				—	301
256a	—	Notturno, E		?Berlin, n.d.	*WRgs; based on 161/5; ? not by Liszt	—	301
		TWO PIANOS					
257	355	Grosses Konzertstück über Mendelssohns Lieder ohne Worte	1834	unpubd	*WRgs	—	242, 301
258	356	Concerto pathétique	before 1857	1866	*Bds; ded. I. von Bronsart; arr. pf, orch, 365, based on 176	—	273, 301–2
		ORGAN					302, 304
259	380	Fantasie und Fuge über den Choral 'Ad nos, ad salutarem undam'	1850	no.4 of Illustrations du Prophète, Leipzig, 1852	†; ded. G. Meyerbeer; arr. pedal pf or pf 4 hands, 624; based on a theme from Le prophète	RS iv/3	302
260	381	Präludium und Fuge über den Namen BACH	1855, rev. 1870	Rotterdam, 1855, rev. version Leipzig, 1870	ded. A. Winterberger; rev. version arr. pf, 529	—	302
261	391	Pio IX ('Der Papsthymnus')	?1863	1865	arr. chorus as Tu es Petrus, added to 3, orch, 361, pf, 530, pf 4 hands, 625	—	
261a	—	Andante religioso		n.d.	†; ded. K. Gille	—	
262	383	Ora pro nobis, litany	1864	1865	ded. Cardinal Prince G. Hohenlohe	—	302
263	388	Resignazione	1877	pubd in Göllerich (1908)	*US-Wc; see also 187a	—	
264	384	Missa pro organo lectarum celebrationi missarum adjumento inserviens	1879	Rome and Leipzig, 1880	*D-WRgs; ded. Princess Sayn-Wittgenstein; based on 20/2 and 38	—	302
265	386	Gebet	1879	n.d.		—	
266	385	Requiem für die Orgel	1883	1885	*US-Wc; based on 12	—	302
267	387	Am Grabe Richard Wagners	1883	unpubd	*NYpm; for pf, 202, harp, str qt, 135	—	
268	390	Zwei Vortragsstücke: 1 Introitus, 2 Trauerode	1884 (no.1), 1870 (no.2)	1887 (no.1), 1890 (both)	no.2 based on 112/1	—	304

SONGS 244, 312–3

(all for 1v, pf: songs existing in multiple versions enumerated separately)

S	R	
269/1	593a	Angiolin dal biondo crin (C. Bocella) 1839 (Berlin, 1843), MSS in private collection of J. & J. Lubrano, S. Lee. Mass., arr. pf, 531/6
269/2	593b	(Berlin, 1856), D-WRgs; B vii/2, LS vi
		Tre sonetti di Petrarca: 1 Pace non trovo, 2 [312] Benedetto sia'l giorno, 3 I vidi in terra angelici costumi
270/1	578a	1844–5 (Berlin, 1846), *D-WRgs, arr. pf, 158; B vii/1
270/2	578b	1854 (Mainz, 1883), *US-Wc, arr. pf. 161/4–6; B [269] vii/3, LS vi
271	566	Il m'aimait tant (D. Gay), c1840 (Mainz, 1843), †, arr. pf, 533; B vii/1
		Am Rhein (Heine), ded. Princess Augusta of Prussia
272/1	567a	c1840 (Berlin, 1843), D-WRgs (photo of autograph in NYp), arr. pf, 531/2; B vii/1
272/2	567b	(Berlin, 1856), *Bory collection; B vii/2
		Die Loreley (Heine)
273/1	591a	Nov 1841 (Berlin, 1843), WRgs, ded. Countess Marie d'Agoult, arr. pf, 531/1
273/2	591b	1854 (Berlin, 1856), arr. 1v, orch, 369, pf. 532; B vii/2
		Die Zelle in Nonnenwerth (F. Lichnowsky)
274/1	618a	before 1842 (Cologne, 1843), †, ded. Countess d'Agoult, arr. pf, 534 and 167
274/2	618b	1860 (1860), *WRgs, ded. E. Genast, arr. vn/vc, pf, 382; B vii/3
		Mignons Lied (Goethe)
275/1	592a	1842 (Berlin, 1843), WRgs, arr. pf, 531/3
275/2	592b	1854 (Berlin, 1856), WRgs; B vii/2
275/3	592c	1860, pubd with 370 (1863), †; B vii/2
276/1	570a	Comment, disaient-ils (Hugo)
276/2	570b	1842 (Berlin, 1844), †, arr. pf, 535; B vii/1 (Berlin, 1859); *WRgs; B vii/2
277	625	Bist du (E. Metschersky), 1843 (Cologne, 1844), rev. c1877–8 (1879), †; B vii/3 Es war ein König in Thule (Goethe)

	S	R
1842 (Berlin, 1844), WRgs, arr. pf, 531/4	278/1	594a
(Berlin, 1856), †; B vii/2	278/2	594b
Der du von dem Himmel bist (Goethe), ded. Princess Augusta of Prussia		
1842 (Berlin, 1843), WRgs, arr. pf, 531/5; B vii/1	279/1	568a
pubd as Invocation (Berlin, 1856), WRgs; B vii/2	279/2	568b
(1860). †; vii/2	279/3	568c
B, *WRgs (frag.); B vii/1	279/4	568d
Freudvoll und leidvoll (Goethe), ded. A. Scheffer		
1844, pubd with 280/2, 292/1, 297/1 and 306/1 (Vienna, 1847), *WRgs, rev. version (1860); B vii/1 [orig.], vii/2 [rev.]	280/1	579a, c
pubd as 280/1, †; B vii/1	280/2	579b
Die Vätergruft (Uhland), 1844 (Berlin, 1860), *WRgs, US-NYpm, arr. 1v, orch, 371; B vii/2 1886, in Stargardt auction catalogue (Marburg, 1976)	281	601
Oh! quand je dors (Hugo)		
1842 (Berlin, 1844), *D-WRgs, arr. pf, 536; B vii/1	282/1	569a
(Berlin, 1859), †, B vii/2	282/2	569b
Enfant, si j'étais roi (Hugo)		
c1844 (Berlin, 1844), *WRgs, arr. pf, 537; B vii/1	283/1	571a
(Berlin, 1859); B vii/2, LS vi	283/2	571b
S'il est un charmant gazon (Hugo)		
c1844 (Berlin, 1844), *Bds, arr. pf, 538; B vii/1	284/1	572a
(Berlin, 1859), *WRgs and F-Pc; B vii/2	284/2	572b
La tombe et la rose (Hugo), c1844 (Berlin, 1844), †, arr. pf, 539; B vii/1	285	573
Gastibelza (Hugo), bolero, c1844 (Berlin, 1844), †, arr. pf, 540; B vii/1	286	574
Du bist wie eine Blume (Heine), c1843 (Cologne, 1844), *D-Mbs, WRgs; B vii/2	287	607
Was Liebe sei (C. von Hagn)		
c1843 (Cologne, 1844), *WRgs, ded. Grand Duchess Sophie of Saxe-Weimar; B vii/1	288/1	575a
c1855, B, WRgs; B vii/2	288/2	575b

Left block

288/3	575c	c1878 (1879), *WRgs*; B vii/3
289	608	Vergiftet sind meine Lieder (Heine), 1842 (Cologne, 1844), **WRgs*; B vii/2, LS vi
		Morgens steh' ich auf und frage (Heine)
290/1	576a	1843 (Cologne, 1844), *WRgs*; B vii/1
290/2	576b	?c1855 (Berlin, 1859), †; B vii/2, LS vi
		Die tote Nachtigall (Kaufmann)
291/1	577a	c1843 (Cologne, 1844), †; B vii/1
291/2	577b	1878 (1879), *private collection of C. Hoche (Stuttgart); B vii/3, LS vi
		Three songs from Wilhelm Tell (Schiller): 1 Der Fischerknabe, 2 Der Hirt, 3 Der Alpenjäger
292/1	582a	1845, pubd as 280/1; B vii/1
292/2	582b	(1859), arr. 1v, orch, 372; B vii/2
		Jeanne d'Arc au bûcher (Dumas)
293/1	586a	1845 (Mainz, 1846), arr. 1v, orch, 373; B vii/3
293/2	586b	1874 (Mainz, 1876)
		Es rauschen die Winde (Rellstab)
294/1	596a	?c1845, rev. before 1856, B; B vii/2
294/2	596b	(1860); B vii/2
295	598	Wo weilt er? (Rellstab), 1844 (1860), **WRgs*; B vii/2
296	606	Ich möchte hingehn (Herwegh), 1845, rev. version (Berlin, 1860); B vii/2, LS vi
		Wer nie sein Brot mit Tränen ass (Goethe)
297/1	609a	?c1845, pubd as 280/1, B vii/2
297/2	609b	c1860 (1860); B vii/3
298	589	O lieb, so lang du lieben kannst (Freiligrath), c1845 (1847), †, arr. 541; B vii/2
299	627	Isten veled [Farewell] (Horvath), 1846–7 (Prague, 1847), rev. version (Leipzig, 1879), †; B vii/3, LS vi
300	585	Le juif errant (Béranger), 1847, unpubd, * and 2 other MSS in *WRgs*
		Kling leise, mein Lied (Nordmann)
301/1	580a	1848, B; B vii/1
301/2	580b	(1860); B vii/2
301a	638a	Oh pourquoi donc (Mme Pavloff), 1843 (Moscow, 1844 as Les pleurs des femmes), *private collection of F. Schnapp (Hamburg), ded. Mme Pavloff; RS iii, appx
301b	638b	En ces lieux (E. Monnier), elegy (Paris, before 1855), †

Right block

302	583	Die Macht der Musik (Duchess Helen of Orleans), 1848–9 (1849), *H-Bl*, ded. Grand Duchess Maria Pavlovna; B vii/1
303	584	Weimars Toten (F. von Schober), dithyramb (1848), †; B vii/1
304	565	Le vieux vagabond (Béranger), before 1849, B, **US-NYp*; B vii/1
		Schwebe, schwebe, blaues Auge (Dingelstedt)
305/1	581a	1845, B, *D-WRgs*; B vii/1
305/2	581b	(1860, **WRgs*; B vii/2
		Über allen Gipfeln ist Ruh (Goethe)
306/1	610a	c1848, pubd as 280/1, †; arr. of 75/1
306/2	610b	(Berlin, 1859), †; B vii/2
306a	—	Quand tu chantes bercée (Hugo), 1843, ed. (Budapest, 1973), **H-Bn*
307	587	Hohe Liebe (Uhland), c1849, pubd with 308 and 298 (1850), **F-Pc, US-Wc*, copy [different] in *D-MZs*, arr. pf, 541; B vii/2
308	588	Gestorben war ich (Uhland), c1845, pubd with 307 and 298 (1850). **F-Pc*, arr. pf, 541; B vii/2
		Ein Fichtenbaum steht einsam (Heine)
309/1	599a	c1845 (1860), **D-WRgs*; B vii/2, LS vi
309/2	599b	1854 (1860), **WRgs*; B vii/2, LS vi
310	600	Ihr Auge ('Nimm einen Strahl der Sonne') (Rellstab), 1843 (1860); B vii/2
311	602	Anfangs wollt' ich fast verzagen (Heine), c1849 (1860). **WRgs*; B vii/2
312	595	Wie singt die Lerche schön (Hoffmann von Fallersleben), ?c1856, pubd as suppl. to *Deutsches-Musen-Almanach* (Würzburg, 1856), **Mbs* (2 copies); B vii/2
313	597	Weimars Volkslied (P. Cornelius), 1857 (Leipzig, 1857), †; based on 87; B vii/2
314	590	Es muss ein Wunderbares sein (Redwitz), 1852 (1859), **WRgs*; B vii/2
315	617	Ich liebe dich (Rückert), 1857 (1860). **US-Wc, D-WRgs*, arr. pf, 542a; B vii/3
316	603	Muttergottes Sträusslein zum Mai-Monate (Müller), 1857 (Berlin, 1859), †; 1 Das Veilchen, 2 Die Schlüsselblumen; B vii/2
317	604	Lasst mich ruhen (Hoffmann von Fallersleben), ?c1858 (Berlin, 1859), †; B vii/2

S	R		
318	605	In Liebeslust (Hoffmann von Fallersleben), ?c1858 (Berlin, 1859), †; B vii/2	
319	611	Ich scheide (Hoffmann von Fallersleben), 1854 (1860), *D-WRgs; B vii/2, LS vi	
320	612	Die drei Zigeuner (Lenau), 1860 (1860), *WRgs, ded. E. Merian-Genast; B vii/3	
321	613	Die stille Wasserrose (Geibel), ?1860 (1860), †; B vii/3	
322	614	Wieder möcht ich dir begegnen (Cornelius), 1860 (1860), †; B vii/3	
323	615	Jugendglück (Pohl), c1860 (1860); B vii/3	
324	616	Blume und Duft (Hebbel), 1854 (1860), *and other MS in D-WRgs; B vii/3	
325	619	Die Fischerstochter (C. Coronini), 1871 (1879), †; B vii/3	
326	623	La perla (Princess Therese von Hohenlohe), 1872 (Rome, n.d.); B vii/3	
327	621	J'ai perdu ma force et ma vie ('Tristesse') (de Musset), 1872 (1879), *US-Wc; B vii/3, LS vi	312
328	621	Ihr Glocken von Marling (E. Kuh), 1874 (1879), † (photo of autograph in Wc), ded. Princess Marie von Hohenlohe; B vii/3, LS vi	
329	622	Und sprich (Biegeleben), 1874 (1879), *H-Ba(mi); ded. Princess Sayn-Wittgenstein; B vii/3	
330	624	Sei still (Nordheim, pseud. of H. von Schorn), 1877 (1879), †; B vii/3, LS vi	
331	628	Gebet (Bodenstedt), ?c1878 (1879), NYpm; B vii/3, LS vi	
332	629	Einst (Bodenstedt), ?c1878 (1879), NYpm; B vii/3, LS vi	
333	630	An Editam (Bodenstedt), ?c1878 (1879), *formerly in Kahnt collection (Leipzig); B vii/3	
334	631	Der Glückliche (Wilbrandt), ?c1878 (1879), †; B vii/3	
335	626	Go not, happy day (Tennyson), 1879, pubd in Tennyson Album (London, n.d.), *Wc; B vii/3, LS vi	312
336	632	Verlassen (G. Michell), 1880 (1880), *H-Bn; B vii/3	312
337	633	Des Tages laute Stimmen schweigen (F. von Saar), 1880, B, * ded. Princess Marie von Hohenlohe; B vii/3, LS vi	

S	R		
338	634	Und wir dachten der Toten (Freiligrath), c1880, B; *D-WRgs, B vii/3, LS vi	312
339	635	A magyarok Istene–Ungarns Gott (Petőfi), 1881 (Budapest, 1881), male vv ad lib, †, arr. pf, 543, pf, left hand, 543a, org, 674, brass, ww, 381a; B vii/3	
340	636	Magyar király-dal–Ungarisches Königslied (Ábrányi), 1883 (Budapest, 1884), *Bn (frag.), arr. pf, 544, pf 4 hands, 626, chorus, orch, 93; B vii/3	
340a	—	Ne brani menya, moy drug (Do not reproach me, my friend') (Tolstoy), 1886 (Moscow, 1958), *USSR-Mcl. ded. Countess Sofia Tolstoy	

OTHER VOCAL WORKS

S	R		
341	640	Ave Maria IV, 1v, org/harmonium/pf, 1881 (Berlin, n.d.), *Bn, arr. pf, 545	
342	642	Le crucifix (Hugo), A, pf/harmonium, 1884 (1884), †; B v/6	
343	643	Sancta Caecilia, A, org/harmonium, after 1880, B, Bl, B v/6	
344	637	O Meer im Abendstrahl (Meissner), S, A, pf/harmonium, c1880 (1883), †, ded. Marie Breidenstein; B vii/3	
345	638	Wartburg-Lieder (J. V. Scheffel), T, 2 Bar, mixed vv, pf/orch, 1872, vocal score (1873), †; 1 Introduction, 2 Wolfram von Eschenbach, 3 Heinrich von Ofterdingen, 4 Walther von der Vogelweide, 5 Der tugendhafte Schreiber, 6 Biterolf und der Schmied von Ruhla, 7 Reimar der Alte; B vii/3	313

RECITATIONS 313

S	R		
346	654	Lenore (Bürger), pf acc., 1858, rev. 1860 (1860), *F-Pc, D-WRgs; B vii/3	
347	655	Vor hundert Jahren (F. Halm), orch acc., 1859, D-WRgs	
348	656	Der traurige Mönch (Lenau), pf acc., 1860 (1872), ded. F. Ritter (née Wagner); B vii/3	264, 313

FOR PIANO AND ORCHESTRA

No.	Entry	Ref.
365	Liszt: Grand solo de concert (176), ?1850, unpubd, *private collection of Lord Londonderry (London), GB-Lbbc	—
366	Schubert: Wandererfantasie d760, before 1852 (Vienna, ?1857-8), arr. 2 pf, 653	459
367	Weber: Polonaise brillante op.72, 1849 (Berlin, ?1851-3), *US-Wc (frag.), ded. Adolf Henselt arr. pf, 455	460

FOR VOICES AND ORCHESTRA

(for 1v, orch, except where stated)

No.	Entry	Ref.
368	Korbay: 2 Songs, 1883, ?unpubd, *SPmoldenhauer: 1 Le matin (Bizet), 2 Gebet (Geibel)	653
369	Liszt: Die Loreley (273/2), 1860 (1863), †	647
370	—: Mignons Lied (275/3), 1860 (1863), *NH	648
371	—: Die Vätergruft (281), 1886 (London and Leipzig, 1886), *NYpm	649
372	—: 3 Songs from Wilhelm Tell (292), c1855 (1872)	645
373	—: Jeanne d'Arc au bûcher (293/1), 1858, rev. 1874 (Mainz, 1877), *D-WRgs, F-Pc	646
374	—: Die drei Zigeuner (320), 1860 (1872), D-WRgs	650
375	Schubert: 6 Songs, 1860: 1 Die junge Nonne, 2 Gretchen am Spinnrade, 3 Lied der Mignon, 4 Erlkönig, all pubd (1863), 5 Der Doppelgänger, unpubd, *WRgs, 6 Abschied, unpubd, MS not traced	651
376	—: Die Allmacht, T/S, male vv, orch, 1871 (1872), *US-Wc	652
377	G. Zichy: Der Zaubersee, ballad, 1884, perf. Vienna, 1885, MS destroyed c1945	451

FOR CHAMBER ENSEMBLE

No.	Entry	Ref.
377a	Liszt: La notte (112/2), vn, pf, 1864-6, unpubd, *Wc	464a
378/1	—: Angelus (163/1) for harmonium, 1877 (Mainz, 1883), *at Editio Musica, Budapest, ded. D. von Bülow]; B ii/6	389
378/2	—: for str qt, 1880 (Mainz, 1883), *H-Ba(mi)	473

No.	Entry	Ref.
349	A holt költő szerelme ('Des toten Dichters Liebe') (Jókai), pf acc., Feb 1874 (Budapest, 1874); B vii/3	657
350	Der blinde Sänger (A. Tolstoy), pf acc., 1875 (Paris, 1878), *D-WRgs, US-Wc, arr. pf, 546; B vii/3	658

ARRANGEMENTS, TRANSCRIPTIONS ETC

FOR ORCHESTRA

No.	Entry	Ref.
351	H. von Bülow: Mazurka-fantasie op.13, 1865 (n.d.), STu	446
352	P. Cornelius: 2nd ov. to Der Barbier von Bagdad, 1877 (1905), D-WRgs	447
353	B. Egressy and F. Erkel: Szózat, Hymnus (Vörösmarty and Kölcsey), 2 patriotic songs, 1873 (Budapest, 1878), *H-Bo, ded. Count Andrássy, arr. pf, 486, pf 4 hands, 628b	448
354	Liszt: Deux légendes (175), 1863, unpubd, *in Stargardt auction catalogue (Marburg, 1975)	440
355	—: Vexilla regis prodeunt (185), 1864, unpubd, *D-WRgs	442
356	—: Festvorspiel (226), 1857 (Stuttgart, 1857), †; B i/11	431
357	—: Huldigungsmarsch (228), 1853, rev. 1857 (Berlin, 1859); B i/11	434
358	—: Vom Fels zum Meer (229), 1860 (Berlin, n.d.), †, ded. Wilhelm I of Prussia; B i/12	435
359	—: [6] Hungarian Rhapsodies (244/14, 12, 6, 2, 5 and 9), 1857-60 (n.d.), collab. F. Doppler, *US-Wc, arr. pf 4 hands, 621	441
360	—: A la Chapelle Sixtine (461), based on works by Allegri and Mozart, unpubd, *D-WRgs	445
361	—: Der Papsthymnus (261), unpubd, *in Stargardt auction catalogue (Marburg, 1975)	443
362	—: Benedictus, from Hungarian Coronation Mass (11), 1875 (1877), ded. A. Kömpel	444
363	Schubert: 4 Marches, from D818, 819, 968b 1859 (Berlin, after 1868), *H-Bn, US-Wc (no.4)	449
364	J. Zarębski: Danses galiciennes, 1881 (Berlin, n.d.), †	450

S	R		
379	470	— Pester Karneval (244/9), pf trio (Mainz, n.d.), *D-MZsch	301
379a	—	— Puszta-Wehmut (246), vn, pf, unpubd, MS in private collection of H. Cardello (New York)	
380	464	Wagner: O du mein holder Abendstern, from Tannhäuser, vc, pf, 1852, ?unpubd, †: see also 444	
381	465	Liszt: Benedictus and Offertorium, from Hungarian Coronation Mass (11), vn, pf, 1869 (1871)	301
381a	—	A magyarok Istene (339), brass, ww, 1882, ?unpubd, *H-Bn	
382	463	— Die Zelle in Nonnenwerth (274/2), vn/vc, pf, unpubd, D-WRgs, † not by Liszt	
383	469	— Die drei Zigeuner (320), vn, pf, 1864 (1896), *formerly in Kahnt collection (Leipzig), ded. E. Remenyi	

FOR PIANO TWO HANDS
(paraphrases, operatic transcriptions etc) — 257, 281-6

S	R		
383a	—	K. Ábrányi: elaboration on Virag dál, 1881, *H-Bn	
384	115	Anon.: Mazurka pour piano composée par un amateur de St Pétersbourg, 1842 (St Petersburg, 1842), ? based on a work by M. Wielhorsky or A. Alabiev	
384a	—	Anon.: Variations on Tiszántuli szép leány, 1846, unpubd, ?*Bn	
385	116	Auber: Grande fantaisie sur la tyrolienne de l'opéra La fiancée, 1829 (Paris and elsewhere, 1829), pubd in Paris as op.1, †, ded. F. Chopin; RS i/1, iv/1 (in different versions)	241, 281
385a	—	— Tyrolean Melody, pubd in Athenaeum musicale (Manchester, 1856), based on theme of 385	
386	117	— Tarantelle di bravura d'après la tarantella de La muette de Portici, 1846 (Vienna, 1847), *US-Wc (frag.), ded. M. Pleyel; RS i/1	283
387	118	— 3 pieces, incl. 2 on themes from La muette de Portici, unpubd, *D-WRgs	
388	125	Beethoven: Capriccio alla turca [from Die Ruinen von Athen], 1846 (Vienna, 1847), *US-Wc (frag.); RS i/1, iv/1	

S	R		
389	126	— Fantasie über Beethovens 'Ruinen von Athen' (Leipzig and Milan, 1865), *D-Bds(frag.), *US-Wc (frag.), based on 122, ded. N. Rubinstein; RS i/1	
390	129	Bellini: Réminiscences des Puritains, 1836 (Mainz, 1837), ded. Princess Belgiojoso; RS ii/2	
391	130	— I puritani, introduction et polonaise, 1840 (Mainz and Paris, c1842), †, polonaise based on 390; RS ii/2	
392	131	— Hexaméron [variations on the March from I puritani], 1837 (Paris, 1837), †, collab. Thalberg, J. P. Pixis, Herz, Czerny and Chopin, ded. Princess Belgiojoso, arr. 2 pf, 654; RS ii/2	258, 269-70
393	132	— Fantaisie sur des motifs favoris de l'opéra La sonnambula, 1839, rev. 1840-41 (1842), †, ded. Princess Augusta of Prussia, arr. pf 4 hands, 627; RS ii/2, iv/1	283
394	133	— Réminiscences de Norma, 1841 (Mainz and elsewhere, 1844), †, ded. M. Pleyel, arr. 2 pf, 655; RS ii/2	283
395	135	Berlioz: L'idée fixe: andante amoroso, ?1833 (Paris and elsewhere, 1833), †, 2nd version Jan 1865, *D-WRgs, ? unpubd	282
396	141	— Bénédiction et serment, deux motifs de Benvenuto Cellini, 1852 (Berlin, 1854), *WRgs, arr. pf 4 hands, 628; RS ii/2	
397	151	Donizetti: Réminiscences de Lucia di Lammermoor [based on the sextet], 1835-6 (Vienna and elsewhere, 1840), pubd as op.13, †, ded. Mme Vanotti; RS ii/1	282
398	152	— Marche et cavatine de Lucie de Lammermoor, 1835-6 (Mainz and elsewhere, 1841), †, arr. pf 4 hands, 628a, †; RS ii/1	282
399	153	— Nuits d'été a Pausilippe, 1838, pubd with 411 (Mainz, 1839), †, ded. Sophie de Medici: 1 Barcajuolo, 2 L'alito di Bice, 3 La torre di Biasone	283
400	154	— Réminiscences de Lucrezia Borgia, 1840: 1 Trio from Act 2, 2 Fantasia on themes from the opera, no.2 first pubd (Vienna, 1841-2), both parts pubd (Paris, 1848), †; RS ii/1, iv/1 [orig. version of no.2]	

401	155	—: Valse à capriccio, sur deux motifs de Lucia et Parisina, 1842 (Vienna, 1842), †, see also 214/3; LS iv, RS ii/1, iv/1	
402	156	—: Marche funèbre de Dom Sébastien, 1844 (Vienna, 1845), †, ? ded. Queen Maria da Gloria of Portugal; RS ii/1	283
403	157	Giuseppe Donizetti: Grande paraphrase de la marche ... pour Sa Majesté le sultan Abdul Medjid-Khan, 1847 (Berlin, 1848), *D-WRgs	
404	159	Duke Ernst of Saxe-Coburg-Gotha: Halloh! Jagdchor und Steyrer [from Tony], 1849 (Leipzig and Paris, 1849), †; RS iii/2	
405	160	F. Erkel: Schwanengesang and March [from Hunyadi László], 1847, unpubd, *WRgs, ded. S. Bohrer	
405a	—	L. Festetics: Variations on Pásztor Lakodalmas, 1858 (Vienna, 1859), †	
406	164	Glinka: Tscherkessenmarsch [from Ruslan and Lyudmila], 1843 (St Petersburg and elsewhere, 1843), †, ded. Count A. Kutuzoff, arr. pf 4 hands, 629; RS ii/2, iv/1	283
407	166	Gounod: Valse de l'opéra Faust (Liège, 1861), †, ded. Baron Alexis Michels; RS iii/2	286
408	167	—: Les sabéennes, berceuse [from La reine de Saba] (Mainz, 1865), †; RS iii/2	
409	169	—: Les adieux, rêverie [from Roméo et Juliette] 1867 (Berlin, 1868); RS iii/2	
409a	170	—: Réminiscences de La juive, 1835 (Berlin, 1836), pubd as op.9, †, ded. C. Kautz; RS ii/1	264, 282
410	219	Mendelssohn: Wedding March and Dance of the Elves [from A Midsummer Night's Dream], 1849–50 (1851), *US-CA (frag.), ded. S. Bohrer; RS ii/2	
411	220	Mercadante: Soirées italiennes, six amusements, 1838, pubd with 399 (Mainz, 1839), †, ded. Archduchess Elisabeth of Austria: 1 La primavera, 2 Il galop, 3 Il pastore svizzero, 4 La serenata del marinaro, 5 Il brindisi, 6 La zingarella spagnola	283
412	221	Meyerbeer: Grande fantaisie sur des thèmes de l'opéra Les Huguenots, 1836 (Berlin, 1837), *D-WRgs, ded. Countess Marie d'Agoult; RS i/2, iv/1 [early version]	
413	222	—: Réminiscences de Robert le diable: Valse infernale, 1841 (Berlin, 1841), †, ded. Princess de Soutzo, arr. pf 4 hands, 630; RS i/2	285
414	222	—: Illustrations du Prophète, 1849–50 (Paris and Leipzig, ?1849–50), *Bds (no.2): 1 Prière, Hymne triomphale, Marche du sacre, 2 Les patineurs, 3 Choeur pastoral, Appel aux armes; RS i/2	
415	224	—: Illustration de L'africaine, 1865 (Berlin, 1866), *US-Wc (no.2), *H-Bn (no.1, frag.), ded. A. Jaëll: 1 Prière des matelots, 2 Marche indienne; RS i/2	249
416	225	—: Le moine, 1841 (Berlin and Paris, 1842), *private collection of J. Bass (New York), ded. Baron Ziegesar	
417	227	M. Mosonyi: Fantaisie sur l'opéra hongrois Szép Ilonka, 1865 (Budapest, 1868), ded. Mosonyi; RS iii/2	
418	228	Mozart: Réminiscences de Don Juan, 1841 (Berlin and elsewhere, 1843), *and other MSS in US-NYp, *Wc (frag.), ded. King Christian VIII of Denmark, arr. 2 pf, 656; RS i/1	283
419	230	Pacini: Divertissement sur la cavatine 'I tuoi frequenti palpiti' [from Niobe], 1835–6 (Paris and elsewhere, 1836), †, ded. Countess Miramont; RS ii/1	264, 282
420	231	Paganini: Grand fantasia de bravoure sur La clochette [on 'La campanella' from the Violin Concerto in B minor op.7], 1831–2 (Vienna and elsewhere, 1834), pubd as op.2, *D-WRgs; B ii/2	242, 281
421	233	Raff: Andante finale and March [from König Alfred], 1853 (Magdeburg, 1853), †, ded. K. Klindworth, arr. pf 4 hands, 631; RS ii/2	285
421a	—	Rossini: Variations sur une marche du Siège de Corinth ['Questo nome qui suono vittoria'], 1830, facs. in FAM, xxiii (1976), 104, *Manhattanville College, Purchase, NY	
422	234	—: La serenata e L'orgia, grande fantaisie sur des motifs des Soirées musicales, 1835–6 (Mainz and Paris, 1837), pubd as op.8 no.1, *H-Bn (frag.), ded J. Montgolfier, see also 424/10–11	282
423	235	—: La pastorella dell'Alpi e Li marinari, 2me fantaisie sur des motifs des Soirées musicales, 1835–6 (Mainz and elsewhere, 1837), pubd as	282

S	R	Description	
424	236	op.8 no.2, †, ded. H. de Musset, see also 424/6 and 12	
		—: [12] Soirées musicales, 1837 [Milan and elsewhere, 1838), †, ded. Countess Julie Samoyloff: 1 La promessa, 2 La regata veneziana, 3 L'invito, 4 La gita in gondola, 5 Il rimprovero, 6 La pastorella dell'Alpi, 7 La partenza, 8 La pesca, 9 La danza, 10 La serenata, 11 L'orgia, 12 Li marinari	282
425	250	Schubert: [3] Mélodies hongroises [based on Divertissement à l'hongroise D818], 1838-9 (Vienna, 1840), *and other MSS in US-Wc, ded. Count Gustav Neipperg: 1 Andante, 2 Marcia, arr. orch in 363, 3 Allegretto	257, 282
426	251	—: [3] Marches, based on D819 and 968b, 1846 (Vienna and elsewhere, 1847), *CA, ded. F. von Koudelka, arr. orch, see 363	
427	252	—: Soirées de Vienne, [9] valses caprices, 1852 (Vienna and elsewhere, 1852-3). *Wc, ded. S. Löwy, cadenza to nos.6, 9, 1883 (Hamburg, 1883)	250, 285
428	258	Sorriano: Feuille morte, elegy, c1845 (Paris, 1845), †	
429	262	Tchaikovsky: Polonaise from Eugene Onegin, Oct 1879 (St Petersburg and Hamburg, 1880). †, ded. K. Klindworth; RS iii/2	286
430	263	J. Végh: Concert Waltz, based on 2ème suite en forme de valse for pf 4 hands (1889). *private collection of E. A. Willmott	
431	264	Verdi: Salve Maria de Jéusalem [I lombardi], 1848 (Mainz and elsewhere, 1848], ded. M. von Kalergis; RS iii/2	
431a	293	—: Concert paraphrase on themes from Ernani, June 1847 rev. 1859, unpubd, *and other MS in D-WRgs	285
432	265	—: Ernani: paraphrase de concert, before 1849, rev. 1859 (1860); *D-WRgs; RS iii/2	285
433	266	—: Miserere du Trovatore, 1859 (1860); RS iii/2	
434	267	—: Rigoletto: paraphrase de concert, 1859 (1860), *WRgs; RS iii/2	
435	268	—: Don Carlos: Coro di festa e marcia funebre, 1867-8 (Milan and elsewhere, 1868), †; RS iii/2	
436	269	—: Aida: Danza sacra e duetto final (Milan and Berlin, 1879), †, ded. T. Raab; RS iii/2	
437	270	—: Agnus Dei de la Messe de Requiem, pf org/harmonium, 1877 (Berlin and Milan, 1879, *private collection of Mrs Jephta Drachman (Stevenson, Maryland) (frag.)	
438	271	—: Réminiscences de Simone Boccanegra, Dec 1882 (Milan, 1883). *GB-Lbm; LS ii, RS iii/2	284, 285, 286, 320
439	272	Wagner: Phantasiestück on themes from Rienzi, 1859 (1861); B arrs./1, RS iii/1	285
440	273	—: Spinning Chorus [from Der fliegende Holländer], 1860 (1861), ded. L. Jungman; B arrs./1, RS iii/1	
441	274	—: Ballad from Der fliegende Holländer, 1872 (Dresden, 1873), *US-Wc; B arrs./1, RS iii/1	
442	275	—: Tannhäuser, ov., 1848 (Dresden, 1849), †; B arrs./1, RS iii/1	262, 263, 285
443	276	—: Pilgrims' Chorus from Tannhäuser, 1861 (1865), rev. based on 676 (Berlin, 1885), †; B arrs./1, RS iii/1	
444	277	—: O du mein holder Abendstern, from Tannhäuser, 1849 (1849), *private collection of Mrs Franklin Geist (New York), H-Bn, ded. Grand Duke Carl Alexander, arr. vc, pf, 380; B arrs./1, RS iii/1	
445	278	—: 2 pieces from Lohengrin and Tannhäuser, 1852 (1853), *D-Bds (no.2), WRgs (no.1), no.1 ded. H. von Bülow 1 Entry of the Guests on the Wartburg, 2 Elsa's Bridal Procession; B arrs./1, RS iii/1	
446	279	—: 2 pieces from Lohengrin, 1854 (1854), †: 1 Festival and Bridal Song, 2 Elsa's Dream and Lohengrin's Rebuke; B arrs./1, RS iii/1	
447	280	—: Isoldens Liebestod [from Tristan und Isolde], 1867 (1868); B arrs./1, RS iii/1	
448	281	—: Am stillen Herd [from Die Meistersinger], 1871 (Berlin, 1871), *at Editio Musica, Budapest, H-Ba(mi); ded. M. von Schleinitz; B arrs./1, RS iii/1	285
449	282	—: Walhall [from Der Ring des Nibelungen], ?c1876 (Mainz, 1876), †; B arrs./1, RS iii/2	

S	R	Work	pp.
450	283	——: Feierlicher Marsch zum heiligen Gral [from Parsifal], 1882 (Mainz, 1883), †; B arrs./1, RS iii/1	
451	284	Weber: Fantasia on themes from Der Freischütz, 1840, unpubd, *D-WRgs	
452	285	——: Leyer und Schwert, April 1848 (Berlin, 1848), *US-Wc, ded. Princess Augusta of Prussia	285
453	286	——: Einsam bin ich, nicht alleine [from La preciosa], 1848 (1848), †, ded. P. Bérard; RS i/1	
454	287	——: Schlummerlied von C. M. von Weber mit Arabesken, 1848 (Leipzig and Paris, 1848), *H-Bn, ded. F. Kroll	
455	460	——: Polonaise brillante, c1851 (Berlin and Vienna, ?1851–3), *US-Wc, ded. A. Henselt, based on 367	
456	292	G. Zichy: Valse d'Adèle (Berlin and Paris, 1877), †	
458	294	Piece based on Italian operatic melodies, unpubd, *D-WRgs	
459	—	deleted, same as 387	
460	295	——: Kavallerie-Geschwindmarsch (Esslingen, 1883), †	

PIANO SCORES, TRANSCRIPTIONS ETC — 243, 281-6

S	R	Work	pp.
461	114	Allegri and Mozart: A la Chapelle Sixtine, 1862 (Leipzig and Milan, 1865), arr. orch, 360, pf 4 hands, 633, org, 658	278
462	119	Bach: 6 Preludes and Fugues for organ, 1842–50 (1852), †	254, 258, 283
463	120	——: Fantasia and Fugue, g, bwv542 (Berlin, 1863), *private collections of Baron von Vietinghoff-Scheel (Berlin) [fantasia] and Dr Lebert (Stuttgart) [fugue]	
464	128	Beethoven: Symphonies, no.5, 1837 (Leipzig and Paris, 1840), ded. J. A. D. Ingres, no.6, 1837 (Leipzig and Paris, 1840), ded. Ingres, no.7, 1837 (Vienna, 1840), others, 1863–4, pubd complete (1865), ded. H. von Bülow, *Breitkopf & Härtel archive, Wiesbaden (nos.1, 2, 5 and 6), D-WRgs (nos.3, 4, 8, 9); B arrs./2-3	243, 257, 282
465	127	——: Septet op.20, 1841 (Leipzig and elsewhere, 1842), †, ded. Grand Duchess Maria Pavlovna, arr. pf 4 hands, 634	283
466	121	——: Adelaide op.46, 1839 (Leipzig and elsewhere, 1840), *US-Wc, ded. Marquise Martellini	283
467	122	——: 6 geistliche Lieder op.48 (Gellert), 1840 (Leipzig and elsewhere, 1841), †, ded. Mlle Zoé de la Rue	283
468	123	——: [6] Lieder von Goethe [from opp.75, 83 and 84] (1849), †: 1 Mignon, 2 Mit einem gemalten Bande, 3 Freudvoll und leidvoll, 4 Es war einmal ein 'König, 5 Wonne der Wehmut, 6 Die Trommel gerühret; RS iv/1 [nos.3 and 6]	
469	124	——: An die ferne Geliebte op.98, 1849 (1850), *D-WRgs	
470	134, 136	Berlioz: Symphonie fantastique, 1833 (Berlin and elsewhere, 1834), 4th movt rev. ?1864–5 (Leipzig, 1866), *WRgs (frag.), *US-Wc (frag.)	242, 281-2
471	137	——: Ouverture des Francs-juges, 1833 (Mainz and Paris, 1845)	242, 281-2
472	138	——: Harold en Italie, with va part, 1837 (Paris and Berlin, 1879), *D-Bds, H-Bn (frag.)	
473	139	——: Marche des pèlerins [from Harold en Italie], 1837, rev. 1862 (1866), D-WRgs	
474	140	——: Ouverture du Roi Lear, 1837, unpubd, *WRgs	
475	142	——: Danse des Sylphes de La damnation de Faust, c1860 (Leipzig and Paris, 1866), †	
476	142a	L. Bertin: Esmeralda, opera, 1837 (Paris, 1837), †	
477	142b	——: Air chanté par Massol [from Esmeralda], 1837 (Paris, 1837), †	
477a	—	——: 3 Pieces from Esmeralda, ?1837 (Paris, 1837), †	
478	143	Bulhakov: Russischer Galopp, 1843 (St Petersburg and Berlin, 1843), *US-Wc	
479	144	Bülow: Dantes Sonett 'Tanto gentile e tanto onesta', 1874 (Berlin, 1875), *Wc	
480	145	Chopin: 6 chants polonais [from op.74], 1847–60 (Berlin, 1860), ded. Princess Marie zu Hohenlohe	250, 285
481	146	Conradi: Le célèbre Zigeunerpolka, c1847 (Berlin, 1849), †	
482	147	Cui: Tarantelle, 1885 (Paris, 1886), *Wc	286
483	148	Dargomïzhsky: Tarantelle, 1879 (St Petersburg and Hamburg, 1880), †, ded. N. Helbig; RS ii/2	286

509	200	——: Gaudeamus igitur (71), *c*1870 (1871), *WRgs* ded. Justizrat Dr Gille; NA xvi	
510	201	——: Marche héroïque (82), *c*1848, unpubd, *D-WRgs*; NA xv	
511	202	——: Geharnischte Lieder (90/4–6) (1861), †; NA xv	
512	179	——: Von der Wiege bis zum Grabe (107), 1881 (Berlin, 1883), †; NA xvii	
513	180	——: Gretchen, 2nd movt of 108, 1877 (1876), *WRgs*; NA xvi	
514	181	——:Der Tanz in der Dorfschenke (110/2), 1859–60 (1862), **US-NYpm*, *D-WRgs*, ded. C. Tausig; NA xv	299
515	182	——: Second Mephisto Waltz (111), 1881 (1881), *H-Bn*, **GB-Lbm* (frag.), ded. C. Saint-Saëns; NA xvii	
516	183	——: Les morts (112/1), 1860, **D-WRgs*, pubd in Göllerich (1908); NA xi	
517	184	——: Le triomphe funèbre du Tasse (112–3), 1866 (1878), **WRgs*, ded. L. Damrosch; NA xvi	
518	185	——: Salve Polonia (113), 1863 (1884),**WRgs*; NA xvii	
519	186	——: 2 polonaises from Stanislaus (688), 1875, unpubd, **D-WRgs*; NA xvii	
520	187	——: Künstlerfestzug zur Schillerfeier (114), 1857–60 (Weimar, Leipzig, 1860); **WRgs*, NA xv	
521	48b	——: Festmarsch zur Goethejubiläumsfeier (115), 1857 (1859); see also 227	
522	51	——: Festmarsch nach Motiven von E. H. zu S.-C.-G. (116), 1857 (1860), *WRgs*; NA xv	
523	55	——: Ungarischer Marsch zur Krönungsfeier in Ofen-Pest (118), 1870, (1871), **WRgs*; NA xvi	
524	54b	——: Ungarischer Sturmmarsch (119), 1875 (Berlin, 1876), *H-Bl*, ded. S. Teleky; NA xvi	
525	188	——: Totentanz (126), *c*1860–65 (1865), **US-Wc* (frag.), ded. H. von Bülow, NA xvi	
526	189	——: Epithalam zu Eduard Reményis Vermählungsfeier (129), ?1872 (Budapest, 1872), †; NA xii	
527	66b	——: Romance oubliée (132), 1880 (Hanover, 1881), **Wc*; LS vii, NA xii	
528	296	——: renumbered as 230a	
529	22	——: Fantasia and Fugue on the Theme B–A–C–H (260), 1871 (1871); *WRgs*, NA v	
530	190	——: L'hymne du pape (261), *c*1864 (Berlin, 1865); †, NA xv	
531	209	——: Buch der Lieder, *c*1843 (Berlin and Milan, 1844), †; 1 Loreley (273/1), 2 Am Rhein (272/1), 3 Mignon (275/1), 4 Es war ein König in Thule (278/1), 5 Der du von dem Himmel bist (279/1), 6 Angiolin dal biondo crin (269/1); NA xv	
532	209	——: Loreley (273/2), 1861 (1862), †; NA xv	
533	203	——: Il m'aimait tant (271), *c*1843 (Mainz, 1843), †; NA xv	
534	213	——: Die Zelle in Nonnenwerth (274), elegy, *c*1843 (Berlin and elsewhere, 1844), †, ded. Prince Felix Lichnowsky; LS vii, NA xvii	
535	204	——: Comment, disaient-ils (276/1), ?1847, unpubd, **D-WRgs*	
536	210	——: Oh, quand je dors (282/1), ?1847, unpubd, **WRgs*	
537	205	——: Enfant, si j'étais roi (283/1), ?1847, unpubd, **WRgs*	
538	206	——: S'il est un charmant gazon (284/1), ?1847, unpubd, **WRgs*	
539	207	——: La tombe et la rose (285), ?1847, unpubd, **WRgs*	
540	208	——: Gastibelza (286), 1847, unpubd, **WRgs*	
541	211	——: Leibesträume, 3 notturnos (307, 308 and 298), *c*1850, no.2 *c*1845 (1850), **WRgs* (nos.2–3); NA xv	250
542	212	——: Weimars Volkslied (87), 1857 (1857); NA xv	
542a	211a	——: Ich liebe dich (315), 1857 **US-Wc*; NA xv	
543	214	——: Ungarns Gott (339), 1881 (Budapest, 1881), **H-Bn*; NA xvii	
543a	—	——: Ungarns Gott (339), arr. left hand alone, 1881 (Budapest, 1881), †, NA xvii	
544	215	——: Ungarisches Königslied (340), 1883 (Budapest, ?1884–5), **H-Bn* (frag.); NA xvii	
545	194	——: Ave Maria IV (341), 1881, ed. (London, 1958), **Bn*; NA xii	
546	216	——: Der blinde Sänger (350), 1878 (Paris and Leipzig, 1881), †; NA xvii	
546a	—	——: O Roma nobilis (54), 1879, unpubd, **D-WRgs*; NA xvii	

S	R		
547	217	Mendelssohn: [7] Lieder, 1840 (Leipzig and elsewhere, 1841), †, ded. Frau Cécile Mendelssohn: 1 Auf Flügeln des Gesanges op.34/2, 2 Sonntagslied op.34/5, 3 Reiselied op.34/6, 4 Neue Liebe op.19a/4, 5 Frühlingslied op.47/3, 6 Winterlied op.19a/3, 7 Suleika op.34/4	283
548	218	—: Wasserfahrt op.50/4, Der Jäger Abschied op.50/2, 1848 (1849), *US-Wc	
549	226	Meyerbeer: Festmarsch zu Schillers 100jähriger Geburtsfeier, 1860 (Paris and Berlin, 1860), *US-Cn (frag.)	
550	229	Mozart: 2 pieces from the Requiem K626, 1865 (Leipzig, 1865), *Wc; 1 Confutatis, 2 Lacrymosa	
551	232	Pezzini: Una stella amica, mazurka (Rome, n.d.), †	
552	237	Rossini: Ouverture de l'opéra Guillaume Tell, 1838 (Mainz and elsewhere, 1842), †; RS ii/1	283
553	238	—: 2 transcriptions, 1847 (Mainz, 1848): 1 Air du Stabat mater ('Cujus animam'), arr. trbn, org, 679, 1v, org, 682, 2 La charité	
554	239	Rubinstein: 2 songs, 1880, †, 1 O! wenn es doch immer so bliebe (1881), ded. Mme Rubinstein, 2 Der Asra (1884)	
554a	239a	—: Introduction to Study, C, unpubd, *Wc	
555	240	Saint-Saëns: Danse macabre op.40, 1876 (Paris and elsewhere, 1876), †, ded. S. Menter	286, 300
556	241	Schubert: Die Rose [D745], 1833 (Leipzig and Paris, 1833), rev. 1833, pubd in Hommage aux dames de Vienne (Vienna, 1838), †, ded. Countess d'Apponyi	
557	242	—: Lob der Tränen, 1837, pubd in Hommage aux dames de Vienne (Vienna, 1838), †	
558	243	—: [12] Lieder, 1835–7 (Vienna and Leipzig, 1838), *Wc (frag.), ded. Countess d'Aragon (nos.1–11), M. d'Agoult, † (no.12): 1 Sei mir gegrüsst, 2 Auf dem Wasser zu singen, 3 Du bist die Ruh, 4 Erlkönig, 5 Meeresstille, 6 Die junge Nonne, 7 Frühlingsglaube, 8 Gretchen am Spinnrade, 9 Ständchen ('Horch, horch! die Lerch'), 10 Rastlose Liebe, 11 Der Wanderer ('Ich komme vom Gebirge her'), 12 Ellens Gesang III	
559	244	—: Gondelfahrer D809, 1838 (Vienna, 1838), *Wc	282
560	245	—: Schwanengesang, 1838–9, no.7 rev. 1865 (Vienna, 1840). H-Bn (no.7), *Bl (no.13), ded Archduchess Sophie: 1 Die Stadt, 2 Das Fischermädchen, 3 Aufenthalt, 4 Am Meer, 5 Abschied, 6 In der Ferne, 7 Ständchen ('Leise flehen'), 8 Ihr Bild, 9 Frühlingssehnsucht, 10 Liebesbotschaft, 11 Der Atlas, 12 Der Doppelgänger, 13 Die Taubenpost, 14 Kriegers Ahnung	
561	246	—: Winterreise, 1839 (Vienna and elsewhere, 1840), *Bl (except nos.4, 6, 7), ded Princess Elenore Schwarzenberg: 1 Gute Nacht, 2 Die Nebensonnen, 3 Mut, 4 Die Post, 5 Erstarrung, 6 Wasserflut, 7 Der Lindenbaum, 8 Der Leiermann, 9 Täuschung, 10 Das Wirtshaus, 11 Der stürmische Morgen, 12 Im Dorfe	282
562	247	—: [8] Geistliche Lieder, 1840 (Leipzig and elsewhere 1841), †; 1 Am Tage Aller Seelen, 2 Himmelsfunken, 3 Die Gestirne, 4 Dem Unendlichen	283
563	248	—: 6 Melodien, 1846 (Berlin and Paris, 1846), †; 1 Lebewohl, 2 Des Mädchens Klage, 3 Das Zügenglöcklein, 4 Trockne Blumen, 5 Ungeduld (1st version), 6 Die Forelle (1st version)	283
564	248	—: Die Forelle, 2nd version, 1846 (Vienna and elsewhere, 1846), Bl; see also 563	
565	249	—: [6] Müllerlieder, 1846 (Vienna, 1847), †, ded. R. Spina: 1 Das Wandern, 2 Der Müller und der Bach, 3 Der Jäger, 4 Die böse Farbe, 5 Wohin?, 6 Ungeduld (2nd version: see also 563)	283
566	253	Schumann: Widmung, 1848 (Leipzig and elsewhere, 1848), *private collection of Queen of Sweden, US-Wc	283
567	255	—: 2 songs (1861): 1 An den Sonnenschein, 2 Dem roten Röslein	
568	256	—: Frühlingsnacht ('Überm Garten durch die Lüfte') (1872)	
569	257	—: Lieder von Robert and Clara Schumann, †; by Robert: 1 Weihnachtslied, 2 Die wandelnde Glocke, 3 Frühlings Ankunft, 4 Des Sennen	

570		Abschied, 5 Er ist's, 6 Nur wer die Sehnsucht kennt, 7 An die Türen will ich schleichen by Clara: 1 Warum willst du andere fragen?, 2 Ich hab' in deinem Auge, 3 Geheimes Flüstern	
571	254	Spohr: Die Rose, romance (Brunswick, 1876), *in Stargardt auction catalogue (Marburg, 1975)	
	259	——: Provenzalisches Lied, 1881 (Berlin, 1882), †	
572	260	Szabady, orchd Massenet: Revive Szegedin, Hungarian march, 1879 (Paris, 1892), *and other MSS in H-SG, ded. A. Gouzien	
573	261	I. Széchényi: Bevezetés és magyar induló-Einleitung und ungarischer Marsch, 1872 (Budapest, 1873), *US-Wc, ded. Széchényi	
573a	—	P. A. Tirindelli: Seconda mazurka variata, Aug 1880 (Turin, n.d.), *in Sotheby's auction catalogue (London, 1969)	
574	288	Weber: Oberon, ov., 1843 (Berlin and Paris, 1847), *US-R; RS i/1	283
575	289	——: Der Freischütz, ov., 1846 (Berlin and Paris, 1847), *R; RS i/1	283
576	290	——: Jubel-Ouvertüre, 1846 (Berlin and Paris, 1847), *R	283
577	291	M. Wielhorsky: Ljubila ja (I love you), romance, 1843 (St Petersburg and Berlin, 1843), *D-WRgs, H-Bn	

PIANO FOUR HANDS

(arrangements of works by Liszt unless otherwise stated)

577a	—	J. Field: Nocturnes nos.1-9, 14, 18 and Nocturne pastorale, E (n.d.)
578	334	Four Pieces from Elisabeth (2), 1862 (1868): 1 Vorspiel, 2 Marsch der Kreuzritter, 3 Sturm, 4 Interludium
579	335	Orchestral pieces from Christus (3), 1866-73 (1873): 1 Hirtenspiel, ? not arr. Liszt, 2 Die heiligen drei Könige, 3 Stabat mater, *US-Wc (no.2)
580	337	Excelsior! [prelude to 6], c1875 (1876)
581	338	Benedictus and Offertorium [from the Hungarian Coronation Mass (11)], 1869 (1871)
582	350	O Lamm Gottes, unschuldig (50/5), 1878-9, unpubd, *D-WRgs
583	339	Via crucis (53), 1878, unpubd, *H-Bn

340	584	Festkantate zur Enthüllung des Beethoven-Denkmals in Bonn (67), 1845 (Mainz, 1846), †
341	585	Pastorale ('Schnitterchor') from 69, 1861 (1861), †
342	586	Gaudeamus igitur (71), 1870-72 (1872), †
343	587	Marche héroïque (82), c1848, unpubd, D-WRgs
344	588	Weimars Volkslied (87), 1857 (1857), †
315	589	Ce qu'on entend sur la montagne (95), 1874 (1875), †
316	590	Tasso (96), c1858 (1859), *Bds
317	591	Les préludes (97), c1858 (1859), †
318	592	Orpheus (98), c1858 (1859), †
319	593	Prometheus (99), c1858 (1862), †
320	594	Mazeppa (100), 1874 (1875), †
321	595	Festklänge (101), ?1854-61 (1861), †
322	596	Hungaria (103), ?1874 (1875), †
—	596a	Héroïde funèbre (102), ?1877 (1878), †
—	596b	Hunnenschlacht (105), ?1877 (1878), †
—	596c	Die Ideale (106), ?1874-7 (1880), †
323	597	Hamlet (104), 1874 (1875), †
324	598	Von der Wiege bis zum Grabe (107), 1881 (Berlin, 1883)
325	599	Two episodes from Lenau's Faust (110), 1861-2 (1862), *F-Pc (frag.), D-WRgs
326	600	Second Mephisto Waltz (111), 1881 (Berlin, 1881), H-Bn
327	601	Les morts (112/1), 1860, unpubd, *and other MS in D-WRgs
328	602	La notte (112/2), 1886, unpubd, *and other MS in WRgs
329	603	Le triomphe funèbre du Tasse (112/3), 1866, rev. 1869, unpubd, *and other MS in WRgs
330	604	Salve Polonia (113), 1863 (1884), *H-Bn (frag.)
331	605	Künstlerfestzug zur Schillerfeier (114), 1859 (1860), †
302	606	Festmarsch zur Goethejubiläumsfeier (115), c1858 (1859), †
303	607	Festmarsch nach Motiven von E. H. zu S.-C.-G. (116), c1859 (1860)
310	608	Rákóczy March (117), 1870 (1871), *private collection of Gregory Benko (New York), Bn
306	609	Ungarischer Marsch zur Krönungsfeier in Ofen-Pest (118), 1870 (1871), †
305	610	Ungarischer Sturmmarsch (119), 1875 (Berlin, 1876), Bl
332	611	Epithalam (129), ?1872 (Budapest, ?1872), †

S	R	
612	333	Elegie (130), 1874 (1875), †
613	307	Weihnachtsbaum (186), 1876 (Berlin, 1882), *Bn*
614	313	Dem Andenken Petőfis (195), 1877 (Budapest, 1877), †
615	298	Grande valse de bravura (209), 1836 (Paris and elsewhere, 1842), †, ded. P. Wolf, ? collab. F. Mockwitz
616	299	Grand galop chromatique (219), 1838 (Leipzig and Paris, 1838), †
617	301	Csárdás macabre (224), 1882, unpubd, *(frag.) and other MS in *D-WRgs*, ded. J. von Végh
618	300	Csárdás obstiné (225/2), ?1884 (Budapest, n.d.)
618a	—	Vom Fels zum Meer (229), unpubd, *private collection of H. Cardello (New York)
619	304	Bülow-Marsch (230), c1883 (Berlin, 1884), †
620	308	Hussitenlied (234), 1840 (Prague, 1841), †
621	309	[6] Hungarian Rhapsodies (244/14, 12, 6, 2, 5, 9), 1874 (Leipzig, 1875), *WRgs (nos.2–6), based on orch version (359)
622	311	Hungarian Rhapsody no.16 (244/16), 1882 (Budapest, 1882), *US-Wc
623	312	Hungarian Rhapsody no.18 (244/18), 1885 (Budapest, 1885)
623a	—	Hungarian Rhapsody no.19 (244/19), 1885–6, ed. (Vienna, n.d.), †
624	314	Fantasia and Fugue on 'Ad nos ad salutarem undam' (259), 1850 (1852), †; RS iv/3
625	336	Der Papsthymnus (261), c1865 (Berlin, 1865), *NYp
626	345	Ungarisches Königslied (340), c1884 (Budapest, c1884), *H-Bn
627	348	Fantasia, on themes from La sonnambula (393), c1852 (1876), †; RS iv/3
628	349	Berlioz: Bénédiction et serment [from Benvenuto Cellini] (396) (Berlin and Brunswick, 1854), †; RS iv/3
628a	—	Donizetti: Marche et cavatine de Lucie de Lammermoor (398), 1835–6 (Mainz and elsewhere, 1841), †; RS iv/3
628b	—	Egressy and Erkel: Szózat und Ungarischer

S	R	
629	351	Hymnus (353) (Budapest, n.d.), †; ? not arr. Liszt
630	352	Glinka: Tscherkessenmarsch (406), 1843 (1843), †; RS iv/3; 2nd version, 1875, ?unpubd, †
631	353	Réminiscences de Robert le diable (413), 1841–3 (Berlin, 1843), †; ? not by Liszt
632	354	Raff: Andante finale and March [from König Alfred] (421), ?1853 (?1853), †
633	346	Schubert: 4 Marches (363), 1879 (Berlin, n.d.), †, ded. W. and L. Thern
634	347	A la Chapelle Sixtine (461), c1865 (1866). †
634a	352a	Beethoven: Septet op.20 (465), ?1841 (?1842), †
634b	296	Mozart: Adagio ('Der welcher wandelt diese Strasse') from Die Zauberflöte, 1875–81, unpubd, *US-Wc
	301	Festpolonaise (230a), 15 Jan 1876, pubd in Göllerich (1908), *Wc, *D-WRgs*

TWO PIANOS

(arrangements of works by Liszt unless otherwise stated)

S	R	
635	357	Ce qu'on entend sur la montagne (95), c1854–7 (1857)
636	358	Tasso (96), c1854–6 (1856)
637	359	Les préludes (97), c1856), *US-PHr (frag.)
638	360	Orpheus (98), c1854–6 (1856), †
639	361	Prometheus (99), 1855–6 (1856)
640	362	Mazeppa (100), 1855 (1856)
641	363	Festklänge (101), c1853–6 (1856)
642	364	Héroïde funèbre (102), c1854–6 (1856)
643	365	Hungaria (103), c1854–61 (1861), *listed in Kinsky
644	366	Hamlet (104), c1858–61 (1861)
645	367	Hunnenschlacht (105), 1857 (1861), *NYpm, ded. W. von Kaulbach
646	368	Die Ideale (106), 1857–8 (1858), †
647	369	Eine Faust-Symphonie (108), 1856, rev. 1860 (1863)
648	370	Eine Symphonie zu Dantes Divina commedia (109), c1856–9 (1859)
649	371	Fantasia, on themes from Beethoven's Die Ruinen von Athen (122), after 1852 (1865), †; RS iv/2
650	372	Piano Concerto no.1 (124), 1853 (Vienna, 1857), †

651 Piano Concerto no.2 (125), 1859 (Mainz, 1862), † 373

652 Totentanz (126), after 1859 (1865) 374

653 Schubert; Wandererfantasie D760 (366), after 1851 (Vienna, 1862), † 375

654 Hexaméron (392), after 1837 (1870), †; RS iv/2 377

655 Réminiscences de Norma (394), after 1841 (Mainz, 1874), †; RS iv/2 378

656 Réminiscences de Don Juan (418), after 1841 (Berlin, 1877), *S-Smf; RS iv/2 379

657 Beethoven: Sym. no.9 (Mainz, c1851), † 376

657a Beethoven: Piano Concertos nos.3–5, 1878–9 (Stuttgart, 1879), *US-Wc (nos.4–5), H-Bl (no.3) —

657b Bülow-Marsch (230), 8 hands, 1884, unpubd, Bl —

ORGAN

658 Allegri and Mozart: Evocation à la Chapelle Sixtine (461), c1862 (1865), *Nydal collection (Stockholm), ded. A. W. Gottschalg 400

659 Arcadelt: Ave Maria (183/2), 1862 (1865), * and ded. A. W. Gottschalg 401

660 Bach: Introduction and Fugue from the cantata Ich hatte viel Bekümmernis BWV21 and Andante from Aus tiefer Not BWV38, 1860 (1862), *US-Wc (frag.) ded. J. G. Töpfer 402

661 —: Adagio from Violin Sonata no.4 BWV1017 (n.d.), †, collab. A. W. Gottschalg 403

662 Chopin: 2 preludes from op.28 [nos.4 and 9] (n.d.), † 404

663 Lassus: Regina coeli laetare, 1865 (n.d.) 405

664 Liszt: Tu es Petrus (3/8), 1867 (n.d.), *F-Pc 391

665 —: S Francesco, prelude to 4, 1880, org/pf, *US-NYpm 392

666 —: Excelsior! [prelude to 6], after 1874 (n.d.), D-WRgs 393

667 —: Offertorium [from the Hungarian Coronation Mass (11)], org/harmonium/pedal pf, after 1867 (n.d.), † 411b

668 —: Slavimo slavno slaveni! (33), 1863, pubd in Jb des Sängerbund Dreizehnlingen (Vienna, 1910–11), *Göllerich collection 397

669 —: 2 Kirchenhymnen (1880), ded. Cardinal 394 302

285

670 Prince G. von Hohenlohe-Schillingsfürst: 1 Salve regina, 1877, *US-Wc, 2 Ave maris stella (34/2), after 1868, *H-Bn 396

671 —: Rosario (56/1–3), 1879, unpubd, *D-WRgs 395

—: Zum Haus des Herren ziehen wir, prelude to 57, org/pf, pubd in Göllerich (1908), *and other MS in H-Bn

672 —: Weimars Volkslied (87), org/harmonium, 1865 (1873) 398

673 —: Variations on a Theme of Bach (180), 1863 (1865), *US-NYp, ded. A. W. Gottschalg 382

674 —: Ungarns Gott (339), c1881 (Budapest, 1882), † 399

674a —: Via Crucis (53), 1878–9, D-WRgs, ?unpubd —

675 Nicolai: Kirchliche Festouvertüre über den Choral 'Ein feste Burg ist unser Gott', 1852 (Vienna, 1852), H-Bn 406

676 Wagner: Pilgrims' Chorus [from Tannhäuser], 1860, rev. 1860, rev. 1862 (Dresden, 1864), arr. pf, 443, *D-WRgs, S-Smf 407

ORGAN AND OTHER INSTRUMENTS

677 Liszt: Hosannah, from 4, org, trbn ad lib, 1862, pubd in Töpferalbum (Leipzig, 1867) 409

678 —: Offertorium and Benedictus, from 11, vn, org, ?1871 (n.d.) 411a

679 Rossini: Cujus animam, from Stabat mater, org, trbn, 1860–70 (Mainz, 1874) 410

VOCAL ARRANGEMENTS

680 Liszt: Ave maris stella (34/2), 1v, pf/harmonium, 1868 (Paris, 1868) 641

681 Liszt: Ave Maria, II (38), 1v, org/harmonium, 1869, B; B v/6 639

682 Rossini: Cujus animam, from Stabat mater, T, org (Mainz, 1874), †; B v/7 643a

683 trad. Serbian: Ein Mädchen sitzt am Meerestrand, 1v, pf, pubd in Orpheon-Album, iv (Stuttgart, n.d.), collab. Prince F. W. C. of Hohenzollern-Hechingen 644a

683a Korbay: Gebet, 1v, org, unpubd, *in Sotheby's —

S	R		p.
684	644	auction catalogue (London, 1969)	
685	644b	Pantaleoni: Barcarole vénitienne, 1v, pf (Leipzig and elsewhere, 1842), †, ded. T. de Bacheracht	
685	644b	M. Pavlovna: Es hat geflammt die ganze Nacht, unpbud, *WRgs	
686	659	Draeseke: Helges Treue (O. Strachwitz), recitation, pf acc., 1860 (1874), ded. B. Davison	

UNFINISHED WORKS

S	R		p.
687	670	Sardanapale (opera, 3, ?Rotondi, after Byron), 1846–51, WRgs (sketches)	311
688	671	Die Legende vom heiligen Stanislaus (oratorio, P. Cornelius, K. E. Edler and ? Princess Sayn-Wittgenstein, after L. Siemienski), 1873–85, *and other MSS in WRgs, *US-Wc (frag.); excerpts: 16, 113 and 519	250, 299, 308, 310
689	672	Singe, wem Gesang gegeben, male vv, c1847, orchd Conradi, D-WRgs	
690	667	Revolutionary Symphony, July 1830, sketches pubd in Raabe (1916), rev. 1848, WRgs (MS and sketches); see also 102	241, 293
691	668	De profundis, inst psalm, pf, orch, 1834–5, *WRgs, ded. Lamennais; see also 126 and 173/4	242, 271, 287
692	669	Violin Concerto, Jan 1860, intended for E. Reményi	301
692a	674	Die vier Jahreszeiten, str qt, c1880, *WRgs	301
692b	—	Piano Sonata, c1835, listed in Schnapp (no.19)	
693	662	Two Hungarian pieces, pf, d, bb, c1840; LS iii	
694	(98)	Fantasia on English Themes, pf, c1840, *WRgs	
695	663	Piano piece, F, ? July 1843, *WRgs	
695a	—	Litanie de Marie, 1847, *WRgs, orig. intended for inclusion in 173; NA ix	
696	661	Fourth Mephisto Waltz, pf, March 1885; LS i	281
697	660	Fantasia on two themes from Le nozze di Figaro, 1842, ed. and completed by F. Busoni (Leipzig, 1912)	283–4
698	666	La mandragore, ballad from Delibes' Jean de Nivelle, pf, after 1880, *WRgs	
699	664	arr. of La notte (112/2), pf, 1864–6, *WRgs; NA xi	
700	665	arr. of Paganini's Carnaval de Venise, pf, *WRgs	
701	673	Den Felsengipfel stieg ich einst hinan, 1v, pf, *Breitkopf & Härtel archive, Wiesbaden	
701a	—	arr. of 151, orch, c1830, *WRgs	

S	R		p.
701b	—	Marie-Poème, 1837, *private collection of J. Pierre, Paris	

DOUBTFUL OR LOST WORKS

CHORAL

S

S			p.
702		Tantum ergo, 1822, mentioned in Ramann, i (1880), 39	305
703		Psalm ii, T, mixed vv, orch, 1851, mentioned in Göllerich (1908), 316	
704		Requiem, on the death of Emperor Maximilian of Mexico, mentioned in Grove 1–4	
705		The Creation, mentioned in Grove 1–4	
706		Benedictus, mixed vv, pf, ed. (New York, 1939), doubtful	
707		arr. of Excelsior!, prelude to 6, Mez/Bar, male vv, pf (Leipzig, n.d.), doubtful	
708		Rinaldo (Goethe), T, male vv, pf, c1848, orchd Conradi, D-WRgs, doubtful	
708a		renumbered as 81a	
708b		Piece for 4 male vv, ?1843–4, companion-piece to 78, mentioned in Schnapp (no.42)	

ORCHESTRAL

S			p.
709		renumbered as 113	
710		Funeral march, mentioned in H. Raff (1901–2), 389	
711		arr. of Csárdás macabre (224), mentioned in Göllerich (1908), 278	
712		arr. of Romance oubliée (132), va, orch (Hanover, n.d.), doubtful	
713		2 pf concs., 1821, mentioned in Ramann, i (1880), 7	
714		renumbered as 126a	
715		Piano Concerto in the Italian Style, mentioned in Göllerich (1908), 281	
716		Grande fantaisie symphonique, a, mentioned in Göllerich (1908), 281	240

CHAMBER

S			p.
717		Trio, 1825, mentioned in Ramann, i (1880), 7	
718		Quintet, 1825, mentioned in Ramann, i, (1880), 7	
719		renumbered as 692a	
720		Allegro moderato, E, vn, pf, mentioned in Göllerich (1908), 280	
721		Prelude, vn, mentioned in Helbig (1907), 74	
722		renumbered as 377a	
723		Tristia, from Vallée d'Obermann (160/6), arr. pf trio, mentioned in Göllerich (1908), 280	
723a		Orpheus (98), harp, harmonium, vn, pf, 1871, unpubd, †, mentioned in letter of 7 March 1871	240

724 Rondo and Fantasia, 1824, mentioned in Ramann, i (1880), 7
725 3 sonatas, 1825, listed in Schnapp (1942), nos.5–7
726 Study, C, mentioned in *Grove 1–4*
726a Valse, E, mentioned in letter of 26 July 1835
727 Prélude omnitonique, mentioned in *Catalogue of the Music Loan Exhibition of the Worshipful Company of Musicians, London, 1904* (London, 1909), 286; †
728 renumbered as 192/5
729 renumbered as 42
730 renumbered as 195
731 Valse élégiaque (Berlin, n.d.), mentioned in *Grove 1–4*
732 renumbered as 215/4
733 renumbered as 233b
734 Ländler, D, *D-WRgs*, doubtful
735 Air cosaque, *WRgs*, doubtful
736 Kerepesi csárdás, mentioned in catalogue of *H-Bn*, doubtful
737 3 morceaux en style de danse ancien hongrois (Budapest, 1850): 1 Maestoso, 2 Tempo di Werbung, 3 Andante ritmico; nos.1–2 by János Liszt, no.3 also doubtful
738 Spanish folksong arrs., mentioned in Conradi and Liszt: *Programme général* (c1850) and Ramann, i (1880), 270
739 arr. of Beethoven's Coriolan, ov., mentioned in La Mara, ed. (1893–1904), i, 66
740 arr. of Beethoven's Egmont, ov., mentioned in Conradi and Liszt
741 arr. of Berlioz's Le carnaval romain, ov., mentioned in Conradi and Liszt
742 arr. of Donizetti's Duettino, mentioned in Pazdírek, ? same as 399
743 arr. of Soldiers' Chorus from Gounod's Faust, mentioned in La Mara, ed. (1898), 344
743a Fantasia on themes from Halévy's Guitarero, perf. Kassel, 1841
744 Paraphrase of Act 4 of Kullak's Dom Sébastien, mentioned in Conradi and Liszt
745 Funeral March, pf, mentioned in La Mara, ed. (1898), 690, ? same as 173/7
746 Andante maestoso (Budapest, n.d.), mentioned in *Grove 1–4*, doubtful
747 Poco adagio [from Missa solemnis (9)], arr. pf, mentioned as for 727; *GB-Lbm*
748 arr. of Mozart's Die Zauberflöte, ov., mentioned in Conradi and Liszt

749 arr. of Radovsky's Preussischer Armeemarsch, mentioned in Thouret's *Katalog der Musikbibliothek im Schlosse zu Berlin*
750 renumbered as 421a
751 Nonetto e Mose, fantasia on themes by Rossini, mentioned in Conradi and Liszt
752 arr. of Rubinstein's Gelb rollt, mentioned in *Grove 1–4*
753 arr. of Schubert's Alfonso und Estrella, Act 1, listed in Schnapp (1942), no.51
754 renumbered as 573a
754a Grand Solo caractéristique à propos d'une chansonette de Panseron, mentioned in letter of 12 Dec 1832
754b Ballade, mentioned in same letter as 754a

755 Sonata, pf 4 hands, mentioned in Ramann, i (1880), 7
756 Mosonyis Grabgeleit (194), arr. pf 4 hands, mentioned in Göllerich (1908), 305
757 Le triomphe funèbre du Tasse (112/3), arr. 2 pf, mentioned in Göllerich (1908), 307
758 The Organ, sym. poem, after Herder, org, mentioned in *Grove 1–4*
759 Consolation, D♭ (172/4), arr. org, pubd as Adagio (Leipzig, n.d.), doubtful
760 Cantico del sol di S Francesco (4), arr. org, mentioned in La Mara, ed. (1893–1904), vii, 327
761 Chopin's Marche funèbre, from op.35, arr. org, vc, pf, mentioned in Habets (1893)

762 Air de Chateaubriand, 1v, pf, mentioned in C. Sayn-Wittgenstein's catalogue of Liszt's works (MS, *D-WRgs*, before 1887)
763 Strophes de Herlossohn, 1v, pf, ibid
764 Kränze pour chant, 1v, pf, ibid
765 Glöckchen (Müller), 1v, pf, ibid [companion-piece to 316/1–2]
765a L'aube naît (Hugo), 1842, mentioned in D. Ollivier, ed. (1933–4), ii, 211
766 Der Papsthymnus (261), arr. S/T, pf (Leipzig, n.d.), doubtful
767 Excelsior!, prelude to 6, arr. 1v, org, *WRgs*, doubtful
768 Der ewige Jude (Schubart), recitation, pf acc., mentioned in *Grove 1–4*

240

WRITINGS

De la fondation Goethe à Weimar (Leipzig, 1851)
Lohengrin et Tannhäuser de R. Wagner (Leipzig, 1851)
F. Chopin (Paris, 1852) — 255-6
Des bohémiens et de leur musique en Hongrie (Paris, 1859) — 256
Über John Fields Nocturne (Leipzig, 1859)
R. Schumanns musikalische Haus- und Lebensregeln (Leipzig, 1860)
Gesammelte Schriften, ed. L. Ramann (Leipzig, 1880-83):
  i: Friedrich Chopin (1852)
  ii/1: Zur Stellung des Künstlers (1835); Über die zukünftige Kirchenmusik (1834); Über Volksausgaben bedeutender Werke (1836); Über Meyerbeers Hugenotten (1837); Thalbergs Grande fantaisie und Caprices (1837) [on op.22 and opp.15 and 19]; R. Schumanns Klavier-Kompositionen (1837) [on opp.5, 11 and 14]; Paganini: ein Nekrolog (1840) — 243, 256, 305
  ii/2: Reisebriefe eines Baccalaureus der Tonkunst (1835-40) [letters to G. Sand, A. Pictet, L. de Ronchaud, M. Schlesinger and M. d'Ortigue]
  iii: Dramaturgische Blätter (1849-56) [incl. articles on well-known 19th-century operas]
  iv: Aus den Annalen des Fortschrittes (1855-9) [essays on Berlioz, R. and C. Schumann, R. Franz, E. Sobolewski and J. Field]
  v: Streifzüge: kritische, polemische und zeithistorische [essays] (1850-58)
  vi: Die Zigeuner und ihre Musik in Ungarn (1859)

BIBLIOGRAPHY

CATALOGUES, BIBLIOGRAPHIES

A. Conradi and F. Liszt: *Programme général des morceaux exécutés par F. Liszt à ses concerts de 1838 à 1848* (MS, D-WRgs, c1850)

F. Liszt: *Thematisches Verzeichnis der Werke von F. Liszt* (Leipzig, 1855)

——: *Thematisches Verzeichnis der Werke, Bearbeitungen und Transkriptionen von F. Liszt* (Leipzig, 1877)

C. Sayn-Wittgenstein: Catalogue of works (MS, *D-WRgs*, before 1887)

L. Friwitzer: 'Chronologisch-systematisches Verzeichnis sämtlicher Tonwerke Franz Liszts', *Musikalische Chronik*, v (Vienna, 1887)

F. Pazdírek: *Universal-Handbuch der Musikliteratur* (Vienna, c1904–10)

A. Göllerich: *Franz Liszt* (Berlin, 1908), 272–331

G. Kinsky: *Musikhistorisches Museum von Wilhelm Heyer in Cöln: Katalog*, iv (Cologne, 1916)

P. Raabe: *Franz Liszt* (Stuttgart, 1931, rev. 2/1968), ii, 241–364; suppl., 7

L. Koch: *Liszt Ferenc bibliografiai kisérlet* [Towards a Liszt bibliography], A fövárosi könyvtar evkönyve, v (Budapest, 1936)

F. Schnapp: 'Verschollene Kompositionen Franz Liszts', *Von deutscher Tonkunst: Festschrift zu Peter Raabes 70. Geburtstag* (Leipzig, 1942), 119–53

C. Suttoni: 'Franz Liszt's Published Correspondence: an Annotated Bibliography', *FAM*, xxvi (1979), 191–234

E. N. Waters: *Liszt Holographs in the Library of Congress* (Washington, DC, 1979)

GENEALOGIES, ICONOGRAPHIES

D. Bartha: *Franz Liszt, 1811–1886: sein Leben in Bildern* (Leipzig, 1936)

R. Bory: *La vie de Liszt par l'image* (Paris, 1936)

W. Füssmann and B. Mátéka: *Franz Liszt: ein Künstlerleben in Wort und Bild* (Langensalza, 1936)

H. E. Wamser: *Abstammung und Familie Liszts* (Eisenstadt, 1936)

E. Ritter von Liszt: *Franz Liszt: Abstammung, Familie, Begebenheiten* (Vienna and Leipzig, 1937)

Z. László and B. Mátéka: *Franz Liszt: sein Leben in zeitgenössischen Bildern* (Kassel, 1969; Eng. trans., 1968)

SPECIALIST PERIODICALS

Zenetudományi tanulmányok, iii (1955) [special Liszt–Bartók issue]

SM, iv (1963) [special Liszt–Bartók issue]

369

Liszt

Journal of the American Liszt Society (Louisville, Kentucky, 1976–)
Liszt Society Journal (London, 1976–)

AUTOBIOGRAPHY, LETTERS TO AND FROM LISZT

F. Hueffer, ed.: *Briefwechsel zwischen Wagner und Liszt* (Leipzig, 1887, 4/1919; Eng. trans., 1888)

La Mara [pseud. of M. Lipsius], ed.: *Franz Liszt's Briefe* (Leipzig, 1893–1902)

C. Bache, ed.: *Letters of Franz Liszt* (London, 1894)

La Mara, ed.: *Briefe hervorragender Zeitgenossen an Franz Liszt* (Leipzig, 1895–1904)

——: *Briefwechsel zwischen Franz Liszt und Hans von Bülow* (Leipzig, 1898)

H. Raff: 'Franz Liszt und Joachim Raff im Spiegel ihrer Briefe', *Die Musik*, i (1901–2)

A. Stern, ed.: *Liszts Briefe an Carl Gille* (Leipzig, 1903)

E. Istel, ed.: 'Elf ungedruckte Briefe Liszts an Schott', *Die Musik*, v (1905–6), 43

'Franz Liszts Briefe an den Fürsten Felix Lichnowsky', *Bayreuther Blätter 1907*, nos.1–3

La Mara, ed.: *Briefwechsel zwischen Franz Liszt und Carl Alexander, Grossherzog von Sachsen* (Leipzig, 1909)

A. W. Gottschalg: *Franz Liszt in Weimar*, ed. C. A. René (Berlin, 1910) [with 48 letters by Liszt]

V. Csapó, ed.: *Liszt Ferenc levelei baró Augusz A.* [Liszt's letters to the Baron Augusz] (Budapest, 1911; Ger. trans., 1911)

J. Kapp: 'Aus Weimars musikalischer Glanzzeit: mit mehreren unveröffentlichen Briefen Franz Liszts', *Die Musik*, xi (1911–12), 223

——, ed.: 'Autobiographisches von Franz Liszt', *Die Musik*, xi (1911–12), 10

N. de Gutmansthal: *Souvenirs de François Liszt: lettres inédites* (Paris, 1913)

La Mara, ed.: *Franz Liszt: Briefe an seine Mutter* (Leipzig, 1918)

R. Bory, ed.: 'Correspondance inédite de Liszt et de la Princesse Marie Sayn-Wittgenstein', *ReM*, ix (1928), 47

——: 'Diverses lettres inédites de Liszt', *Schweizerisches Jb für Musikwissenschaft*, iii (1928), 5

A. Orel, ed.: 'Unbekannte Briefe von Franz Liszt', *Jb der Österreichischen Leo-Gesellschaft 1930*

F. Schnapp, ed.: *Liszts Testament: aus dem Französischen ins Deutsche übertragen* (Weimar, 1931)

D. Ollivier, ed.: *Correspondance de Liszt et de la comtesse d'Agoult 1833–1840* (Paris, 1933–4)

Bibliography

F. Schnapp, ed.: 'Briefe Franz Liszts und seiner Familie', *Die Musik*, xxviii (1935–6), 662

D. Ollivier, ed.: *Correspondance de Liszt et de sa fille Mme Emile Ollivier* (Paris, 1936)

V. Kisselev, ed.: 'Nezdannïye pis'ma F. Lista' [Unpublished letters of Liszt], *SovM* vii (1951), 77

H. E. Hugo, ed.: *The Letters of Franz Liszt to Marie zu Sayn-Wittgenstein* (Cambridge, Mass., 1953)

M. Prahacs, ed.: *Franz Liszt: Briefe aus ungarischen Sammlungen 1835–1886* (Kassel, 1969)

W. R. Tyler and E. N. Waters, eds.: *Letters of Franz Liszt to Olga von Meyendorff* (Washington, DC, 1979)

OTHER CORRESPONDENCE, RECOLLECTIONS, MEMOIRS

G. Sand: *Lettres d'un voyageur* (Paris, 1837, numerous subsequent edns.)

——: *Histoire de ma vie*, x (Paris, 1856, numerous subsequent edns.)

D. Stern [pseud. of M. d'Agoult]: *Mes souvenirs (1806–33)* (Paris, 1877, numerous subsequent edns.)

A. Fay: *Music Study in Germany* (Chicago, 1881, 4/1893/*R*1965)

A. Habets, ed.: *Borodine et Liszt* (Paris, 1893; Eng. trans., 1895)

C. Hallé: *Life and Letters* (London, 1896)

J. Vianna da Motta: *Einige Beobachtungen über Franz Liszt* (Munich, 1898)

W. Weissheimer: *Erlebnisse mit Richard Wagner* (Stuttgart, 1898)

E. Sauer: *Meine Welt* (Stuttgart, 1901)

A. von Schorn: *Zwei Menschenalter: Erinnerungen und Briefe* (Berlin, 1901)

La Mara: *Aus der Glanzzeit der Weimarer Altenburg* (Leipzig, 1906)

C. Wagner: *Franz Liszt: ein Gedenkblatt von seiner Tochter* (Munich, 1911)

K. von Schlözer: *Römische Briefe* (Berlin and Leipzig, 1913, 16/1926)

R. Wagner: *Briefe an Hans von Bülow* (Jena, 1916)

La Mara: *An der Schwelle des Jenseits* (Leipzig, 1925)

V. Boissier: *Liszt pédagogue* (Paris, 1927/*R*1976)

D. Stern: *Mémoires 1833–1854*, ed. D. Ollivier (Paris, 1927)

M. Herwegh: *Au printemps des dieux* (Paris, 1930)

——: *Au banquet des dieux* (Paris, 1931)

A. Apponyi: *Erlebnisse und Ergebnisse* (Berlin, 1933)

M. Herwegh: *Au soir des dieux* (Paris, 1933)

D. Ollivier: *Liszt et ses enfants* (Paris, 1936) [incl. 1 letter by Liszt]

J. Vier: *François Liszt, l'artiste – le clerc: documents inédits* (Paris, 1950)

371

P. Benary: 'Liszt und Wagner in Briefen der Fürstin von Wittgenstein', *Mf*, xii (1959), 468

H. Weilguny: *Das Liszthaus in Weimar* (Weimar, 1963) [incl. Borodin's recollections of Liszt]

W. Jerger: *Franz Liszts Klavierunterricht von 1884–1886 dargestellt an den Tagebuchaufzeichnungen von August Göllerich* (Regensburg, 1975)

M. Gregor-Dellin and D. Mack, eds.: *Cosima Wagner: die Tagebücher 1869–1877* (Munich and Zurich, 1976–7; Eng. trans., 1978–9)

M. P. Eckhardt and C. Knotik: *Franz Liszt und sein Kreis in Briefen und Dokumenten aus den Beständen des Burgenländischen Landesmuseums* (Eisenstadt, 1983)

BIOGRAPHICAL STUDIES, LIFE AND WORKS

J. W. Christern: *Franz Liszt, nach seinem Leben und Werke, aus authentischen Berichten dargestellt* (Hamburg, 1841)

L. Rellstab: *Franz Liszt: Beurteilungen–Bericht–Lebensskizze* (Berlin, 1842, 2/1861)

G. Schilling: *Franz Liszt: sein Leben und Wirken aus nächster Beschauung* (Stuttgart, 1844)

R. L. de Beaufort: *Franz Liszt: the Story of his Life* (Boston and London, 1866, 2/1910) [incl. N. Helbig: 'F. Liszt in Rome', 207]

La Mara: *Musikalische Studienköpfe*, i (Leipzig, 1868), 233–75

J. Schubert: *Franz Liszts Biographie* (Leipzig, 1871)

L. Ramann: *Franz Liszt als Künstler und Mensch*, i–iii (Leipzig, 1880–94; vol.i, Eng. trans., 1882)

O. Lessmann: *Franz Liszt: eine Charakterstudie* (Berlin, 1881)

L. Nohl: *Franz Liszt* (Leipzig, 1882–8)

R. Pohl: *Franz Liszt*, Gesammelte Schriften über Musik und Musiker, ii (Leipzig, 1883)

B. Vogel: *Franz Liszt: Abriss seines Lebens und Würdigung seiner Werke* (Leipzig, 1888)

La Mara: *Classisches und Romantisches aus der Tonwelt* (Leipzig, 1892)

O. Lüning: *Franz Liszt, ein Apostel der Ideale* (Zurich, 1896)

E. Reuss: *Franz Liszt* (Leipzig, 1898)

R. Louis: *Franz Liszt*, Vorkämpfer des Jahrhunderts, ii (Berlin, 1900)

C. Bache: *Brother Musicians* (London, 1901)

M. D. Calvocoressi: *Franz Liszt: biographie critique* (Paris, 1905)

A. Göllerich: *Franz Liszt* (Berlin, 1908)

J. Kapp: *Franz Liszt* (Berlin, 1909, 20/1924)

——: *Liszt-Brevier* (Leipzig, 1910)

——: *R. Wagner und Franz Liszt: eine Freundschaft* (Berlin, 1910)

J. G. Prod'homme: *Franz Liszt* (Paris, 1910)

Bibliography

J. Chantavoine: *Liszt* (Paris, 1911, 6/1950)

A. Hervey: *Franz Liszt and his Music* (London, 1911)

J. G. Huneker: *Franz Liszt* (New York and London, 1911)

A. Sallès: *Le centenaire de Liszt: Liszt à Lyon* (Paris, 1911)

H. Thode: *Franz Liszt* (Heidelberg, 1911)

A. von Schorn: *Das nachklassische Weimar* (Weimar, 1911–12)

P. Bekker: *Franz Liszt* (Bielefeld, 1912)

G. R. Kruse: 'August Conradi', *Die Musik*, xii/4 (1912–13), 3

G. Hollitzer: *Liszt Ferenc és a weimari irodalmi élet* [Liszt and the literary life of Weimar], Német philologiai dolgozatok, vi (Budapest, 1913)

J. Kapp: 'Hector Berlioz über F. Liszt', *Der Merker*, iv (1913), 86

——: 'Franz Liszt und Robert Schumann', *Die Musik*, xiii (1913–14), 67

B. Schrade: *Franz Liszt* (Berlin, 1914)

P. Raabe: *Grossherzog Karl Alexander und Liszt* (Leipzig, 1918)

La Mara: *Liszt und die Frauen* (Leipzig, 2/1919)

L. Bourgues and A. Dénéréaz: *La musique et vie intérieure: essai d'une histoire psychologique de l'art universel* (Paris, 1921)

R. Bory: *Une retraite romantique en Suisse: Liszt et la Comtesse d'Agoult* (Geneva, 1923, 2/1930)

K. Grunsky: *Franz Liszt* (Leipzig, 1924)

H. J. Moser: *Geschichte der deutschen Musik vom Auftreten Beethovens bis zur Gegenwart* (Stuttgart, 1924)

G. de Pourtalès: *La vie de Franz Liszt* (Paris, 1925, 2/1950; Eng. trans., 1926)

W. Wallace: *Liszt, Wagner and the Princess* (London, 1927)

C. van Wessem: *Franz Liszt* (The Hague, 1927)

E. Mesa: *Liszt: su vida y sus obras* (Paris, 1929)

G. de Pourtalès: *Liszt et Chopin* (Paris, 1929)

P. Raabe: *Franz Liszt* (Stuttgart, 1931, rev. 2/1968)

F. Corder: *Ferenc (François) Liszt* (London, 2/1933/R1976)

E. Newman: *The Life of Richard Wagner* (London, 1933–47)

C. Gray: 'Franz Liszt', *The Heritage of Music*, ii, ed. H. J. Foss (London, 1934)

E. Newman: *The Man Liszt* (London, 1934, 2/1970)

S. Sitwell: *Liszt* (London, 1934/R1965, rev. 3/1966)

H. Engel: *Franz Liszt* (Potsdam, 1936)

E. Haraszti: 'Liszt à Paris: quelques documents inédits', *ReM* (1936), no.165, p.241

A. Hevesy: *Liszt, ou le roi Lear de la musique* (Paris, 1936)

R. Hill: *Liszt* (London, 1936, 2/1949)

L. Nowak: *Franz Liszt* (Innsbruck, 1936)

B. Ollivier: *Liszt, le musicien passionné* (Paris, 1936)

E. Haraszti: 'Deux franciscains: Adam et Franz Liszt', *ReM* (1937), no.174, p.269

M. Tibaldi Chiesa: *Vita romantica di Liszt* (Milan, 1937, 2/1941)

P. Raabe: *Wege zu Liszt* (Regensburg, 1944)

C. Gray: *Contingencies* (London, 1947)

A. M. Pols: *Franz Liszt* (Bloemendaal, 1951)

I. Revesz: 'Liszt es Lamennais', *Zenetudományi tanulmányok*, i (1953), 115

T. Marix-Spire: *Les romantiques et la musique: le cas George Sand* (Paris, 1954)

H. Searle: *The Music of Liszt* (London, 1954, rev. 2/1966)

E. Haraszti: 'Pierre Louis Dietsch und seine Opfer', *Mf*, viii (1955), 39

J. Vier: *La Comtesse d'Agoult et son temps* (Paris, 1955–63)

B. Voelcker: *Franz Liszt, der grosse Mensch* (Weimar, 1955)

W. Beckett: *Liszt* (London, 1956, 2/1963)

I. Csekey: *Liszt Ferenc Baranyában* [Liszt in Baranya] (Fünfkirchen, 1956)

Y. Mil'shteyn: *F. List* (Moscow, 1956, rev. 2/1971; Hung. trans., 1965)

W. Serauky: 'Robert Schumann in seinem Verhältnis zu Ludwig van Beethoven und Franz Liszt', *Robert Schumann: aus Anlass seines 100. Todestages* (Leipzig, 1956), 68

B. Szabolcsi: *Liszt Ferenc estéje* [The twilight of Franz Liszt] (Budapest, 1956; Eng. trans., 1959)

G. Z. Törnbom: *Liszt* (Stockholm, 1956)

C. Haldane: *The Galley Slaves of Love* (London, 1957)

I. Csekey: 'Franz Liszt in Pécs (Fünfkirchen)', *Mf*, xi (1958), 69

H. Engel: 'Liszt, Franz', *MGG*

C. Rostand: *Liszt* (Paris, 1960; Eng. trans., 1972)

P. Rehberg and G. Nestler: *Franz Liszt* (Zurich, 1961)

K. Hamburger: *Liszt Ferenc* (Budapest, 1966; Ger. trans., 1973)

E. Haraszti: *Franz Liszt* (Paris, 1967)

A. Walker, ed.: *Franz Liszt: the Man and his Music* (London, 1970, 2/1976)

A. Walker: *Liszt* (London, 1971)

N. Gerson: *George Sand* (London, 1973)

E. Perényi: *Liszt* (New York and London, 1975)

B. A. Brombert: *Cristina: Portraits of a Princess* (New York, 1977)

K. Hamburger, ed.: *Franz Liszt: Beiträge von ungarischen autoren* (Leipzig, 1978)

E. K. Horvath: *Franz Liszt: Kindheit (1811–1827)* (Eisenstadt, 1978)

P. Rehberg: *Liszt: die Geschichte seines Lebens, Schaffens und Wirkens* (Munich, 1978)

M. Eckhardt: 'Liszt à Marseille', *SM*, xxiv (1982), 163

M. Kovács: 'Documents sur Liszt en Belgique', *SM*, xxiv (1982), 157

Bibliography

A. Walker: *Franz Liszt: the Virtuoso Years 1811–1847* (New York, 1983)

ORCHESTRAL MUSIC, CONCERTOS

R. Wagner: *Ein Brief über Liszts symphonische Dichtungen* (Leipzig, 1857)

F. Draeseke: 'Liszts symphonische Dichtungen', *Anregungen für Kunst und Wissenschaft*, ii (1857); iii (1858)

——: 'Liszts Dante-Symphonie', *NZM*, xxvii (1860), 193, 201, 213, 221

A. Heuss: 'Eine motivisch-thematische Studie über Liszts sinfonische Dichtung "Ce qu'on entend sur la montagne" ', *ZIMG*, xiii (1911–12), 10

P. Raabe: *Die Entstehungsgeschichte der ersten Orchesterwerke Franz Liszts* (diss., U. of Jena, 1916)

H. Engel: *Die Entwicklung des deutschen Klavierkonzerts von Mozart bis Liszt* (Leipzig, 1927)

J. Weber: *Die symphonischen Dichtungen Franz Liszts* (diss., U. of Vienna, 1928)

J. Bergfeld: *Die formale Struktur der 'Symphonischen Dichtungen' Franz Liszts* (diss., U. of Berlin, 1931)

T. Stengel: *Die Entwicklung des Klavierkonzerts von Liszt bis zur Gegenwart* (diss., U. of Berlin, 1931)

E. Haraszti: 'Les origines de l'orchestration de Franz Liszt', *RdM*, xxxi (1952), 81

——: 'Genèse des Préludes de Liszt', *RdM*, xxxii (1953), 111

G. Króo: 'Gemeinsame Formprobleme in den Klavierkonzerten von Schumann und Liszt', *Robert Schumann: aus Anlass seines 100. Todestages* (Leipzig, 1956), 136

A. Main: 'Liszt after Lamartine: "Les Préludes" ', *ML*, lx (1979), 133

PIANO MUSIC, LISZT AS PIANIST

F. Busoni: 'Die Ausgaben der Liszt'schen Klavierwerke', *AMz*, xxvii (1900); repr. in F. Busoni: *Von der Einheit der Musik* (Berlin, 1922)

H. Arminski: *Die ungarischen Phantasien von Franz Liszt* (diss., U. of Vienna, 1929)

H. Dobiey: *Die Klaviertechnik des jungen Franz Liszts* (diss., U. of Berlin, 1932)

W. Howard: *Franz Liszt: Rhapsodie Nr.5, ein Kapitel Romantik* (Berlin, 1932)

I. Philipp: *La technique de Liszt* (Paris, 1932)

R. Kókai: *Franz Liszt in seinen frühen Klavierwerken* (diss., U. of Freiburg, 1933)

H. Landau: *Die Neuerungen der Klaviertechnik bei Franz Liszt* (diss., U. of Vienna, 1933)

R. Klasinc: *Die konzertante Klaviersatztechnik seit Liszt* (diss., U. of Vienna, 1934)

W. Rüsch: *Franz Liszts Années de pèlerinage: Beiträge zur Geschichte seiner Persönlichkeit und seines Stiles* (diss., U. of Zurich, 1934)

H. Westerby: *Liszt, Composer, and his Piano Works* (London, 1934)

D. Presser: *Studien zu den Opern- und Liedbearbeitungen Franz Liszts* (diss., U. of Cologne, 1953)

I. Kecskeméti: 'Two Liszt Discoveries – 1: an Unknown Piano Piece', *MT*, cxv (1974), 646

R. M. Longyear: 'The Text of Liszt's B minor Sonata', *MQ*, lx (1974), 435

N. B. Reich: 'Liszt's Variations on the March from Rossini's Siège de Corinthe', *FAM*, xxiii (1976), 102

B. Ott: *Liszt et la pédagogie du piano* (Issy-les-Moulineaux, 1978)

S. Winklhofer: *Liszt's Sonata in B Minor: a Study of Autograph Sources and Documents* (diss., U. of Michigan, Ann Arbor, 1980)

A. Main: 'Liszt's *Lyon*: Music and the Social Conscience', *19th Century Music*, iv (1980–81), 228

OTHER STUDIES

E. Reuss: 'Liszts Lieder', *Bayreuther Blätter 1906*, nos.7–12

La Mara: 'Franz Liszt und sein unvollendetes Stanislaus-Oratorium', *Österreichischer Rundschau* (15 Oct 1911)

E. Segnitz: *Franz Liszts Kirchenmusik* (Langensalza, 1911)

P. Roberts: *Etudes sur Boieldieu, Chopin et Liszt* (Rouen, 1913)

J. Wenz: *Franz Liszt als Liederkomponist* (diss., U. of Frankfurt am Main, 1921)

H. Sambeth: *Die gregorianischen Melodien in den Werken Franz Liszts* (diss., U. of Münster, 1923)

G. Galston: *Studienbuch: Franz Liszt* (Munich, 1926)

A. Schering: 'Über Liszts Persönlichkeit und Kunst', *JbMP 1926*, 31

W. Danckert: 'Liszt als Vorläufer des musikalischen Impressionismus', *Die Musik*, xxi (1928–9), 341

J. Heinrichs: *Über den Sinn der Lisztschen Programmusik* (diss., U. of Bonn, 1929)

E. Haraszti: 'La question tzigane-hongroise', *IMSCR, i Liège 1930*, 140

Z. Gárdonyi: *Die ungarischen Stileigentümlichkeiten in den musikalischen Werken Franz Liszts* (diss., U. of Berlin, 1931)

W. Howard: *Liszts Bearbeitung des Cujus animam aus dem Stabat Mater von Rossini* (Berlin, 1935)

B. Bartók: *Liszt problémák* (Budapest, 1936)

Z. Gárdonyi: *Liszt Ferenc magyar stilusa/Le style hongrois de François Liszt* (Budapest, 1936)

A. Molnár: *Liszt Ferenc alkotásai az esztétika tükrében* [Liszt's compositions in the light of aesthetics] (Budapest, 1936)

Bibliography

T. Bayer: 'Franz Liszt in der Dichtung', *Deutsche Musikkultur*, i (1936–7), 226, 285

E. Haraszti: 'Le problème Liszt', *AcM*, ix (1937), 123; x (1938), 32

M. Cooper: 'Liszt as a Song Writer', *ML*, xix (1938), 171

B. de Miramon Fitz-James: 'Liszt et La divine comédie', *RdM*, xix (1938), 81

E. Major: *Liszt Ferenc és a magyar zenetörtenet* [Liszt and Hungarian music history], Magyar zenei dolgozatok, iv–i (Budapest, 1940)

A. Toth: *Liszt Ferenc a magyar zene utjan* [Liszt along the path of Hungarian music] (Budapest, 1941)

A. Einstein: *Greatness in Music* (New York, 1941)

——: 'Franz Liszt, écrivain et penseur', *RdM*, xxiii (1944), 12; Eng. trans., *MQ*, xxxiii (1947), 410

R. Raffalt: *Über die Problematik der Programmusik* (diss., U. of Tübingen, 1949)

B. Szabolcsi: *A XIX. század magyar romantikus zenéje* [Hungarian Romantic music in the 19th century] (Budapest, 1951)

E. Haraszti: 'Un romantique déguisé en tzigane', *RBM*, vii (1953), 26

F. Noske: *La mélodie française de Berlioz à Duparc* (Amsterdam and Paris, 1954; Eng. trans., rev., 1970)

W. Felix: 'Die Reformideen Franz Liszts', *Festchrift Richard Münnich* (Leipzig, 1957), 104

G. Kraft: 'Franz Liszt und die nationalen Schulen in Europa', *Festschrift Richard Münnich* (Leipzig, 1957), 85

B. Hansen: *Variationen und Varianten in den musikalischen Werken Franz Liszts* (diss., U. of Hamburg, 1959)

Z. Hrabussay: *Unbekannte Manuskripte von Franz Liszt in der Slowakei*, Musikwissenschaftliche Studien, iv (Bratislava, 1960)

I. Kecskeméti: 'Unbekannte Eigenschrift der XVIII. Rhapsodie von Franz Liszt', *SM*, iii (1962), 173

Z. Gárdonyi: 'Neue Tonleiter- und Sequenztypen in Liszts Frühwerken', *SM*, xi (1969), 169

I. Kecskeméti: 'Two Liszt Discoveries – 2: an Unknown Song', *MT*, cxv (1974), 743

W. J. Dart: 'Revisions and Reworkings in the Lieder of Franz Liszt', *SMA* (1975), no.9, p.41

L. Bárdos: *Liszt Ferenc a jövő zenésze* [Liszt the innovator] (Budapest, 1976)

D. Legány: *Liszt Ferenc Magyarországos 1869–1873/Ferenc Liszt in Hungary 1869–73* (Budapest, 1976)

S. Winklhofer: 'Liszt, Marie d'Agoult and the "Dante" Sonata', *19th Century Music*, i (1977–8), 15

E. G. Heinemann: *Franz Liszts Auseinandersetzung mit der geistlichen Musik* (Munich, 1978)

Liszt Studien II: Eisenstadt 1978

377

D. Torkewitz: *Harmonisches Denken im Frühwerk Franz Liszts* (Munich, 1978)

A. Walker: 'Schumann, Liszt and the C Major Fantasie, op.17: a Declining Relationship', *ML*, lx (1979), 156

R.Stevenson:'LisztatMadridandLisbon:1844–45', *MQ*,lxv(1979),493

P. A. Autexier: 'Actualité de la recherche Lisztienne en Hongrie', *RdM*, lxvii (1981), 80

E. F. Jensen: 'Liszt, Nerval, and *Faust*', *19th Century Music*, v (1981–2), 151

R. P. Locke: 'Liszt's Saint-Simonian Adventure', *19th Century Music*, iv (1980–81), 209

R. L. Todd: 'Liszt, Fantasy and Fugue for Organ on 'Ad nos, Ad Salutarem Undam'' ', *19th Century Music*, iv (1980–81), 250

A. Walker: 'Liszt and the Schubert Song Transcriptions', *MQ*, lxvii (1981), 50

Index

Index

Index

Index

Index

387

Revue et gazette musicale de Paris, 255
Revue musicale, 14
Rhine, river, 109, 160, 244
Rhineland, 106
Richarz, Dr, 172, 173
Richter, J. P. F. [Jean Paul], 102, 104, 192
——, *Flegeljahre*, 111, 176
Rietz, Julius, 154
Riga, 137
Rome, 248, 249, 250, 253, 255, 294, 299, 306
——, Oratorio della Madonna del Rosario, 249
——, S Maria dell' Anima, 250
Rosen, Gisbert, 104, 106
Rosenthal, Moriz, 249
Rossini, Gioachino, 14, 43, 257, *268*, 282
——, *Guillaume Tell* Overture (transcr. Liszt), 283
——, *La gazza ladra*, 106
——, *Otello*, 269
——, *Soirées musicales*, 282
Rothschild, Baron and Baroness James de, 15, 24, 242
Rotterdam, 171
Rouget de Lisle, Claude-Joseph: *La Marseillaise*, 176, 293
Rubinstein, Anton, 63, 249
Rückert, (Johann Michael) Friedrich, 132, 146, 164, 182
——, *Minnespiel* (from *Liebesfrühling*), 148, 182
——, *Neujahrslied*, 150–51, 157
——, *Verzweifle nicht im Schmerzenstal*, 148, 164
Rudel, Gottlob [Schumann's guardian], 104, 106, 108
Russia, 3, 137, 139, 244, 246, 283

Saint-Cricq, Caroline de, 240, 248, 285
——, father of, 240
St Francis of Assissi, 320

St Petersburg (now Leningrad), 139
Saint-Saëns, Camille, 63, 249, 289, 300
——, *Danse macabre* (arr. Liszt), 286, 300
Salieri, Antonio, 238
Sand, George [Dudevant (née Dupin), Baroness Aurore], 19, 20, 21, 22, *23*, 25, 26–7, 33, 242, 268
——, grandmother of, 20
——, *Le contrebandier*, 269
——, *Lucrezia Floriani*, 20
Sauer, Emil von, 249
Saxony, 99
Sayn-Wittgenstein, Princess Carolyne, 246, 247, 248, 249, 252, 255, 256, 271, 279, 285, 288, 289, 292, 298, 310, 321
——, *Causes intérieures de la faiblesse de l'Eglise en 1870*, 252
Sayn-Wittgenstein, Prince Nicholas, 246, 249
Scarlatti, Domenico, 243
——, Sonatas, 258
Scheveningen, 165
Schiller, Friedrich von, 181, 295, 312
——, *Die Braut von Messina*, 158
——, *Die Ideale*, 295
Schilling, Gustav, 135
Schirmer [music publishers], 64
Schlesinger [music publisher], 64
Schnabel [Schumann's grandfather], 99
Schneeberg, nr. Zwickau, 113, 115, 126, 137
Schneider, Friedrich: *Weltgericht*, 99
Schoenberg, Arnold, 63
Scholz, Bernhard, 248
Schönefeld, nr. Leipzig, 131
Schöpff, F. W. T. [Wielfried von der Neun], 154
Schubart, Christian Friedrich Daniel: *Ideen zu einer Aesthetik der Tonkunst*, 101
Schubert, Ferdinand, 125, 176
Schubert, Franz, 24, 49, 102, 104,

Index

Index